Divided Environments

What are the implications of climate change for twenty-first-century conflict and security? Rising temperatures, it is often said, will bring increased drought, more famine, heightened social vulnerability and large-scale political and violent conflict; indeed, many claim that this future is already with us. *Divided Environments*, however, shows that this is mistaken. Focusing especially on the links between climate change, water and security, and drawing on detailed evidence from Israel–Palestine, Syria, Sudan and elsewhere, it shows both that mainstream environmental security narratives are misleading and that the actual security implications of climate change are very different from how they are often imagined. Addressing themes as wide-ranging as the politics of droughts, the contradictions of capitalist development and the role of racism in environmental change, while simultaneously articulating an original 'international political ecology' approach to the study of socio-environmental conflicts, *Divided Environments* offers a new and important interpretation of our planetary future.

Jan Selby is Professor of Politics and International Relations at the University of Sheffield.

Gabrielle Daoust is Assistant Professor in the Department of Global and International Studies, University of Northern British Columbia.

Clemens Hoffmann is Lecturer in International Politics in the Division of History, Heritage and Politics, University of Stirling.

Divided Environments
An International Political Ecology of Climate Change, Water and Security

Jan Selby
University of Sheffield

Gabrielle Daoust
University of Northern British Columbia

Clemens Hoffmann
University of Stirling

CAMBRIDGE
UNIVERSITY PRESS

University Printing House, Cambridge CB2 8BS, United Kingdom

One Liberty Plaza, 20th Floor, New York, NY 10006, USA

477 Williamstown Road, Port Melbourne, VIC 3207, Australia

314–321, 3rd Floor, Plot 3, Splendor Forum, Jasola District Centre, New Delhi – 110025, India

103 Penang Road, #05–06/07, Visioncrest Commercial, Singapore 238467

Cambridge University Press is part of the University of Cambridge.

It furthers the University's mission by disseminating knowledge in the pursuit of education, learning, and research at the highest international levels of excellence.

www.cambridge.org
Information on this title: www.cambridge.org/9781009098021
DOI: 10.1017/9781009106801

© Jan Selby, Gabrielle Daoust and Clemens Hoffmann 2022

This publication is in copyright. Subject to statutory exception and to the provisions of relevant collective licensing agreements, no reproduction of any part may take place without the written permission of Cambridge University Press.

First published 2022

A catalogue record for this publication is available from the British Library.

Library of Congress Cataloging-in-Publication Data
Names: Selby, Jan, 1972– author. | Daoust, Gabrielle, 1984– author. | Hoffmann, Clemens, 1978– author.
Title: Divided environments : an international political ecology of climate change, water and security / Jan Selby, Gabrielle Daoust, Clemens Hoffmann.
Description: Cambridge ; New York, NY : Cambridge University Press, 2022. | Includes bibliographical references.
Identifiers: LCCN 2022007319 (print) | LCCN 2022007320 (ebook) | ISBN 9781009098021 (hardback) | ISBN 9781009107600 (paperback) | ISBN 9781009106801 (ebook)
Subjects: LCSH: Climatic changes – International cooperation. | Human security – International cooperation. | Security, International – Environmental aspects. | Environmental policy – International cooperation. | BISAC: POLITICAL SCIENCE / International Relations / General
Classification: LCC QC903 .S475 2022 (print) | LCC QC903 (ebook) | DDC 363.738/74526–dc23/eng20220503
LC record available at https://lccn.loc.gov/2022007319
LC ebook record available at https://lccn.loc.gov/2022007320

ISBN 978-1-009-09802-1 Hardback
ISBN 978-1-009-10760-0 Paperback

Cambridge University Press has no responsibility for the persistence or accuracy of URLs for external or third-party internet websites referred to in this publication and does not guarantee that any content on such websites is, or will remain, accurate or appropriate.

'What are they after?'

'People say they came to look for water.'

'You – what did you hear them say?'

'There were too many people around them – I only heard a word here and there' ...

They all looked at the strange scene before them as if it were a dream or a hallucination, but the line of neatly parked tractors, and the deadly silence that reigned over the new land – for the wadi now looked like part of the desert beyond, except for the hills and the heaps of ruined trees – convinced them that it was real: a cruel, wicked sight that resembled death.

Beside the ruins on the broad plain they sat, determined to stay and wait ...

The order came from afar, in a harsh but muffled voice, like a cry from a cave. 'Fire!'

Cities of Salt, by Abdelrahman Munif, trans. P. Theroux
(Vintage, 1989), 26, 110, 614.

Unbiased at least he was when he arrived on his mission,
Having never set eyes on the land he was called to partition
Between two peoples fanatically at odds,
With their different diets and incompatible gods.
'Time,' they had briefed him in London, 'is short. It's too late
For mutual reconciliation or rational debate:
The only solution now lies in separation.
The Viceroy thinks, as you will see from his letter,
That the less you are seen in his company the better,
So we've arranged to provide you with other accommodation.
We can give you four judges, two Moslem and two Hindu,
To consult with, but the final decision must rest with you.'

Shut up in a lonely mansion, with police night and day
Patrolling the gardens to keep the assassins away,
He got down to work, to the task of settling the fate
Of millions. The maps at his disposal were out of date
And the Census Returns almost certainly incorrect,
But there was no time to check them, no time to inspect
Contested areas. The weather was frightfully hot,
And a bout of dysentery kept him constantly on the trot,
But in seven weeks it was done, the frontiers decided,
A continent for better or worse divided.

The next day he sailed for England, where he could quickly forget
The case, as a good lawyer must. Return he would not,
Afraid, as he told his Club, that he might get shot.

'Partition', by W. H. Auden (1966)

Contents

List of Figures and Tables	*page* x
List of Maps	xi
Preface	xvii
List of Abbreviations	xxi

1 Introduction   1
Looming Climate Instability?   1
Our Approach and Argument in Brief   9
Questions of Method   10
An International Political Ecology   18
Climate Change, Water and (In)Security   23
Five Divided Environments   26
Organisation of the Book   30

2 Geography versus Demography   33
The Question of Scarcity   33
The Water Stress Paradox   35
Water Is Not Life   40
Where Are the Water Wars?   46
Eco-Determinism: The Zombie Theory of Politics   54
The Politics of Scarcity Narratives   60
A Revenge of Geography?   63

3 Drought   69
More Drought, More Conflict?   69
Darfur 2003–5: The 'First Climate Change War'?   71
The Syrian Civil War: Drought, Migration and Conflict Revisited   77
Lake Chad: Who Needs Evidence?   83
A Brief History of Drought   88
Everybody Loves a Good Drought   95
A Revenge of Uncertainty?   100

4 Others   103
A Destroyer of Trees   103
The Global Environmental Colour Line   105
A Bifurcated State on the Nile   114

viii   Contents

Liberalism and its Contradictions in Cyprus	120
Palestine: A Setter-Colonial Encounter	125
Identity and Alterity in an Age of Climate Change	131

5 Hydraulics — 137

From Greenock to Tennessee	137
Engineering Ethno-Nationalism in Israel and Cyprus	141
River Development and Beyond in Syria and Sudan	149
The Five Contradictions of Hydraulic Development	158
Resurgent Hydraulic Politics	165

6 Frontiers — 169

Water and the Capitalist Resource Frontier	169
Abundance and Violence on Sudan's Multiple Frontiers	172
Water Apartheid in the West Bank and Gaza	181
The Rise and Fall of Hasakah	190
The Lake Chad Basin: Nigeria's Outstanding Internal Frontier	196
Resistance and Resilience	201
The New Frontier Lands of Climate Change	205

7 War — 209

Where is the Bloody Business?	209
Infrastructure Destruction and its Consequences	213
The Antinomies of Displacement	219
Expropriation	226
Institution-Building and Development	232
War and Climate Change Vulnerability	236

8 Peace — 240

The Promises of Environmental Peacebuilding	240
Is Peace Really Blue?	242
Liberal Functionalism	246
The Oslo Process: Dissimulating Decolonisation	252
Naivasha and After: Self-Determination and Militarisation	258
The Annan Plan: Dissimulating Unification	263
Peace and Climate Change Vulnerability	266

9 Transformations and Circulations — 270

Nexus, Nexus, Nexus	270
The International Political Ecology of Virtual Water	273
Rebound in the Negev	282
Fossil Fuels, Fossil Capital and Water Security	286
Capital into Water, Water into Capital	291
Capitalism over Climate	294

10 Conclusions — 300

Reimagining Environmental Security	300
Our Divided Environments Revisited	301
Ten Theses on Water (In)Security	306

Theorising Ecological Politics	314
The Coming Landscape of Climate Change and Conflict	317
Index	325

Figures and Tables

Figures

2.1 Renewable water resource availability per person,
for selected countries, 1962–2017   *page* 37
2.2 Agriculture's worldwide contribution to GDP,
employment and exports, 1962–2019   52
6.1 Gaza is Manhattan   189

Tables

2.1 Falkenmark's Water Stress Index, for selected countries, 2017   36
2.2 SDG Indicator 6.4.2: water stress, for selected countries, 2017   37

Maps

1 Divided Cyprus *page* xii
2 Israel, the West Bank and Gaza xiii
3 The Lake Chad region xiv
4 Sudan and South Sudan xv
5 Syria xvi

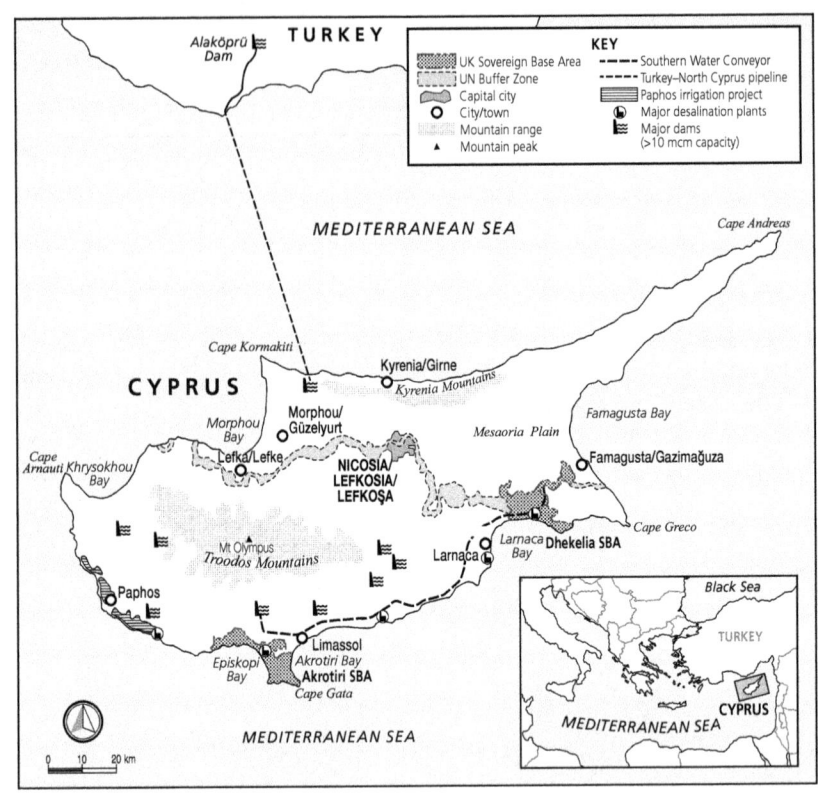

Map 1 Divided Cyprus

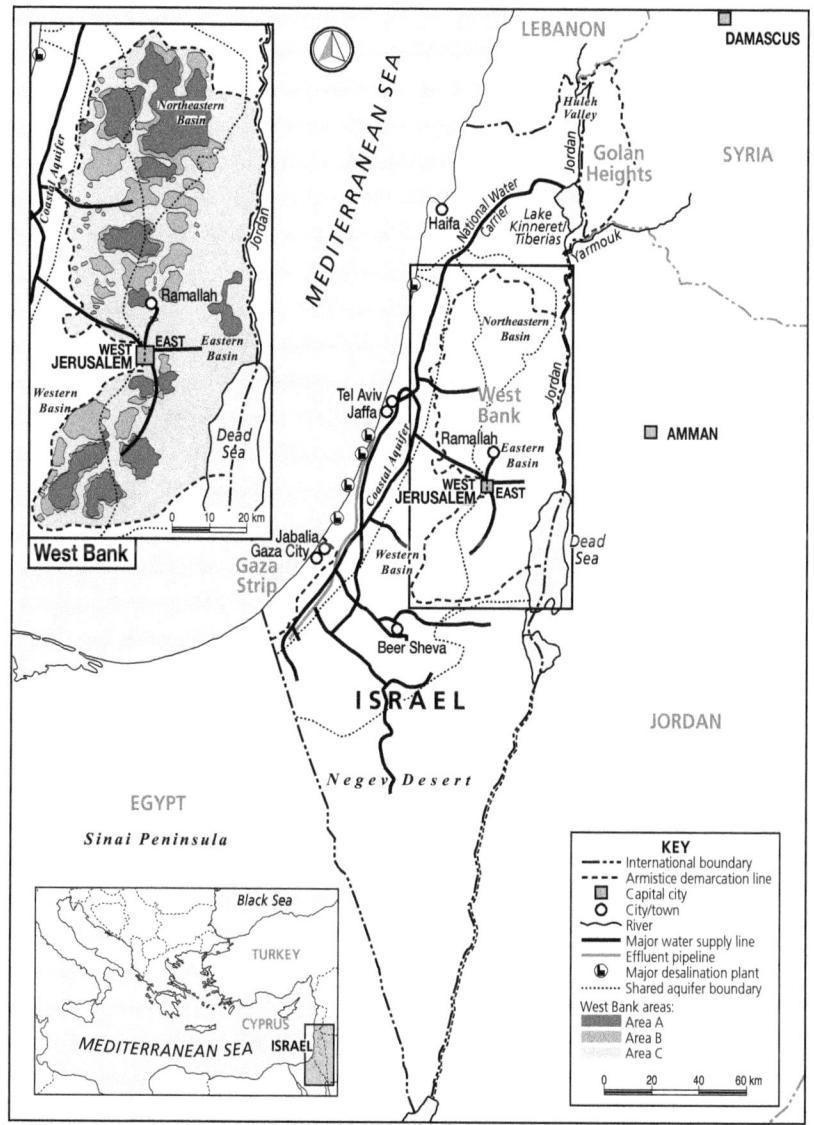

Map 2 Israel, the West Bank and Gaza

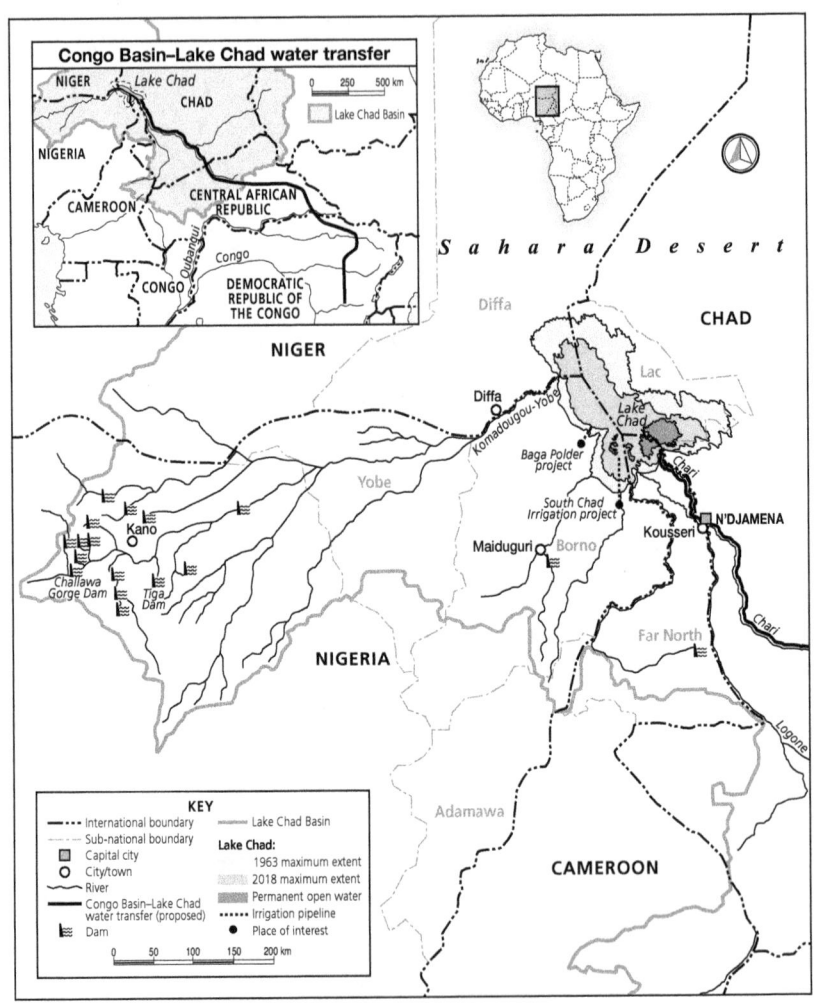

Map 3 The Lake Chad region

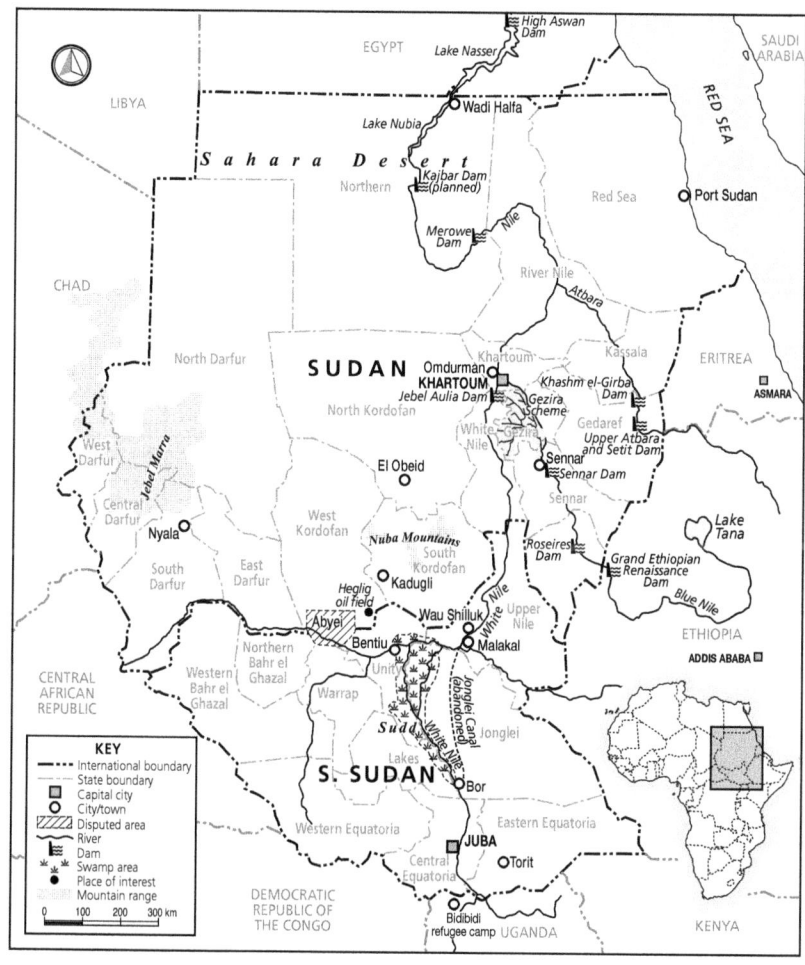

Map 4 Sudan and South Sudan

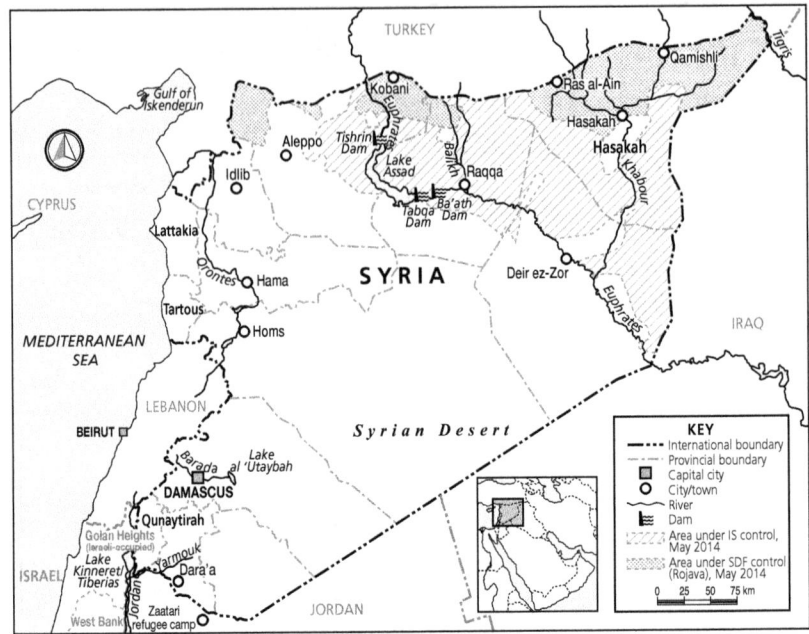

Map 5 Syria. Note that this map shows Rojava and Islamic State territory as of May 2014; since then, the extent of these areas has changed multiple times.

Preface

If this book began anywhere, it was with a September 2008 email from Giorgos Kallis and Christos Zografos about a proposed three-year research consortium on climate change, water and security. The Climate Change, Hydro-Conflicts and Human Security (or CLICO) consortium, funded by the EU's Framework 7 programme and led by Giorgos and Christos, brought together researchers on climate, water politics and environmental security to explore the implications of climate change for water resources and security in the Mediterranean, North Africa and the Middle East. Such programmes are often more trouble than they're worth, bedevilled by the EU's predilection for research collaboration and integration above all else. Thankfully, Giorgos and Christos' politics meant that they managed CLICO with an exceptionally light touch, letting many flowers bloom. More than a dozen years on, this book is one belated result.

The initial idea behind this project, as proposed to the EU, was for a comparative analysis of water–climate–conflict linkages in three partitioned societies: Cyprus, Israel–Palestine and Sudan. The plan was for a book comprising three case study chapters sandwiched between broader theoretical reflections. But this was always too ambitious given the project timescale. And then a whole host of other things got in the way, not least several new jobs, two new babies, a failed coup d'état, relocations between countries, the standard teaching and administrative demands, periodic research assessment requirements and multiple other research projects. The consequence was that Selby and Hoffmann barely worked on the book during the five years after CLICO had concluded, in December 2012. Only in late 2017, after Selby had completed a separate stream of research on climate change and the Syrian civil war, was it decided that the book should be rethought and restructured, this time with a thematic structure and Syria added as a fourth 'divided environment'. Daoust joined the team in October 2018, leading to more sustained work on the manuscript plus the addition of Lake Chad as a fifth case, with the bulk of the manuscript being written between 2019 and 2021.

The text of *Divided Environments* is entirely new, though various sections draw on our previous individual and co-authored publications, summarising some while expanding on others. The final text was for the most part conceived, framed and written by Selby, mostly while sitting, during lockdown, at a picnic table next to the compost heap of his tiny Brighton garden: hence, where there are major analytical, theoretical or narrative flaws, these are in all likelihood his doing. By contrast, responsibility for the analysis of cases was divided. Hoffmann led and undertook most of the research on Cyprus. Daoust did the same for Lake Chad. Selby led most of the research on Israel–Palestine and Syria. Each author has led different segments of the research on Sudan–South Sudan. Above all, what follows is the product of innumerable conversations and exchanges to identify cross-cutting patterns, divergent trajectories, narrative inconsistencies and unanswered puzzles within and across our five cases. It is in this latter respect, above all, that the writing of this book has involved particularly close collaboration.

Our analysis is underpinned by extensive primary research which reaches back to 2010, some of it even earlier. This primary research, which cuts across all of our cases, has involved a combination of elite interviews (mainly with local water managers and policymakers, and international donor and NGO officials); informal discussions with farmers, aid recipients and community leaders; the organisation of local stakeholder workshops; participant observation in a wide range of expert forums; archival analysis of water negotiation files; and discourse analysis of media and policy representations of specific water and climate issues. Research was undertaken in Cyprus from 2010 to 2012 (by Hoffmann and Selby); in South Sudan during 2011–12 (by Hoffmann) and 2014–15 (by Daoust, as part of prior doctoral research); in Khartoum and Omdurman in 2013 (by Hoffmann); in N'Djamena in 2020 (by Daoust); and in Israel–Palestine on multiple occasions (by Selby). In addition, we have benefitted from a good many Skype and more recently Zoom interviews and from regular email correspondence and conference discussions with both local and international experts. Lastly, we wish to stress that much of our analysis is underpinned by close interrogations of existing primary and secondary sources – sources which often misrepresent water and climate security issues.

Our debts are many: over the course of the last decade, scores of people have provided inspiration, answered irritating questions, been co-authors, offered research support or helped shape the journal articles that underpin much of what follows. On climate and water politics, and political ecology more broadly, we particularly wish to mention Neil Adger, Jeremy Allouche, Jessica Barnes, Tor Benjaminsen, Andrea

Brock, Halvard Buhaug, Simon Dalby, Alex De Waal, Rosaleen Duffy, James Fairhead, Earl Gammon, Emma Hakala, Betsy Hartmann, Amber Huff, Tobias Ide, Jeremy Lind, Michael Mason, Lyla Mehta, Filippo Menga, Naho Mirumachi, Daanish Mustafa, Lars Otto Naess, Pete Newell, Dave Ockwell, Paul Rogers, Elizabeth Shove, Ayesha Siddiqi, Jeannie Sowers, Eric Swyngedouw, Judith Verweijen and Steffan Wyn-Jones, and to thank Lokendra Karki for his research assistance.

On Cyprus, we particularly wish to acknowledge Hakan Djuma, Gozen Elkiran, Iacovos Iacovides, Hayriye Kahveci, Dilek Latif, Julian Saurin and Nicos Trimikliniotis. On Israel–Palestine, Tony Allan, Shaddad Attili, Fuad Bateh, Gidon Bromberg, Natasha Carmi, Camilla Corradin, Ciarán Ó Cuinn, Muna Dajani, Rebhi El-Sheikh, Itay Fischhendler, Amira Hass, Jad Isaac, Peter Laban, Steve McCaffrey, Ruhan Nagra, Jacob Norris, Yoav Peled, the late Dave Phillips, Greg Shapland, Michael Talhami, Abdelrahman Tamimi, Ahmad Yaqubi and Mark Zeitoun have been particularly important interlocutors. The Palestine Liberation Organization's Negotiation Support Unit is acknowledged for providing Selby with access to Palestinian bilateral water negotiation files, as is the UN Food and Agricultural Organization for granting Selby the opportunity to participate in a strategic evaluation of its West Bank and Gaza programme. Vanessa Iaria, Bhabani Nayak, Cassandra Poyiadjis-Osler and Dan Watson all provided valuable research assistance at various points.

On Lake Chad, especial thanks to Maarten Gischler, Géraud Magrin, Amiera Sawas and the organisers of the 18th Colloque Méga-Tchad, Zakinet Dangbet and Christine Raimond. On Sudan and South Sudan, thanks to Yacob Arsano, Atta el-Battahani, Luka Biong Deng, Ana Elisa Cascao and Alan Nicol. On Syria, thanks to Selby's collaborators Omar Dahi, Christiane Fröhlich and Mike Hulme; to Myriam Ababsa, Geraldine Chaterlard, Marwa Daoudy, Lina Eklund, Danijel Orešić and Jodi Tejel for their help with particular points of detail and interpretation; and to Mark de Jong, Katy Joyce, Sofia Kesidou, Wassim Naboulsi and Gerardo Torres Contreras for their research assistance. Thanks to Shane Brighton, Adriana Bruggeman, Clemens Messerschmid, Ethemcan Turhan, Bob Vitalis and Louise Wise for their acute observations on draft chapters, to John Hall for the maps and to John Haslam at Cambridge University Press for being willing to take this book on.

The research for and writing of this book was undertaken across the universities of Sussex (Selby until 2020, Hoffmann 2010–12, Daoust 2018–20), Bilkent (Hoffmann 2012–16), Stirling (Hoffmann from 2016), Sheffield (Selby from 2020) and St Francis Xavier (Daoust from 2020). At Sussex, we would particularly like to thank Saul Becker,

Richard Black, Mike Collyer, Andrea Cornwall, Synne Dyvik, Stefan Elbe, Kit Eves, Medeni Fordham, Buzz Harrison, Beate Jahn, Sam Knafo, Richard Lane, Kamran Matin, Mario Novelli, Patricia Owens, Justin Rosenberg, Anna Stavrianakis, Benno Teschke, Cindy Weber, Rorden Wilkinson and Joanna Wood for all their support and, more broadly, to acknowledge former colleagues in the Department of International Relations and School of Global Studies for being so fantastic. At Bilkent, thanks to Ali Bilgiç, Pınar Bilgin, Can Mutlu and Erinç Yeldan, and at Stirling, Andrea Baumeister, Paul Cairney, Megan Dee, Sally Foster, Sian Jones, Sossie Kasbarian, Peter Lynch, Matias Margulis, Emma McLeod, Holger Nehring, Richard Orham, Bashir Saade, Andrea Schapper and Hannes Stephan. At Sheffield, thanks especially to Ruth Blakeley and Andy Hindmoor. And at St Francis Xavier, particular thanks to Youngwon Cho.

In addition, we want to acknowledge those whose work has been a major influence but is not adequately cited in what follows. Space constraints have required us to condense our footnotes considerably, including cutting all explanatory text and indirect references (as well as grouping references together). Our apologies where the result is that influences are not properly recognised.

Last but not least, Jan Selby would like to thank Alison for her solidarity, love and general good sense. I know that this book has dragged on for far too long: I promise not to attempt another one like it again.

Abbreviations

ADP	Agricultural Development Project
AfDB	African Development Bank
bcm/y	billion cubic metres per year
CBS	Central Bureau of Statistics
CCS	Center for Climate and Security
cm/cap/y	cubic metres per capita per year
CO_2	carbon dioxide
COP	Conference of the Parties
CRED	Centre for Research on the Epidemiology of Disasters
DIU	Dams Implementation Unit (Sudan)
DSI	State Hydraulic Works (Turkey)
ECSC	European Coal and Steel Community
EM-DAT	Emergency Events Database
FAO	Food and Agricultural Organization of the United Nations
G7	Group of Seven
GDP	gross domestic product
GERD	Grand Ethiopian Renaissance Dam
GIZ	German Agency for International Cooperation
HRW	Human Rights Watch
ICG	International Crisis Group
IDF	Israel Defense Forces
IDMC	Internal Displacement Monitoring Centre
IDP	internally displaced person
ILO	International Labour Organization
IMF	International Monetary Fund
IPCC	Intergovernmental Panel on Climate Change
IR	International Relations
IS	Islamic State
IWA	Israeli Water Authority
JICA	Japan International Cooperation Agency
JNF	Jewish National Fund

List of Abbreviations

JWC	Joint Water Committee (Israeli–Palestinian)
LCBC	Lake Chad Basin Commission
l/c/d	litres per capita per day
LGAs	Local Government Areas (Nigeria)
mcm/y	million cubic metres per year
MFC	Mechanised Farming Corporation (Sudan)
NASA	National Aeronautics and Space Administration
NGO	non-governmental organisation
PA	Palestinian Authority
PLO	Palestine Liberation Organization
ppm	parts per million
PRIO	Peace Research Institute Oslo
PWA	Palestinian Water Authority
RCP	Representative Concentration Pathway
RoC	Republic of Cyprus
SCIP	South Chad Irrigation Project (Nigeria)
SDF	Syrian Democratic Forces
SDG	Sustainable Development Goal
SFG	Strategic Foresight Group
SPLM/A	Sudan People's Liberation Movement/Army
TVA	Tennessee Valley Authority
UAR	United Arab Republic
UN	United Nations
UNCCD	United Nations Convention to Combat Desertification
UNCCP	United Nations Conciliation Commission for Palestine
UNDP	United Nations Development Programme
UN-ECOSOC	United Nations Economic and Social Council
UNEP	United Nations Environment Programme
UNESCO	United Nations Educational, Scientific and Cultural Organization
UN-ESCWA	United Nations Economic and Social Commission for Western Asia
UNHCR	United Nations High Commissioner for Refugees
UNHRC	United Nations Human Rights Council
UNICEF	United Nations Children's Fund
UN-OCHA	United Nations Office for the Coordination of Humanitarian Affairs
UNRWA	United Nations Relief and Works Agency for Palestine Refugees
UNSCO	United Nations Special Coordinator for the Middle East Peace Process
USDA	United States Department of Agriculture

US ODNI	United States Office of the Director of National Intelligence
WASH	water, sanitation and hygiene
WDD	Water Development Department (Republic of Cyprus)
WEF	World Economic Forum
WFP	World Food Programme
WHO	World Health Organization
XR	Extinction Rebellion

1 Introduction

Looming Climate Instability?

That anthropogenic climate change is one of the foremost twenty-first-century global security challenges is a view now firmly, if rather superficially, ensconced within Western liberal public and policy discourse. National security strategies have depicted it as 'an urgent and growing threat' and possibly 'the greatest challenge' there is to global stability, potentially presaging a 'breakdown of the rules-based international system' and a 're-emergence of major inter-state conflict'. Foreign ministers have labelled it 'perhaps the twenty-first century's biggest foreign policy challenge' and 'the world's most fearsome weapon of mass destruction' and claimed that 'the threat that a changing climate presents to ... international peace and security cannot be underestimated'. Climate change ministers have argued that 'we need to be ready for a world where climate instability drives political instability' and that a 'world where climate change goes unchallenged will be a Hobbesian world, where life for far more people is "nasty, brutish, and short"'. The United States Congress and Pentagon have both described climate change as a threat to US national security. Successive United Nations (UN) Secretary Generals have called climate change 'the defining threat of our time' and 'the pre-eminent geopolitical and economic issue of the twenty-first century'. Activist movements from Extinction Rebellion (XR) to Greenpeace have characterised it as 'an unprecedented global emergency' that puts us 'in a life or death situation of our own making' and as 'the world's biggest threat ... ranked close to weapons of mass destruction in terms of potential impact' (indeed, one of the co-founders of XR has claimed that climate change is already 'turning whole regions of the world into death zones' and that a climate change–induced 'global holocaust ... is already underway'). And figures from Barack Obama to Russell Brand, among many others, have suggested that climate change is a bigger threat than terrorism.[1]

---

[1] The White House, *National Security Strategy 2015* (2015), 12; UK Cabinet Office, *The National Security Strategy of the United Kingdom: Security in an Interdependent World* (2008),

Indeed, such views have a conspicuously diverse array of proponents. Western militaries and defence planners, national security think tanks, intelligence agencies, UN and Bretton Woods organisations, state development agencies, humanitarian and development NGOs, environmental campaigners, mainstream liberal media, eco-socialist commentators and even authoritarian Southern governments: all have in one way or another, and for one reason or another, argued that climate change has sweeping implications for conflict and security. Climate change deniers have often ridiculed claims to this effect, along with more basic evidence on the extent and causes of global warming. Many non-Western and Southern governments, most notably China, India, Russia and Brazil, have also been sceptical, arguing, among other things, that the UN Security Council is not the appropriate venue for addressing the challenge of climate change. And various academics have also expressed doubts about the links between climate change, conflict and security, as detailed below. Yet for all this, the breadth of the contemporary Western public and policy consensus on the question of climate security is striking. On this issue, both the American military machine and its fiercest critics can in broad terms agree, as can neo-liberal economists and their anti-capitalist opponents. Climate security discourse is a space where John Kerry and Naomi Klein, Prince Charles and the Syrian state, George Monbiot and the World Bank, Friends of the Earth and the US Central Intelligence Agency all converge.[2]

18–19; W. Hague, 'The diplomacy of climate change', Speech to Council on Foreign Relations, New York (27/09/2010); S. Denyer, 'Kerry calls climate change a weapon of mass destruction, derides sceptics', *Washington Post* (16/02/2014); M. Wallström, Statement at the UN Security Council Debate on Climate-Related Security Risks (11/07/2018); E. Davey, Speech to a Climate and Resource Security Dialogue for the 21st Century conference, London (22/03/2012); C. Huhne, 'The geopolitics of climate change', Speech to Future Maritime Operations conference, Royal United Services Institute, London (07/07/2011); US Congress, *National Defense Authorisation Act for Fiscal Year 2018*, HR2810, Section 335; Department of Defense, *Report on Effects of a Changing Climate to the Department of Defense* (2019), 2; A. Guterres, 'Remarks at the High-Level Event at COP 23' (15/11/2017); B. Ki-moon, 'Opening remarks to UN Climate Change Summit Plenary' (22/09/2009); Extinction Rebellion, 'The Emergency', https://rebellion.earth/the-truth/the-emergency/; K. Naidoo, 'Nature does not negotiate: climate catastrophe is with us now!', Greenpeace (08/12/2014); C. Baynes, 'Extinction Rebellion founder told he is not welcome in movement after Holocaust comments', *Independent* (21/11/2019); 'Obama: The Vox conversation, part two', *Vox.com* (09/02/2015); R. Leber, 'Obama is right: climate change kills more people than terrorism', *New Republic* (11/02/2015); R. Brand, 'ISIS versus climate change – which kills more?', *The Trews* (05/03/2015), https://www.youtube.com/watch?v=Zrr5BvrAo-Y.
[2] T. R. Bromund, 'Climate change is not a national security threat', The Heritage Foundation (04/06/2015); P. J. Michaels and C. A. Preble, 'Does climate change actually fuel terrorism?', The Cato Institute (18/11/2015); UN Security Council, *6587th Meeting* (S/PV.6587) (20/07/2011), 7–9, 12–13, 18–20; UN Security Council, *8451st Meeting* (S/PV.8451) (25/01/2019), 15–17, 42–3, 61–2; Denyer, 'Kerry calls climate change

For proponents of this climate security orthodoxy, the implications of climate change for global security are abundantly clear: that through its impacts on both short-term environmental shocks and long-term trends, climate change will exacerbate resource pressures and scarcities and in turn feed increased resource competition, economic and social vulnerability, migration and displacement, and civil and political conflict at multiple sites and scales – all aided and abetted by existing patterns of poverty and fragility. The central concern of climate security discourse, in other words, is with climate-induced resource scarcity crises and their consequences, which are typically envisaged as taking place in, and as emanating from, the developing world. The 2010 US National Security Strategy, for example, characterised the changes likely to be 'wrought by a warming planet' as 'new conflicts over refugees and resources; new suffering from drought and famine; catastrophic natural disasters; and the degradation of land across the globe'. UK Foreign Secretary Margaret Beckett introduced the first ever UN Security Council debate on the subject by asserting that an 'unstable climate will exacerbate some of the core drivers of conflict, such as migratory pressures and competition for resources'. 'What makes wars start?' she asked, before answering: '[f]ights over water. Changing patterns of rainfall. Fights over food production, land use.' And during a follow-up Security Council meeting, UN Secretary General Ban Ki-moon couched the issue as follows:

We must make no mistake. The facts are clear. Climate change is real and it is accelerating in a dangerous manner. It not only exacerbates threats to international peace and security, it is a threat to international peace and security ... Competition between communities and countries for scarce resources, especially water, is increasing, exacerbating old security dilemmas and creating new ones. Environmental refugees are reshaping the human geography of the planet, a trend that will only increase as deserts advance, forests are felled and sea-levels rise. Mega-crises may well become the new normal. Those are all threats to human security, as well as to international peace and security.[3]

---

a weapon of mass destruction'; J. Box and N. Klein, 'Why a climate deal is the best hope for peace', *New Yorker* (18/11/2015); R. Mills, 'Charles: Syria's war linked to climate change', *Sky News* (23/11/2015); ICG, *Popular Protest in North Africa and the Middle East (VI): The Syrian People's Slow-Motion Revolution* (2011), 23; G. Monbiot, 'How fossil fuel burning nearly wiped out life on Earth – 250m years ago', *Guardian* (27/05/2015); World Bank, *Turn Down the Heat: Confronting the New Climate Normal* (2014); C. Bennett, 'Failure to act on climate change means an even bigger refugee crisis', *Guardian* (07/09/2015); D. R. Coates, *Worldwide Threat Assessment of the US Intelligence Community*, Statement to Senate Select Committee on Intelligence (29/01/2019), 23.
[3] The White House, *National Security Strategy 2010* (2010), 47; UN Security Council, *5663rd Meeting* (S/PV.5663) (17/04/2007); P. Reynolds, 'Security Council takes on global warming', *BBC News* (18/04/2007); UN Security Council, *6587th Meeting*, 2.

In all of these formulations and many others besides, it is *resource scarcities* and their socio-economic, especially migration, consequences which are viewed as the key 'intervening variables' between global climate change and worsening instability.

There is, on one level, very good reason for these concerns. Global average temperatures are already more than 1°C above pre-industrial levels (and temperatures over land around 1.5°C higher). The level of atmospheric carbon dioxide (CO_2) is not just rising but doing so at an ever-accelerating rate (during the 1960s, atmospheric CO_2 was rising at below 0.8 parts per million (ppm) annually; by the 1990s, this growth rate had become 1.5 ppm; by May 2019, the atmospheric CO_2 level was 3.5 ppm higher than twelve months previously). Global greenhouse gas emissions will continue rising through to at least 2030 even if all countries' 2015 Paris Agreement commitments are fully implemented. And, of course, they are not being. As a result, our Anthropocene planet is currently on track to have warmed by 1.5°C sometime during the 2030s or 2040s and by 2°C – the internationally accepted target for avoiding 'dangerous climate change' – not long after that. It is projected that, even with full implementation of the Paris Agreement, the Earth will have warmed by between 2.6 and 3.2°C by 2100. Unless worldwide mitigation policies and implementation efforts are significantly expanded, the warming will be greater still. And, in some regions, average temperature rises are likely to be even higher than these global figures suggest.[4]

Climate change of this magnitude and velocity will undoubtedly have wide-ranging environmental, economic, political and humanitarian consequences. Global heating will transform regional climates and ecosystems. Heat death risks will soar. Precipitation may shift considerably, with some regions becoming hotter and drier, others hotter and wetter. Most forms of extreme weather event will become both more frequent and more extreme. Sea levels will rise – albeit unlikely by more than one metre this century, and with sea levels not fully stabilising for several

---

[4] IPCC, *Climate Change and Land: An IPCC Special Report on Climate Change, Desertification, Land Degradation, Sustainable Land Management, Food Security, and Greenhouse Gas Fluxes in Terrestrial Ecosystems* (2019), 44; US National Oceanic and Atmospheric Administration, Earth Systems Research Laboratory, 'Trends in atmospheric carbon dioxide', www.esrl.noaa.gov/gmd/ccgg/trends/data.html; Scripps Institution of Oceanography, 'Carbon dioxide levels hit record peak in May', Keeling Curve blog (04/06/2019); J. G. J. Olivier et al., Trends in Global CO_2 and Total Greenhouse Gas Emissions: 2017 Report (PBL Netherlands Environmental Assessment Agency, 2017); International Energy Agency, *Global Energy and CO_2 Status Report 2017* (2018); IPCC, *Global Warming of 1.5°C: An IPCC Special Report on the Impacts of Global Warming of 1.5°C above Pre-Industrial Levels and Related Global Greenhouse Gas Emission Pathways*, Summary for Policymakers (2018), 4; J. Rogelj et al., 'Paris agreement climate proposals need a boost to keep warming well below 2°C', *Nature*, 534 (2016), 631–9.

millennia.⁵ And population distribution and food production patterns will inevitably be hugely affected and will have to change. It is frankly inconceivable, given all of this, that climate change will not have significant ramifications for patterns of conflict, insecurity and instability. And in these respects it is neither surprising nor particularly troubling that there is such a wide liberal public and policy consensus on the question of climate security.

In other respects, however, the value of climate security discourse is far from clear. The precise *meanings* of climate security claims are, to start with, often inconstant or ambiguous. Sometimes their reference points are 'national' or 'global' security, while at other times their focus is the likely impacts of climate change on 'human security' or even 'ecological security' – which are different matters altogether. Sometimes climate impacts are discussed in determinist and mono-causal terms, while elsewhere climate change is portrayed as but one 'contributory factor' to conflict among numerous others – with the question of how many others (tens? hundreds? thousands?) usually being left open and unaddressed. The *assumptions* underpinning climate security thinking are often questionable, including those about the nature of human–environment relations and about the causes of conflict, instability and insecurity. Moreover, the *purposes* – the aims and agendas – guiding climate security discourse also warrant interrogation. The framing of climate change as a security challenge – its discursive 'securitisation' – has no doubt been motivated above all by a desire to highlight the urgency of the climate change challenge, and through that to help push the issue up assorted social, political and international policy agendas. But other agendas have also often been in play: military interests in identifying new rationales for intervention; economic interests associated with new 'crisis response' technologies; donor and NGO preferences for depoliticised framings of socio-ecological crises; and more. To adapt Robert Cox's pithy phrase, climate security discourse is 'always for someone and for some purpose' – and not all of these purposes are benign. What's more, even allowing for the best of intentions, there remain questions about the *impacts* of climate security discourse on efforts to reduce greenhouse gas emissions. A good case can be made that, far from supporting mitigation efforts, the language of climate 'threats', 'chaos', 'emergency' and 'catastrophe' feeds feelings of helplessness and fatalism and may even provide an excuse for inaction.⁶

⁵ C. Mora et al., 'Global risk of deadly heat', *Nature Climate Change*, 7 (2017), 501–6; P. U. Clark et al., 'Consequences of twenty-first century policy for multi-millennial climate and sea-level change', *Nature Climate Change*, 6 (2016), 360–9.
⁶ M. McDonald, 'Discourses of climate security', *Political Geography*, 33 (2013), 42–51; R. Cox, 'Social forces, states and world orders: beyond international relations theory',

Most important, the scholarly *evidence* on the links between climate change and security is weak and divided, and when it departs from dominant policy framings is routinely ignored. For the most part, scientific research has played a formative if far from straightforward role in pushing forward national and international action on climate change. Within climate security discourse, by contrast, it has been defence planners and their scenario reports which have been most influential. Thus, the first major climate security study, commissioned by the Pentagon's leading futurologist, contained very little evidence but nonetheless envisaged large-scale military confrontations over natural resources, a 'flood of refugees' arriving in the United States from the Caribbean (by 2012!) and civil war in China plus the 'near collapse' of the European Union (EU) (by 2025). Likewise, the single most influential report on the subject, a 2007 study authored by a dozen retired three- and four-star US generals and admirals, concluded that climate change will act as a 'threat multiplier for instability in some of the most volatile regions of the world' – though, once again, with only the barest of evidence bases.[7] Only in the wake of these early military-led reports has a significant body of actual research on the subject been conducted. While any discourse always has multiple origins, climate security discourse has clearly been led and shaped more by policy and defence actors, most prominently the US military establishment and its offshoots, than by any weight of scientific evidence.

On the evidence itself, researchers are deeply, and often bitterly, divided – in a manner that cuts across epistemologies and methods. Some quantitative studies have identified striking historical relationships between the climate, weather and conflict, and from that developed projections about the potential conflict and security impacts of climate change. A widely read 2009 study by Marshall Burke and colleagues, for example, identified strong correlations between temperature variations and battle deaths in Africa, and on this basis predicted that by 2030, an additional 393,000 lives may be lost each year across Africa because of global warming. Another more recent study, by Anouch Missirian and Wolfram Schlenker, has claimed to find associations between asylum applications received by the EU and weather fluctuations in source countries, and on this basis suggested that by the end of the century these

---

Millennium, 10:2 (1981), 128; K. M. Norgaard, *Living in Denial: Climate Change, Emotions, and Everyday Life* (MIT Press, 2011).

[7] P. Schwartz and D. Randall, *An Abrupt Climate Change Scenario and Its Implications for United States National Security* (California Institute of Technology, 2003), 17; Center for Naval Analysis Military Advisory Board, *National Security and the Threat of Climate Change* (2007), 44–5.

applications may have increased by as much as 188 per cent. Both studies, however, have been sharply critiqued, including by fellow quantitative researchers.[8] More broadly, while most quantitative studies do identify some manner of connection between specific environmental and conflict variables, a large minority do not, and among those that do, the findings are consistently contradictory. Of recent studies on the impacts of rainfall variability in Africa, for example, some find *low* rainfall to be associated with increased conflict but others *high* rainfall; still others find high rainfall to be associated with *reduced* conflict; at least one study finds that droughts aid democratic transitions; others conclude that precipitation extremes, of either sign, are associated with increased conflict; and numerous studies have found no meaningful correlations, either in Africa specifically or beyond. Just as striking, even review essays on quantitative scholarship on climate security have not been able to come to common conclusions about the extent of agreement on the subject.[9]

Moreover, qualitative researchers have, if anything, been even more split. Some have agreed with and perhaps even gone beyond the policy orthodoxy, with one leading genocide studies scholar concluding that climate change will probably be 'the biggest trigger of genocide in the twenty-first century' and many others foreseeing 'climate wars' and 'climate chaos'. And yet, on the other hand, many qualitative researchers have been profoundly sceptical and critical of climate security thinking. Case study analyses have repeatedly disputed claims about particular conflicts – the civil wars in Darfur and Syria, for example – and the role of climate change therein. Likewise, discourse analyses of climate security narratives have consistently argued, on a range of

---

[8] M. Burke et al., 'Warming increases the risk of civil war in Africa', *Proceedings of the National Academy of Sciences*, 106:49 (2009), 20670–4; A. Missirian and W. Schlenker, 'Asylum applications respond to temperature fluctuations', *Science*, 358:6370 (22/12/2017), 1610–14; H. Buhaug, 'Climate not to blame for African civil wars', *Proceedings of the National Academy of Sciences*, 107:38 (2010), 16477–8; A. Bojanowski, 'Asyl-studie entsetzt wissenschaftler', *Der Spiegel* (22/12/2017).

[9] J. Selby, 'Positivist climate conflict research: a critique', *Geopolitics*, 19:4 (2014), 829–56 provides further detail on these disparate findings. See also I. Salehyan, 'From climate change to conflict? No consensus yet', *Journal of Peace Research*, 45:3 (2008), 315–26; J. Scheffran et al., 'Disentangling the climate–conflict nexus: empirical and theoretical assessment of vulnerabilities and pathways', *Review of European Studies*, 4:5 (2012); O. M. Theisen et al., 'Is climate change a driver of armed conflict?', *Climatic Change*, 117:3 (2013), 613–25; S. M. Hsiang and M. Burke, 'Climate, conflict, and social stability: what does the evidence say?', *Climatic Change*, 123:1 (2013), 39–55; I. Salehyan, 'Climate change and conflict: making sense of disparate findings', *Political Geography*, 14 (2014), 1–5; M. Burke et al., 'Climate and conflict', *Annual Review of Economics*, 7 (2015), 577–617; H. Buhaug, 'Climate–conflict research: some reflections on the way forward', *Wiley Interdisciplinary Reviews: Climate Change*, 6:3 (2015), 269–75; J. Busby, 'Taking stock: the field of climate and security', *Current Climate Change Reports*, 4 (2018), 338–46.

historical, evidential and normative grounds, against attempts to link climate change and security.[10]

The treatment of climate security issues within the reports of the Intergovernmental Panel on Climate Change (IPCC) broadly reflects these disagreements and uncertainties. The IPCC's Third Assessment Report of 2001 conformed clearly to the popular and policy orthodoxy, foreseeing a 'destabilization of international order by environmental refugees' and the 'emergence of conflicts as a result of multiple climate change impacts' – even though little by way of supporting evidence was provided. The IPCC's 2007 report was rather different in tone but still suggested that, in Africa at least, 'climate change may become a contributing factor to conflicts in the future, particularly those concerning resource scarcity, for example, scarcity of water'. By contrast, the IPCC's Fifth Assessment Report of 2014 was both far more strongly evidence-based and much more equivocal, concluding that 'collectively the research does not conclude that there is a strong positive relationship between warming and armed conflict' and that '[c]onfident statements about the effects of future changes in climate on armed conflict are not possible'. While it is sometimes claimed that there now exists agreement that climate is a 'risk factor' in conflict, the measure of agreement is in truth exceedingly thin. Unlike within Western policy and media circles, among researchers there is at present no consensus on the question of climate security.[11]

---

[10] J. Zimmerer, 'Foreword', special issue on climate change, environmental violence and genocide, *International Journal of Human Rights*, 18:3 (2014), 263; G. Dyer, *Climate Wars: The Fight for Survival as the World Overheats* (Oneworld, 2008); J. Mazo, *Climate Conflict: How Global Warming Threatens Security and What to Do About It* (International Institute for Strategic Studies, 2010); H. Welzer, *Climate Wars: What People Will Be Killed for in the Twenty-First Century*, trans. P. Camiller (Polity, 2012); A. Alvarez, *Unstable Ground: Climate Change, Conflict and Genocide* (Rowman and Littlefield, 2017). Discourse critiques are discussed later in this chapter; the evidence on Darfur and Syria is interrogated in Chapter 3.

[11] J. B. Smith et al., 'Vulnerability to climate change and reasons for concern: a synthesis', in J. J. McCarthy et al. (eds.), *Climate Change 2001: Impacts, Adaptation and Vulnerability. Contribution of Working Group II to the Third Assessment Report of the Intergovernmental Panel on Climate Change* (Cambridge University Press, 2001), 950; M. Boko et al., 'Africa', in M. L. Parry et al. (eds.), *Climate Change 2007: Impacts, Adaptation and Vulnerability. Contribution of Working Group II to the Fourth Assessment Report of the Intergovernmental Panel on Climate Change* (Cambridge University Press, 2007), 443; W. N. Adger et al., 'Human security', in C. B. Field et al. (eds.), *Climate Change 2014: Impacts, Adaptation, and Vulnerability. Part A: Global and Sectoral Aspects. Contribution of Working Group II to the Fifth Assessment Report of the Intergovernmental Panel on Climate Change* (Cambridge University Press, 2014), 772–3; R. Nordås and N. P. Gleditsch, 'Climate change and conflict', *Political Geography*, 26:6 (2007), 627–38; R. Nordås and N. P. Gleditsch, 'IPCC and the climate–conflict nexus', paper presented at the International Studies Association annual convention 2009; N. P. Gleditsch and R. Nordås, 'Conflicting messages? The IPCC on conflict and human security', *Political*

Our Approach and Argument in Brief

This book is intended as a contribution to this simultaneously orthodoxy-dominated and scientifically contested intellectual and political terrain. It asks whether the public and policy climate security orthodoxy is well founded or is built instead on foundations of sand. It considers whether climate change– and resource scarcity–induced civil or inter-state conflicts are on the cards or not. It explores what sort of dynamics of conflict, instability and insecurity climate change and the responses to it might bequeath. And it reflects, in passing, on the diverse purposes, interests and agendas served by climate security discourse.

Our approach to these issues is qualitative and loosely comparative, focused on a specific aspect of the climate change challenge as well as on specific geopolitical spaces – and through that, seeking to tease out broader, more general conclusions. Our analysis focuses on water as a particularly crucial site of, and 'intervening variable' in, the claimed climate change–security relationship. It explores these water, climate change and security dynamics in relation to five contemporary 'divided environments': Israel–Palestine, Syria, Cyprus, Sudan–South Sudan and the Lake Chad region. It investigates not just the future but also the past and present links between climate, water and conflict within these five geographical spaces, and seeks to identify and explain similarities and differences across them. It is also theoretically informed, specifically by the tradition of political ecology and by what we label an 'international political ecology' approach to our subject matter, and takes periodic forays into theoretical debates on human–environment relations, conflict and security, international relations and the nature of our global capitalist order. Through all this, the book is intended to serve as a contribution not just to research on climate change and security but also to research on water politics and environmental security, and within the broad field of political ecology.

Our central argument, developed right through the book, is that the conflict and security implications of climate change are very different from those typically imagined within the climate security orthodoxy. Comparing across our five cases, we show that there exists no correlation between environmental resource scarcities on the one hand and water- and climate-related conflicts, vulnerabilities and insecurities on the other. Comparing across time, we show that in many respects the risks of climate, water and environmentally induced chaos are in historical decline and that this dynamic is unlikely to go into reverse in the

Geography, 43 (2014), 82–90; K. Mach et al., 'Climate as a risk factor for armed conflict', *Nature*, 571 (2019), 193–7.

foreseeable future, even under conditions of accelerating human-induced global warming. We demonstrate that while environment-related insecurities, vulnerabilities and conflicts are unfortunately all too real, these are much more determined by political and economic forces and power relations – by processes of state-building, war-making and development – than by environmentally defined resource scarcities, and that this is unlikely to change anytime soon either. We argue, by extension, that the conflict implications of climate change relate less to resource scarcities than to how climate change may transform, contribute to or legitimate new projects of state- and nation-building, development, appropriation and dispossession. We contend that the resource- and environment-centrism of most scholarship on climate security, and within some critical scholarship too, is both analytically and politically problematic. We argue for a very different, political ecology–informed approach to reflecting on the conflict and security implications of climate change – one which both recognises the irreducibly political character of contemporary environmental insecurities and views climate change, and the challenges it poses, as much more than a climatic or environmental problem. And lastly, as a theoretical contribution to political ecology, we stress the importance of international structures and relations within all these processes – adopting, illustrating and arguing for an 'international political ecology' approach to the study of environmental conflict, crisis and insecurity.

We are at risk of getting ahead of ourselves, however. For, before wading in too deep, we need first to explain and justify our approach, methods and premises. The remainder of this introductory chapter seeks to do just that. How, we need to ask, have others sought to investigate the conflict and security implications of climate change? What approaches have they adopted, and how have these approaches fared? What options are available to us? What methods are appropriate? Or, in short, how can we possibly know?

Questions of Method

There can be few objects of analysis where this 'how can we know?' question – this question of method – is more daunting than in the case of the claimed links between climate change and security. For, with the exception of the possible direct connection between high temperatures and aggressive behaviour, any climate change effects on patterns of conflict and instability would only be indirect, mediated via impacts on assorted environmental conditions and socio-economic structures and dynamics. Indeed, there are innumerable possible causal pathways between climate change and security, many of which involve long chains of intervening variables. These hypothetical causal chains would mostly be spatially extremely complex, typically connecting distant locales and operating at multiple scales. With the important exception of impacts from short-term environmental shocks like floods, they would also often be temporally complex, with gradual changes in climatic and environmental conditions potentially taking years, or even decades, to feed through into impacts on patterns of conflict. Questions inevitably arise at each step of claimed causal chains over the extent to which specific effects can be attributed to their hypothesised climate change–induced cause. For example, it has sometimes been suggested that droughts in Russia and China during 2010 were a central cause of the 2011 'Arab Spring', via their impacts on local wheat yields and, in turn, global wheat prices.[12] But this immediately suggests a whole cascade of questions: about whether the 2010 droughts were products of more than natural variability; about the extent to which the increased global wheat prices were caused by these droughts versus other factors (such as increased global biofuels production and the effects of investment strategies on international commodity markets); about whether and how global wheat prices translated into domestic wheat and bread prices within the Arab world; and about whether and how local food price rises contributed to the Tunisian, Egyptian and other uprisings, given their very many other causes. Establishing firm evidence of significant causal connections between climate change and instability is, in short, incredibly challenging.

Making matters more complicated still is the fact that climate security discourse is principally about the future, about threats and risks which are viewed as likely to deepen as the planet warms – but which, in the final analysis, are unknowable and unknown. Even on the direct climatic consequences of human carbon emissions the uncertainties are huge. For, while climate scientists have no doubt about the basic physics of the greenhouse effect, and thus little doubt about the future direction of global temperature changes, beyond this the uncertainties proliferate. In its Fifth Assessment Report, for example, the IPCC concluded that global mean surface temperatures would likely rise by between 0.3°C and 0.7°C during the period 2016–35 (relative to 1986–2005) – but 'likely' was defined here only as a probability of more than 66 per cent, and in any case the IPCC only had 'medium confidence' in this assessment (these uncertainties, it is worth noting, were less a function of different emissions scenarios than of the divergent assumptions embedded in different

---

[12] T. Sternberg, 'Chinese drought, bread and the Arab spring', *Applied Geography*, 34 (2012), 519–24.

climate models).¹³ Moreover, on other issues – precipitation, drought incidence, other extreme weather events and more – the future uncertainties are still greater, as discussed in Chapters 2 and 3.

However, this is nothing when compared to the uncertainties which stalk any analysis of social or political futures. Political forecasts, in particular, often prove spectacularly wrong. Kenneth Waltz, the pre-eminent figure in late twentieth-century International Relations (IR), asked, in his major work of 1979, 'Who is likely to be around 100 years from now – the United States, the Soviet Union, France, Egypt, Thailand, and Uganda? Or Ford, IBM, Shell, Unilever, and Massey-Fergusson?', before answering, 'I would bet on the states, perhaps even on Uganda.' Along with Waltz, the entire discipline of IR failed to predict the end of the Cold War, just as Middle East area studies failed to foresee the Arab Spring, and most economists the global financial crash of 2007.¹⁴ Indeed, political and conflict forecasting is such a hazardous business that most scholarly research on these subjects is historical and interpretive, not predictive. And if this applies to conflict generally, then it applies especially to the conflict and security implications of global climate change – given that, unlike the former, the latter is historically unprecedented.

Broadly speaking, these daunting epistemological challenges have been navigated by climate security researchers in one of five ways. One approach has simply been to ignore them in favour of unreflexively *environment-centric narratives* which read human history – past, present and future alike – as determined by environmental and climatic forces alone. Many of the early defence planning studies on climate security fall into this category, typically portraying the 'fates of societies and civilisations' as 'intimately connected to', and indeed universally shaped by, climate; and on this basis projecting or imagining a future of global warming–driven societal 'collapse'. But a surprising number of academic studies are in essence little different. Raphael Reuveny's finding that climate change–induced migration is likely to cause conflict and 'may foster a fertile atmosphere for global terrorism', for instance, is premised on little more than a list of historical episodes of 'environmental migrations', which is advanced without any analysis of the relative contributions of environmental and non-environmental factors to

---

[13] IPCC, *Climate Change 2013: The Physical Science Basis. Contribution of Working Group I to the Fifth Assessment Report of the Intergovernmental Panel on Climate Change* (Cambridge University Press, 2013), 20, 36, 956.

[14] K. Waltz, *Theory of International Politics* (Addison-Wesley, 1979), 95; R. N. Lebow and T. Risse-Kappen (eds.), *International Relations Theory and the End of the Cold War* (Columbia University Press, 1996); G. Gause III, 'Why Middle East Studies missed the Arab Spring: the myth of authoritarian stability', *Foreign Affairs*, 20 (2014); P. Krugman, 'How did economists get it so wrong?', *New York Times* (06/09/2009).

the migrations in question – a gaping analytical flaw which did not prevent publication in the first major peer-reviewed collection on the subject. Equally, Jürgen Zimmerer's recent assertions that 'environmental violence is among the main driving forces of collective violence', and that 'climate change will dramatically increase the likelihood of genocide' and probably be the 'biggest trigger of genocide in the twenty-first century' are simply that: ungrounded assertions, devoid of any analysis of the non-environmental causes of violence, or of the social, economic and political pathways through which genocide might be 'triggered' by climate change. We will consider such eco-determinist reasoning more fully in Chapter 2. But suffice to say for now that just as human history has not been wholly determined by environmental factors, so humankind's future cannot be 'reduced to climate' either.[15] Environment-centric narratives may proffer superficially powerful accounts of looming climate chaos, but they achieve this only through the most blatant and untenable reductionism – to the extent that they are of negligible scientific value, and little value either as guides to the future.

A related approach – which we label *methodological environment–centrism* – involves focusing on environmental pathways to conflict not as an ontological position, as in the above examples, but rather for reasons of method. The classic exemplar here is the 'process-tracing' approach to environmental security taken by Thomas Homer-Dixon, the pre-eminent theorist of the subject, which aims only to identify causal linkages between environmental and conflict variables, and explicitly repudiates the goal of explaining conflict, or particular conflicts, overall. Similar, though, is the approach taken by the Washington DC–based think tank the Center for Climate and Security (CCS), which focuses on identifying and highlighting climatic causes of conflict, while simultaneously acknowledging that there are always innumerable other 'contributory factors' at play. Now, this is perfectly reasonable, in so far as it goes. However, what it is important to recognise is that this approach provides no logical basis for drawing conclusions about the importance of climatic and environmental factors and pathways, relative to other political, economic and social causes of conflict – for knowing whether climate change is a 'significant factor' behind conflicts, or just one of a thousand or even a million others. Moreover, in practice there is regular slippage between the methodological and ontological variants of environment-centrism. Homer-Dixon,

---

[15] Mazo, *Climate Conflict*, 43; R. Reuveny, 'Climate change-induced migration and violent conflict', *Political Geography*, 26:6 (2007), 656–73; J. Zimmerer, 'Climate change, environmental violence and genocide', *International Journal of Human Rights*, 18:3 (2014), 265; Zimmerer, 'Foreword', 263; M. Hulme, 'Reducing the future to climate: a story of climate determinism and reductionism', *Osiris*, 26:1 (2011), 245–66.

for example, has argued that the frequency of violent conflicts 'will probably jump sharply ... as scarcities rapidly worsen' and that climate change will produce 'insurgencies, genocide, guerrilla attacks, gang warfare and global terrorism', even though his methodology provides no basis for these conclusions, and even though he insists on 'the impossibility of discriminating among the relative power of causes'. Equally, for all its claimed sensitivity to non-environmental factors, CCS regularly lurches into climate reductionism.[16] Although typically presented as departing from environmental determinism, this approach really does not deviate from it – and indeed provides no resources whatsoever, let alone a coherent analytical or methodological framework, for examining the likely significance of climate change within the overall landscape of twenty-first-century global politics.

A very different approach has been pursued within *large-N quantitative studies*: here the aim has been to test for statistical correlations between particular historical environmental and conflict variables, and on this basis to draw inferences about the conflict implications of global climate change specifically. A large body of such quantitative work has been produced over the last decade or so, as already indicated. Building upon a large and ever-expanding body of quantitative research on conflict, this research has undoubtedly brought methodological rigour to the study of climate security, especially when compared to the narrative approaches discussed above. And crucially, it does not 'reduce the future to climate', instead providing a method for exploring the difference that climatic variations make to patterns of economy, politics, society and conflict, without assuming that climatic or environmental variables are their sole or primary determinants.

However, this quantitative research programme has major shortcomings. First, many of the historical correlations identified within it are highly questionable, shaped as much by unreliable and frequently contradictory datasets, and by arbitrary or untenable modelling and data boundary assumptions, as by anything else. Consider the two quantitative

---

[16] T. Homer-Dixon 'Strategies for studying causation in complex ecological–political systems', *The Journal of Environment and Development*, 5:2 (1996), 132–48; C. Werrell and F. Femia, 'Let's not say climate change causes war. But let's not also ignore the real security risks', Center for Climate and Security blog (05/12/2015), https://climateandsecurity.org/2015/12/lets-not-say-climate-change-causes-war-but-lets-also-not-ignore-the-real-risks/; T. Homer-Dixon, 'Environmental scarcities and violent conflicts: evidence from cases', *International Security*, 19:1 (1994), 39; T. Homer-Dixon, 'Terror in the weather forecast', *New York Times* (27/04/2007); T. Homer-Dixon, 'Cause and effect', Making Sense of Sudan blog (02/08/2007), https://africanarguments.org/2007/08/cause-and-effect/; J. Selby et al., 'Climate change and the Syrian civil war revisited', *Political Geography*, 60 (2017), 232–44; J. Selby et al., 'Climate change and the Syrian civil war revisited: a rejoinder', *Political Geography*, 60 (2017), 253–5.

findings mentioned briefly at the beginning of this chapter. Burke and colleagues' prediction that climate change may cause 393,000 extra battle deaths in Africa by 2030 was rooted in a finding that, between 1981 and 2002, a 1°C temperature rise in Africa was associated with a 4.5 per cent increase in the incidence of civil war. Yet this claimed relationship only holds true for the 1981–2002 period, not post-2000, as the authors themselves acknowledged; and their 1981–2002 findings are highly dependent on just six historical conflicts, all of which were sparked by foreign interventions which could not have been caused by local temperature anomalies (discounting these specific conflicts makes their claimed temperature–civil war relationship all but disappear). Equally, the coming increase in asylum applications projected by Missirian and Schlenker is rooted in claimed associations between countries with an 'optimal temperature range for agriculture', defined as around 20°C, and asylum applications received by the EU between 2000 and 2014. However, this finding is disproportionately affected by a small number of states, most notably Iraq, where the suggested link between asylum applications and the climate is utterly coincidental (Iraqi applications rose between 2004 and 2007 and fell between 2010 and 2013, with temperatures following a similar pattern – but there is no reason to think that the former pattern was driven by the latter, rather than by the US-led invasion and its aftermath). Moreover, their statistical finding is essentially that most migration to the EU is from relatively hot countries – a fact which, though no doubt true, tells us nothing about the causes of this migration, or about whether and to what extent future temperature increases will lead it to increase. Altogether, this suggests that the correlations identified within quantitative climate conflict research are often little more than statistical artefacts.[17]

It is regularly suggested that more research – more fine-grained data and methods! – will ultimately bring greater clarity. But fifteen years of quantitative climate security scholarship have brought nothing of the sort, as we have seen. And conflict research as a whole arguably provides far fewer firm answers than is often thought or implied.[18] Given this and the data and modelling problems touched on above, as well as the huge

---

[17] Burke et al., 'Warming increases the risk of civil war'; M. Burke et al., 'Climate robustly linked to African civil war', *Proceedings of the National Academy of Sciences*, 107:51 (2010), E185; H. Buhaug et al., *Sensitivity Analysis of Climate Variability and Civil War* (PRIO, 2010); Missirian and Schlenker, 'Asylum applications', Fig. S5, Table S8; Bojanowski, 'Asyl-studie entsetzt wissenschaftler'; J. Selby and G. Daoust, *Rapid Evidence Assessment on the Impacts of Climate Change on Migration Patterns* (UK Foreign, Commonwealth and Development Office, 2021); Selby, 'Positivist climate conflict research'.
[18] Nordås and Gleditsch, 'Climate change and conflict'; Buhaug, 'Climate–conflict research'; H. Hegre and N. Sambanis, 'Sensitivity analysis of empirical results on civil

number of causal pathways through which environmental changes might conceivably affect economic, political and presumably conflict dynamics, it seems unlikely that a consensus will ever emerge on the correlates of climate and conflict, let alone a clear causality.

Moreover, even if such a consensus did emerge it is doubtful how much this would tell us about the conflict implications of global anthropogenic climate change. The latter is a historically unprecedented development which, in addition to promising worldwide changes in temperature and rainfall patterns, poses profound challenges to contemporary modes of economic reproduction, social organisation and political power, requires a wide range of policy responses, and is already a subject of fierce political disagreement. Global climate change is a social–political as much as an environmental phenomenon. We therefore need to ask: does it really make sense to develop predictions about the conflict and security implications of this unprecedented human-induced global challenge through positivist-quantitative analyses of historical variations in the weather?[19] Our assessment is that it does not.

At the other end of the epistemological spectrum from this quantitative work, the approach taken with *discourse critiques* has been to express concern about – and often to completely reject – climate security narratives on the grounds of both the purposes and agendas assumed to be motivating them and their anticipated political implications and consequences. From a diverse range of theoretical starting points – constructivist, post-structuralist and post-colonial, above all – the common premise of these critiques has been that narratives are not merely attempts to represent reality, but are also interested in and productive of it. So viewed, climate chaos 'imaginaries' and 'securitisations' are troubling and potentially dangerous, whether because they reinforce colonial or Northern stereotypes, divert attention from, and culpability for, more directly human causes of conflict, or produce new rationales for resource expropriation, state policing or external military intervention.[20] Seen thus,

---

war onset', *Journal of Conflict Resolution*, 50:4 (2006), 508–35; N. Sambanis, 'What is a civil war? conceptual and empirical complexities of an operational definition', *Journal of Conflict Resolution* 48:6 (2004) 814–58; C. Cramer, 'Homo economicus goes to war: methodological individualism, rational choice and the political economy of war', *World Development*, 30:11 (2002), 1845–64.

[19] N. P. Gleditsch, 'Whither the weather? Climate change and conflict', *Journal of Peace Research*, 49:1 (2012), 7.

[20] M. J. Trombetta, 'Environmental security and climate change: analysing the discourse', *Cambridge Review of International Affairs*, 21:4 (2009), 585–602; B. Hartmann, 'Rethinking climate refugees and climate conflict: rhetoric, reality, and the politics of policy discourse', *Journal of International Development*, 22:2 (2010), 233–46; M. Carr, 'Slouching towards dystopia: the new military futurism', *Race and Class*, 51:3 (2010), 13–32; McDonald, 'Discourses of climate security'; E. Swyngedouw, 'Apocalypse now!

climate security narratives are not neutral representations but instruments of, or unwitting participants in, operations of power.

In our view, such perspectives have much to commend them, sharing what is best in quantitative analyses – above all, their anti-reductionism and, in their own way, methodological rigour – while also departing from them in important respects. Discourse critiques are in our view rightly suspicious of easy policy consensuses, as well as the commitment to prediction which characterises so much climate security talk. The focus on the ways in which discursive and policy responses to climate change may themselves contribute to or legitimise appropriation, interventions and violence is also, in our view, important – a useful counter to quantitative and environment-centric researchers' exclusive focus on climate impacts. Discourse critiques are not just critiques, but provide tools for analysing the causal and constitutive impacts of climate change–related representations and narratives on patterns of politics, conflict and insecurity – something which quantitative methods have no resources to do.

Yet for all this, there are undoubted limitations to such discourse-centrism. By understanding climate security primarily through the prism of discourse, discourse analyses effectively assign the environment and climate change only secondary roles in analysis. In Latourian terms, they fail to approach environment and society 'symmetrically', being comfortable exploring how the former is represented and constituted by the latter, but shying away from discussion of the reverse.[21] Moreover, the essentially critical orientation of discourse critiques means that, though they have much to say on the limitations of climate security talk, they offer only the most limited explanatory or predictive assessments. Discourse critiques ultimately provide few answers to the question of what role climate change might play in twenty-first-century world order and global security – other than to insist, however rightly, that any impacts will be mediated via discourse.

The above discussion suggests a series of methodological requirements: for a method that is sensitive to both the material and the discursive dimensions of climate security; that considers not just causal pathways between climate and instability, but also how these pathways intersect with other 'non-environmental' factors; that considers the

---

Fear and doomsday pleasures', *Capitalism Nature Socialism*, 24:1 (2013), 9–18; G. Bettini, 'Climate barbarians at the gate? A critique of apocalyptic narratives on "climate refugees"', *Geoforum*, 45 (2013), 63–72; B. Hartmann, 'Converging on disaster: climate security and the Malthusian anticipatory regime for Africa', *Geopolitics*, 19:4 (2014), 757–83; J. Warner and I. Boas, 'Securitization of climate change: how invoking global dangers for instrumental ends can backfire', *Environment and Planning C: Politics and Space*, 37:8 (2019), 1471–88.

[21] B. Latour, *We Have Never Been Modern* (Harvard University Press, 1993).

conflict and security implications of both global climatic and associated environmental changes, and the social and political responses to them; and that provides some sort of basis, however qualified, for both generalisation and prediction. Our premise is that the *political ecology* tradition can furnish just such a method.

An International Political Ecology

Political ecology, in Piers Blaikie and Harold Brookfield's oft-used if schematic formulation, 'combines the concerns of ecology and a broadly defined political economy'. Its central thesis, it may be said, is that ecological transformations and crises are always political in both cause and consequence. Less schematically, however, political ecology combines these twin concerns not just by analysing their interactions – by treating 'politics' and 'ecology' as connected but essentially separate spheres – but by approaching them as a dialectical unity, wherein ecology is viewed as internal to society, and politics and power are simultaneously understood as internal to all major contemporary ecological transformations and crises. The underlying premise here is that the very distinction between non-human 'ecology' and the 'environment' on the one hand and human 'politics' on the other is – like the distinction between 'politics' and 'economics' – a modern epistemological construction which can obscure as much as it reveals. There of course exist countless natural environmental objects and processes which predate and have not been created by humans, from the natural course of the Amazon River to the atmospheric heat-trapping properties of CO_2. Yet 'nature' no longer exists as an asocial domain separate from human praxis; and, conversely, 'politics' is always rooted, however indirectly, in the exploitation, transformation, circulation and control of the fruits of the Earth. Political ecology, as we understand it, is thus not merely concerned with the political causes and consequences of environmental change, but is the study of the 'metabolic relations' – the patterned and uneven flows of commodities, capital, carbon, bodies, ideas, waste and more – through which both modern political life and our planet's socialised nature are constituted.[22]

While many different approaches to political ecology have been articulated ('regional', 'feminist', 'Third World', 'critical' and so on), taken as

---

[22] P. Blaikie and H. Brookfield (eds.), *Land Degradation and Society* (Routledge, 1987), 17; P. Warde et al., *The Environment: A History of the Idea* (John Hopkins University Press, 2018); E. M. Wood, 'The separation of the economic and the political in capitalism', *New Left Review*, 127 (1981), 66–95; K. Marx, *Capital: A Critique of Political Economy, Vol. I*, trans. B. Fowkes (Penguin, 1990 [1867]), 283.

a whole political ecology has principally been inspired by historical materialism, with important additional influences coming from Foucault, new materialism and post-colonialism.[23] Our approach in this book can be characterised in these terms too. Hence our analysis is materialist, focused on the extraction, destruction, transformation, distribution and appropriation of nature through human and mechanical labour, and on those social relations, including class, racial, gender and other hierarchies, which are the corollaries of these socio-ecological processes. Our approach is also historical, attentive to both general historical patterns, trends, ruptures and continuities – from legacies of colonialism to ever-rising greenhouse gas emissions – and the specific historical dynamics through which these processes have been articulated at particular times and in particular places. We view the contemporary global social order as, in essence, capitalist – that is, as a system of generalised market dependence where all are compelled to enter the market and where the demands of competition and profit-maximisation dictate relentless expansion, commodification, technological innovation and the unprecedented exploitation and degradation of nature. As per the Marxist tradition, we approach economic development as an inherently political, conflict-ridden and often violent process, involving complex admixtures of accumulation and dispossession, incorporation and marginalisation. We view the state both as a key agent of, and as a crucial arena for contestation over, these developmental and distributive processes. Along the lines of the discourse analyses discussed above, we analyse representations as inherently political – as both politically interested and politically consequential – especially through the support that they have historically lent to colonial and modern state power. And in keeping with most research in political ecology, our analysis is rooted in critical realist philosophical premises.[24]

Understood thus, what ultimately distinguishes political ecology from the other approaches discussed above is its anti-reductionism. Unlike environment-centrism, political ecology does not reduce the future to pressures emanating from environmental change. It takes nature and the environment seriously, including by emphasising the effectivity and power – or what actor network theorists term, however hyperbolically, the

---

[23] Blaikie and Brookfield, *Land Degradation*; D. Rocheleau et al. (eds.), *Feminist Political Ecology: Global Issues and Local Experiences* (Routledge, 1996); R. Bryant and S. Bailey, *Third World Political Ecology: An Introduction* (Routledge, 1997); T. Forsyth, *Critical Political Ecology: The Politics of Environmental Science* (Routledge, 2003).

[24] A. Sayer, *Method in Social Science: A Realist Approach*, 2nd ed. (Routledge, 1992); E. M. Wood, *The Origin of Capitalism: A Longer View* (Verso, 2002); J. B. Foster et al., *The Ecological Rift: Capitalism's War on the Earth* (Monthly Review Press, 2010); A. Malm, *The Progress of This Storm: Nature and Society in a Warming World* (Verso, 2018).

'agency' – of such things as droughts, floods, groundwater depletion, mosquitoes and rising atmospheric carbon. But it locates all such environmental objects and developments in socio-historical context, exploring how they have been generated through specific histories and political economies, and insisting that, just as the past has not been shaped by ecological forces alone, so the future will not be either. Equally, by contrast with discourse analyses, political ecology does not reduce social relations to representation. It freely acknowledges the historical and likely future importance of specific (environmental and social) narratives in the making and remaking of polities and societies. But it nonetheless insists that 'social life is essentially practical', shaped by work, energy and material exchanges and capabilities, and that representations are only as influential as socio-political circumstances permit.[25] Moreover, unlike quantitative analyses, political ecology does not reduce social life to numerals, let alone to correlations or regressions. It makes frequent use of statistical evidence, as we also do in the chapters to follow in this book. But it nonetheless insists that quantitative models cannot capture the irreducible complexity, historicity or spatial specificity of environment–politics relations, and that a much more qualitative form of analysis is therefore required – including when reflecting on the future.

So understood, political ecology–informed research has already made a broad range of contributions to understanding the conflict and security implications of climate change. Political ecologists have provided close ethnographic studies of how communities manage and respond, including politically, in the face of climatic variability and vulnerability and environmental change. They have advanced direct critiques of supposedly textbook cases of climate-induced conflict. They have explored the already significant conflict and security impacts of government-led climate adaptation and mitigation strategies – of what in broader terms is often characterised as 'green grabbing' – including the global neo-liberal political and economic structures implicated in them. In a reversal of the standard focus of climate security research, they have also interrogated the role of security institutions – most notably the US military – in the making of both climate change and responses to it. Certain eco-Marxist scholars have sought to imagine, and advocate, a future in which revolutionary global political conflict and change, including perhaps the establishment of green dictatorships, becomes necessary to resolving the climate crisis (as well as providing a route to a more equal and less exploitative non-capitalist world order). At the boundary between the critical geopolitics and political ecology traditions, others have sought to

[25] K. Marx, 'Theses on Feuerbach', in *Early Writings*, ed. L. Colletti (Penguin, 1975), 422.

rethink the nature and meaning of 'security' under conditions of the Anthropocene. And beyond climate security specifically, political ecology-informed research has had a tremendous amount to say on the politics of natural resources, and on the patterns of insecurity, marginalisation and violence relating to them – much of which is directly relevant to thinking through the conflict and security implications of planetary heating.[26]

The present volume seeks to contribute to this existing body of political ecology research in two main ways. On the one hand, substantively, we seek simply to offer a holistic and multi-case study informed analysis of the security implications of climate change from a political ecology perspective – something which, for all the wealth of research on the subject, has not yet been attempted. But in addition, theoretically, we adopt and seek to suggest the broad contours of a new approach to political ecology, in which consideration of 'the international' is appropriately integrated into its core subject matter – what we conceive of as an 'international political ecology' approach to the study of socio-ecological crisis.

By way of explanation: political ecology as a field and approach is essentially an offshoot of research in human geography and anthropology; inevitably it thus reflects these two disciplines' priorities as well as oversights. Its hallmarks have long been an attentiveness to the local – as captured, for instance, in Piers Blaikie's characterisation of political ecology as inherently 'place-based' in focus and method – combined with a critical attitude towards the state, and an implicit dependency or world systems theory–informed understanding of how environmental crises and vulnerabilities are shaped by worldwide capitalist structures and imperatives. By contrast, an 'international political ecology', as we conceive it, should approach the international as a key constitutive feature of modern

---

[26] T. A. Benjaminsen et al., 'Does climate change drive land-use conflicts in the Sahel?', *Journal of Peace Research*, 49:1 (2012), 97–111; T. A. Benjaminsen and B. Ba, 'Why do pastoralists in Mali join jihadist groups? A political ecological explanation', *Journal of Peasant Studies*, 46:1 (2019), 1–20; H. Verhoeven, 'Climate change, conflict and development in Sudan: global neo-Malthusian narratives and local power struggles', *Development and Change*, 42:3 (2011), 679–707; J. Fairhead et al., 'Green grabbing: a new appropriation of nature', *Journal of Peasant Studies*, 39:2 (2012), 237–61; P. Bigger and B. D. Neimark, 'Weaponizing nature: the geopolitical ecology of the US Navy's biofuels program', *Political Geography*, 60 (2017), 13–22; O. Belcher et al., 'Hidden carbon costs of "everywhere war": logistics, geopolitical ecology, and the carbon bootprint of the US military', *Transactions of the Institute of British Geographers*, 45:1 (2020), 65–80; G. Mann and J. Wainwright, *Climate Leviathan: A Political Theory of Our Planetary Future* (Verso, 2018); S. Dalby, *Security and Environmental Change* (Polity, 2009); S. Dalby, *Anthropocene Geopolitics: Globalization, Security, Sustainability* (University of Ottawa Press, 2020); P. Le Billon, 'The political ecology of war: natural resources and armed conflicts', *Political Geography*, 20:5 (2001), 561–84; N. Peluso and M. Watts (eds.), *Violent Environments* (Cornell University Press, 2001).

world politics and, in turn, as a primary cause of environmental degradation and environment-related vulnerabilities. The idea of the 'international', here, points towards a number of things: to what Justin Rosenberg has theorised as the coexistence of a 'multiplicity' of interacting states and societies; to the importance of those borders and borderland spaces separating them; to the political and economic relations – of competition, emulation, alliance-building, colonisation, war-making and more – between them; to the specificity of different national economic, political and developmental capacities and trajectories; and, not least, to the constitutive impacts of international relations on 'internal' social processes.[27] Our premise is that these features of the international are, both individually and in combination, crucial to understanding contemporary patterns of environment-related conflict, security and insecurity.

To be clear, our proposition is not that the international is the scalar category above all others, or even the most important among them; global capitalist structures and globalising processes exist too, as do North–South hierarchies, local specificities and multiple, intersecting axes of division including along lines of race, gender, class, language, region and religion. Unlike within much IR theory, we do not wish to privilege the international – but instead, merely to bring it more fully back into the analytical mix. Likewise, we do not view the international as a transhistorical and asocial realm, but rather as a modern social and historical construct; we consider this true both of its general form – organised around the supposedly Westphalian principles of absolute sovereignty and inter-state anarchy – and of the particular states, or 'nation-states', which currently constitute it. We thus seek to explore the role of the international both historically and intersectionally, that is, with a view to how international divisions and hierarchies intersect with other axes of domination and subordination, and in recognition of their complex, cumulative impacts on both ecology and society.[28]

An international political ecology, we suggest, should attend both to the consequences of international relations for the environment and

---

[27] P. Blaikie, *The Political Economy of Soil Erosion in Developing Countries* (Longman, 1985), ch. 5; Blaikie and Brookfield, *Land Degradation*; Rocheleau et al., *Feminist Political Ecology*; R. Peet et al. (eds.), *Global Political Ecology* (Routledge, 2011); J. Rosenberg, 'Basic problems in the theory of uneven and combined development. Part II: unevenness and political multiplicity', *Cambridge Review of International Affairs*, 20:1 (2010), 165–89; J. Rosenberg, 'International Relations in the prison of political science', *International Relations*, 30:2 (2016), 127–53.

[28] B. Teschke, *The Myth of 1648: Class, Geopolitics and the Making of Modern International Relations* (Verso, 2003); K. Crenshaw, 'Demarginalizing the intersection of race and sex: a black feminist critique of antidiscrimination doctrine, feminist theory and antiracist politics', *University of Chicago Legal Forum*, 1 (1989), 139–67; F. Sultana, 'Political ecology 1: from margins to center', *Progress in Human Geography*, 45:1 (2021), 156–65.

environment-related insecurities – to the consequences of multiplicity, boundaries, competition, hierarchies, enmities, alliances and nationally specific development pathways – and, conversely, to the diverse ways in which the appropriation, transformation and circulation of nature and its resources is complicit in processes of nation-building and state-building and the constitution of geopolitical orders. It should seek to do this while recognising that the international is but one important scale or dimension of politics among others. Indeed, if anything, an international political ecology should valorise neither the 'international' nor 'politics' nor even 'ecology' but instead history, by exploring the historical transformations which have been visited on both politics and nature under conditions of capitalist modernity – and the implications of these historical transformations for thinking about the future. While a hard ask, it is such an approach that we seek to adopt in the pages to follow.

Climate Change, Water and (In)Security

How, though, might these abstract formulations be translated into a realisable project? As already discussed, the question of the links between climate change and conflict and security poses acute methodological challenges, relating principally to the huge number of possible causal pathways between the two, and to the inherent unpredictability of many aspects of social and political life, particularly historically unprecedented ones. Moreover, whatever we might wish, a commitment to political ecology does not magic away these methodological challenges. Hence at minimum we require both a more limited and precise focus or object of analysis – since 'climate change and security' in general is way too fuzzy and unwieldy a topic – and a political ecology-appropriate method for studying the future.

On the first requirement our answer is water – or more precisely the complex, multi-directional relations between climate change, water and (in)security. There are two reasons why water serves as such an ideal focus for us. In the first place, no other environmental resource has been so regularly identified within climate security discourse as a likely factor – a potential 'intervening variable' – in climate change–related instability. The quotations cited at the beginning of the chapter from Barack Obama, Margaret Beckett and Ban Ki-moon are all illustrative of this, with their common emphases on how climate change–induced water shocks and scarcities will increase levels of competition, displacement and, in turn, social, political and violent conflict. Mass media coverage suggests likewise, focusing as it so often does on a simultaneously climate- and water-related phenomenon, drought. Moreover, academic research points in a similar direction: IPCC assessment reports have focused on climate change–related 'water wars' more than any other issue, while within recent quantitative climate security research, impacts on and through rainfall have been one of two main focuses (the other being impacts on and through temperature variations). One retired US admiral even goes so far as to claim that 'from a national security perspective, climate change is all about the water'. While this is overstated, if global climate change is to become a 'threat multiplier' for instability then water will undoubtedly be central to this.[29]

Second – and turning to the second main methodological challenge noted above – water in many ways provides an ideal analogue for exploring the future conflict and security implications of climate change. Recall that quantitative-positivist methods are distinctly limited as a tool for understanding the conflict and security implications of global anthropogenic climate change, simply because the latter has never occurred before; there exists no database of prior instances of it that might provide a basis for inductive reasoning or prediction. Hence a predictively inclined student of climate security has no option but to reason by analogy – that is, to identify and analyse appropriate historical or contemporary analogues, and then to apply any findings about them to climate security specifically. Though rarely acknowledged, this is precisely what most quantitative climate conflict researchers do: they treat short-term weather variations as analogues for human-induced global warming and on this basis apply findings about the former to the latter. Likewise, if we, as authors, want to advance predictions, however qualified, about the future security implications of climate change, we have no option but to engage in historical, analogical reasoning. Hence the question which faces us is this: given that there are so many conceivable analogues for climate chaos – not just weather changes, but everything from weapons of mass destruction and international terrorism through to the end of the transatlantic slave trade – what should our analogue be?[30]

Our answer is water mainly because of the striking parallels that exist between the contemporary climate security orthodoxy and long-established narratives around water and conflict. Water has long been regarded within many liberal policy and academic circles as a likely contributor to instability and violence, just as climate change is today (we

---

[29] Nordås and Gleditsch, 'IPCC and the climate-conflict nexus'; Selby, 'Positivist climate conflict research', 832; D. Titley, 'Global warming a threat to national security', *Cognoscenti* (20/02/2013).

[30] Selby, 'Positivist climate conflict research', 840–5; W. Nuttall and D. Manz, 'A new energy security paradigm for the twenty-first century', *Technological Forecasting and Social Change*, 75:8 (2008), 1256–7.

detail this further in Chapter 2). Indeed, 'water wars' have long been held as paradigmatic within environmental security thinking: river water is 'the renewable resource most likely to stimulate interstate war', claimed Homer-Dixon in his major work on the subject.[31] Many of the standard features of water security discourse – the language used, the theories deployed and the predictions made – now recur within contemporary climate security thinking. Many of the major proponents are also the same. In addition, water politics has been extensively investigated – probably more so than any other resource politics issue, and from a wide variety of different perspectives – thus furnishing us with a hefty weight of empirical and historical material for analysis and critique. For all these reasons, the politics of water provides an ideal analogue for, and study aid to, exploring the conflict and security implications of climate change.

Two final points need to be made about the thematic scope of the analysis to follow. First, our overall focus is the connections between climate change, water and *(in)security*, particularly but not limited to insecurities associated with *conflict*. We need to be clear about these terms. We use the term 'conflict' to denote both its political and violent forms, not restricting our analysis to armed conflict alone; we consider the term to apply to the full range of scales, from the international to the household, though in practice we concentrate mainly on large-scale intergroup conflicts for reasons set out below. As for 'insecurity', we use this as a loose umbrella for the diverse forms of suffering, pain, vulnerability, marginalisation, displacement, dispossession, denial of basic needs and violence, and/or risks to this effect, that are experienced or faced by individuals and collectivities; as we understand it, this includes everything from direct and immediate physical harms to what Johann Galtung called 'structural violence' and what Rob Nixon characterises as 'slow violence', acts of harm that can take years or decades to unfold. By the same token, we understand 'security' as referring, at least on one level, to freedom from such threats and harms, 'security' in this sense being a normative good and aspiration. Yet we also assume, to complicate matters, that actions taken in the name of security – against some particular threat or to protect some specific group of people – often create insecurities, whether for those outside the group being secured or indeed for those within it; and that 'security' and 'insecurity' are thus transitive phenomena, things which people do to each other. We thus attend to many different forms and causes of water-related conflict and insecurity. We analyse discourses and policies in pursuit of 'water security' and 'climate

---

[31] T. Homer-Dixon, *Environment, Scarcity and Violence* (Princeton University Press, 1999), 179.

security' with an eye to how they may themselves contribute to insecurities. And, above all, we approach this whole terrain as materialists, our central object of analysis being human suffering and conflict, rather than – as in constructivist critical security studies – security politics understood as arising from and centring on speech acts.[32]

Second, we seek to analyse climate change, water, conflict and (in)security alike in a holistic and symmetrical fashion. Thus we consider not only the conflict and security implications of climate change's environmental impacts, but also the conflict implications of adaptation and mitigation efforts; and we examine not just climate change's material impacts, but also those associated with discourse on it. Similarly, on water, while we do attend to the politics surrounding major named water resources, our overall approach is to explore its conflict and security implications right across the 'hydro-social cycle' – from rain, swamps and desert margins, right through to dams, pipelines, agro-industry, metering and sewage.[33] And we examine not only how water and climate change matter for conflict and security but also, conversely, how the latter – and more broadly politics – determine patterns of water- and climate-related degradation, distress and vulnerability. Our overall objective, in sum, is to investigate the relations between climate change, water and (in)security in their full multi-directional complexity, including by exploring the socio-ecological processes through which each is internally and dialectically constituted. Such an approach not only follows from the political ecology premises outlined above; in our view, it is also a *sine qua non* for developing a rounded assessment of the conflict and security implications of global climate change.

Five Divided Environments

Our empirical strategy for exploring these complex relations between water, climate change and (in)security centres on particular political geographical spaces – each of which have long histories of both ethno-nationalist violence and division, and conflict or insecurity relating to environmental resources. These spaces are 'divided environments' in two senses: each of them is politically and territorially divided; and their

---

[32] J. Galtung, 'Violence, peace, and peace research', *Journal of Peace Research*, 6:3 (1969), 167–91; R. Nixon, *Slow Violence and the Environmentalism of the Poor* (Harvard University Press, 2011); K. Booth, 'Security and emancipation', *Review of International Studies*, 17:4 (1991), 313–26; B. Buzan et al., *Security: A New Framework for Analysis* (Lynne Rienner, 1998).

[33] J. Linton and J. Budds, 'The hydrosocial cycle: defining and mobilizing a relational-dialectical approach to water', *Geoforum*, 57 (2014), 170–80.

'natural environments', most notably water, have been repeatedly contested and divided as well. The divided environments in question are Israel–Palestine, Syria, Cyprus, Sudan–South Sudan and the Lake Chad region.

These five cases have been selected with a view to both commonalities and differences. All five are sites of significant political conflict structured around rival ethno-national identities and militarised state projects. All five have experienced large-scale internal or inter-societal violence and at least one full-scale war – Israel–Palestine and Sudan have been sites of repeated wars; Syria, at the time of writing, is in the tenth year of its horrific conflict; and the Lake Chad crisis has lasted even longer – as well as forced displacement, internal colonisation and a range of human rights abuses. All five are territorially divided in one way or another, whether as a result of war (Israel–Palestine, Cyprus, Syria), de jure post-colonial partition (Sudan) or arbitrary boundary-drawing by colonial powers (Lake Chad). They are geographically connected in a giant arc from the Mediterranean shore of West Asia to northern Africa and the Sahel. Their contemporary politics still bears the scars of either British or French colonial rule. Much of their areas comprise arid or semi-arid drylands. And each of them is home to severe, though contrasting, water problems – some of which have been sources or sites of political discord, and many of which are predicted to worsen as the planet warms. Our cases touch on some of the rivers that feature most regularly in water wars discourse (the Euphrates, the Jordan and the Nile) as well as the three most oft-cited examples of supposed climate change-induced armed conflict (Darfur, Syria and Lake Chad). The commonalities between our cases cut across history, geography, hydrology and politics.

In other respects, however, it is the differences between these five divided environments that are the more striking. They are hugely different in scale: Cyprus may be the third largest island in the Mediterranean, but on a clear day its entire coastline can be scanned from Mount Olympus in the Troodos Mountains; by contrast, Sudan pre-2011 was roughly the size of western Europe, with a territory of almost 2.5 million km^2, 268 times that of Cyprus. Our cases are socio-economically diverse: Israel has a highly educated society and a high-tech dominated economy, ranked 22nd, just above Korea, in the most recent Human Development Index; while, for all their oil exports, South Sudan and Chad are ranked 186th and 187th respectively, near the foot of this global league. They vary enormously in state power and capacity – from Israel, with its nuclear arsenal and panoptic administrative and surveillance systems, to South Sudan, which was dubbed a 'failed state' on independence in 2011, only to regress from that beginning into on–off civil war. They vary in their types of government – from the Republic of Cyprus' EU member parliamentary democracy, through to Israel's 'ethnic democracy' (or 'ethnocracy') and the military–authoritarian regimes of Damascus, N'Djamena and Khartoum.[34] They vary in the nature of their 'dividedness' – with the majority of our cases having once been unitary political territories but the Lake Chad region not, lying at the junction of Chad, Cameroon, Nigeria and Niger. They cross climatic and ecological zones – from the Mediterranean climates of Cyprus, western Syria and Israel–Palestine, through to the desert lands of the Negev, eastern Syria and the Sahara, to the Sahel and tropical savannah further south. And most important in the context of this book, our five cases are characterised by a great diversity of water politics and conflicts.

The value of exploring such diverse cases lies partly in providing a basis for generalisation, as per John Stuart Mill's 'method of difference'.[35] That said, this book is comparative only in the loosest of senses: it is organised thematically rather than around case studies, and the individual thematic chapters typically discuss only some of the cases in any detail (either because the theme in question does not speak particularly to the case, or thanks to a surfeit of empirical material). As such, while our analysis includes detailed and original readings of particular aspects of Israeli–Palestinian, Syrian, Cypriot, Sudanese and Lake Chad water and climate politics, it is not, and does not seek to be, comprehensive in its treatment of these cases.

There are, we acknowledge, possible shortcomings to this empirical strategy and selection of cases. For one, a focus on conflict-ridden and violent contexts rather than peaceful ones has sometimes been criticised by climate security researchers as involving a dependent variable sampling bias, wherein the links between climate and instability are systematically overstated out of inattentiveness to the 'vastly more ubiquitous and continuing condition of peace'.[36] Similarly, some may wonder whether a focus on large-scale ethno-national divisions is appropriate given that, as even mainstream environmental security discourse recognises, environment-related conflicts are often highly localised and do not necessary follow national political identities or state boundaries. Neither of these points is, in our view, without merit. Yet there are also definite

---

[34] UNDP, *Human Development Report 2019* (2019); D. Howden, 'A failed state before it's born? Inside the capital of the world's next nation', *Independent* (07/01/2011); S. Smooha, 'Minority status in an ethnic democracy: the status of the Arab minority in Israel', *Ethnic and Racial Studies*, 13:3 (1990), 389–413; O. Yiftachel, *Ethnocracy: Land and Identity Politics in Israel/Palestine* (University of Pennsylvania Press, 2006).
[35] J. S. Mill, *A System of Logic* (Longmans, Green, Reader, and Dyer, 1872), book 3, ch. 8.
[36] C. Adams et al., 'Sampling bias in climate–conflict research', *Nature Climate Change*, 8 (2018), 200.

advantages to focusing on national-level conflicts. Mainstream public and policy narratives on climate and water security focus mostly on just such cases – on the role of water in Nile basin politics, on the contribution of drought to civil war onset in Syria or Lake Chad – making it crucial that at least some critical academic analysis does likewise. Equally, sampling bias is only really problematic if it results in exaggeration (or underestimation) of climate–conflict linkages: if it does not do this, and still less where findings contradict climate chaos narratives, then the problem disappears. Moreover, while it is absolutely right that many environmental and other conflicts are highly localised, they are typically also to a significant degree shaped by country-specific national-level – and international – political and economic dynamics. Most importantly, although state-defined spaces provide us with our five cases, our analysis operates across scales, and thus discusses many locally specific incidents and dynamics and many axes of division beyond those revolving around ethno-national identities or the nation-state.

More positively, our title *Divided Environments* is also intended as a nod towards Nancy Peluso and Michael Watts' important collection *Violent Environments* – and to indicate our debt to its critical and political ecology-informed engagement with mainstream environmental security narratives. All too often, studies of the environment and security place their primary emphases on the ways in which politics and conflict are determined, or are soon going to become determined, by environmental limits and forces. We wish to insist, by contrast, that it is historically configured human agency in the division – and also the exploitation, transformation, destruction, appropriation, distribution and commodification – of environmental spaces and resources that lies at the core of the environmental insecurity problematique. The poem and extracts which preface this book are chosen because they point in this direction. More typical for works on environmental conflict would be to quote from Percy Shelley's 'Ozymandias', and to invoke its imagined, romanticised and frankly Orientalist portrayal of 'bare' and 'lifeless' ruins in the desert where 'nothing beside remains' as a metaphor for eco-led political collapse.[37] We turn instead to Abdelrahman Munif's *Cities of Salt* for its exploration of how places, people and water are transformed and of the displacement, bewilderment, exploitation, resentments and violence that so often ensue on frontiers of capitalist extractivism. We invoke W. H. Auden's 'Partition', for its part, given its focus on the role of the political – on the centrality of

---

[37] Peluso and Watts, *Violent Environments*; J. Diamond, *Collapse: How Societies Chose to Fail or Succeed* (Viking, 2005), vii; B. Fagan, *The Great Warming: Climate Change and the Rise and Fall of Civilizations* (Bloomsbury, 2008), ix; M. Reisner, *Cadillac Desert: The American West and Its Disappearing Water*, rev. ed. (Penguin, 1986).

political structures, decisions and agency – in the drawing of boundaries and the dividing of environments, and in 'settling the fate of millions'. We turn to Munif and Auden together because they are both concerned with coloniality, the one presenting a subaltern reading of the United States' quest for world oil, the other an elite-centred if melancholic take on Britain's departure from India. We invoke them, moreover, because they both end with gunshot, albeit in different registers. Whereas Auden's colonial administrator is merely afraid of being shot, anticipating threats to his life and walling himself off for his own protection, for the Bedouin of Munif's thinly disguised Saudi Arabia the word 'Fire!' and the ensuing carnage are actualities, the culmination of long-term processes of displacement, dispossession and development. Between Auden and Munif, the analysis that follows is in many respects an exploration of these two faces of violence and (in)security. There exists, as we show, an enormous gulf between Northern and elite fears of anticipated 'security threats', and the historical, present-day and likely future realities of marginalisation, exploitation, domination and conflict – or, in short, 'insecurity' – across the global periphery. And this, we hold, applies to the questions of climate and water security just as it does to most other areas.

Organisation of the Book

In sum, this book operates on three levels. On a first, it is concerned with the implications of global anthropogenic climate change for twenty-first-century world politics and its landscape of conflict and insecurity: this is our primary research puzzle and question. On a second, it is about the past and present of the relations between water and (in)security, it being the study of these relations which dominates the pages to follow and which furnishes us with our evidence base for reflecting on climate change and the future. And third, it is intended as a contribution to research in political ecology – both substantively, in providing a political ecology–informed multi-case interpretation of some of the most pressing resource security challenges worldwide; and theoretically, in calling for and illustrating an 'international political ecology' approach to understanding them.

The analysis is structured, as already mentioned, around themes rather than our five cases. Each of the eight substantive chapters explores one such theme – a particular aspect, or site, of the water–conflict problematique. Each combines general theoretical discussion with case study–informed insights and comparison. Each also focuses first and foremost on the past and present, only in conclusion turning systematically to the future, and future climate change. Each of the chapters follow this structure and hence can be read, if need be, as stand-alone explorations of a theme.

Hopefully, however, this book will not principally be read in this way – for it is also structured as a single unfolding argument which starts by developing a series of critiques, and from there turns to exposition. Thus the next two and a bit chapters are essentially critiques of water and climate security orthodoxies; the focus here is on questions of geography and demography (Chapter 2) and climatic variability (Chapter 3) – and in both of them on the language of scarcity, on Malthusian or 'eco-determinist' reasoning, on claims that water scarcity and climate change are already contributing to large-scale violence and on showing that such claims are thin, routinely overstated and ethico-politically problematic. Thereafter, by contrast, the book turns to exposition, developing a series of arguments about the past and present relations between water, conflict and insecurity, and their likely future relations under conditions of accelerating climate change. We start this reconstruction by considering questions of identity and alterity (Chapter 4). We then explore processes of material transformation, expropriation and marginalisation across space, first in relation to core processes of hydraulic development and state-building (Chapter 5), and next in relation to territorial frontiers (Chapter 6). In two chapters that are also conceived as a pair, we then turn to the impacts of war (Chapter 7) and peace (Chapter 8) on water and climate-related transformations and insecurities. And in the final substantive chapter, we consider how international circulations of food, energy and capital underpin patterns of water and climate security (Chapter 9).

Across the chapters, we explore not just the impacts of water on conflict and security, but simultaneously the reverse causality: namely, how water resources and water-related inequalities, vulnerabilities and conflicts have been caused and configured by political strategies, violence, divisions and hierarchies. What we show throughout is that, both historically and still today, the political ecology of water – that is, the patterns of production, transformation, consumption, distribution, degradation, scarcity, insecurity and conflict relating to it – has been shaped much more by political and political–economic forces, structures and divides, than vice versa. Moreover, water, we show, is becoming less, not more, economically and politically important. Viewed thus, water as a resource is in our view unlikely to become a significant cause of or contributor to conflict and insecurity, even in an era of accelerating climate change. This does not mean, we insist, that the whole issue is inflated and overblown. We do not doubt the existence of profound water security challenges. We do not question the reality of human-induced global heating, or the urgent need to tackle it. We consider it close to certain that both water issues and climate change will, in future, have profound and wide-ranging consequences for conflict and security. What we do believe, however, is that these consequences will be rather different from how they are usually imagined. This book seeks to make this case.

2   Geography versus Demography

The Question of Scarcity

No idea is more central to mainstream water, climate and environmental security discourse than the idea of 'scarcity'. Scarcity, it is typically assumed, leads to competition, marginalisation, migration, insecurity and conflict, and in an era of accelerating climate change will likely do so with ever wider and more acute consequences. Such is the basic premise underpinning most environmental security thinking, whether the word 'scarcity' is deployed or not. Thus Thomas Homer-Dixon's major explanatory claim on the subject is that 'environmental scarcity causes violent conflict'. Michael Klare maintains that global limits on freshwater availability combined with rising demand will, by 2050, produce 'intense competition for this essential substance in all but a few well-watered areas of the planet'. The Environmental Justice Foundation foresees a world of 'scarce futures'. Margaret Beckett, as UK foreign secretary, claimed that '[w]hen people are exposed to the stresses caused by overpopulation, resource scarcity, environmental degradation, as they feel the security upon which they and their families depend progressively slipping away, so we see the slide down the spectrum from stability to instability'. And the first major defence planning study of climate security risks put the issue thus:

> Abrupt climate change is likely to stretch carrying capacity well beyond its already precarious limits. And there's a natural tendency or need for carrying capacity to become realigned. As abrupt climate change lowers the world's carrying capacity aggressive wars are likely to be fought over food, water, and energy. Deaths from war as well as starvation and disease will decrease population size, which overtime, will re-balance with carrying capacity.[1]

---

[1] Homer-Dixon, 'Environmental scarcities', 39; Homer-Dixon, *Environment, Scarcity and Violence*, 93; M. Klare, 'The new geography of conflict', *Foreign Affairs*, 80:3 (2001), 57; Environmental Justice Foundation, *The Gathering Storm: Climate Change, Security and Conflict* (2014), 14; M. Beckett, 'Foreign policy and climate security', Speech at the British Embassy, Berlin (23/10/2006); Schwartz and Randall, *Abrupt Climate Change Scenario*, 15.

There is, no doubt, an intuitively common sensical quality to much scarcity reasoning. Who, after all, can doubt that shortages of water, food and other essentials can have the most devastating of effects, and could increasingly do so in the future? And yet scarcity discourse is at the same time extremely slippery. One moment scarcity is an absolute; but then the next it is relational, only existing relative to cultural perceptions or socially constructed needs. One moment it is defined spatially; the next, temporally. One moment its causes are physical; but then the next they are economic, and sometimes institutional. One moment it refers to individuals, families and communities not having enough of whatever resource is under consideration; but then the next it refers to there not being enough of that resource – which as we know from Amartya Sen's famous dictum on famine are two entirely different things. One moment it – whatever 'it' is taken to be – is depicted as foundational to social life, the underlying cause both of economic exchange and of 'the utter inevitability of politics'; but then the next it becomes a much more limited phenomenon, referring just to specific groups or times or spaces. Scratch the surface of scarcity talk and all manner of meanings suddenly appear.[2]

In practice, however, two particular ideas and images of scarcity dominate contemporary water, climate and environmental security discourse. One is that of scarcities resulting from short-term environmental shocks, especially droughts, which are thought likely to become both more common and more extreme under the influence of global climate change. And the other is appropriately captured by Timothy Mitchell in his study of Egypt-in-the-world, *Rule of Experts*:

> Open almost any study of Egypt produced by an American or international development agency and you are likely to find it starting with the same simple image. The question of Egypt's economic development is almost invariably introduced as a problem of geography versus demography, pictured by describing the narrow valley of the Nile River, surrounded by desert, crowded with rapidly multiplying millions of inhabitants.[3]

Geography, over-population and drought: these are the unholy trinity which, within mainstream academic and policy discourse, are viewed as the principal causes of environmental scarcity and attendant conflicts and insecurities.

---

[2] A. Sen, *Poverty and Famines: An Essay on Entitlement and Deprivation* (Clarendon, 1981), 1; W. Ophuls, *Ecology and the Politics of Scarcity* (W.H. Freeman and Co, 1977), 8; J. Selby and C. Hoffmann, 'Beyond scarcity: rethinking water, climate change and conflict in the Sudans', *Global Environmental Change*, 29 (2014), 361–2.

[3] T. Mitchell, *Rule of Experts: Egypt, Techno-Politics and Modernity* (University of California Press, 2002), 209.

This chapter considers the first two parts of this trinity – geography and demography – since they are standardly bracketed together, and are also typically viewed as establishing the foundation and contexts out of which other scarcity-related crises arise; drought is a case in point, and so is the focus of the chapter to follow. The chapter shows that, contrary to the orthodoxy identified by Mitchell, the tension between local geographical constraints and demographic pressures is not the central or underlying cause of contemporary water-related insecurities, and that there are good structural reasons for this, rooted in the logics of global capitalism. We argue, moreover, that the 'eco-determinist' premises underpinning this orthodoxy are both substantively misleading and normatively questionable. And we argue, on these grounds, that climate change-induced scarcities are in and of themselves unlikely to become a major cause of conflict. As an introductory foray into our subject, the chapter inevitably leaves many threads hanging: it does not, for instance, discuss the politics and purposes of climate security discourse (initial consideration of which is deferred until the next chapter; systematic analysis must wait until the conclusion). The chapter's central purpose is simply to identify some of the major substantive weaknesses in eco-determinist and environmental security reasoning, and through that to show that what Robert Kaplan has called a 'revenge of geography' is in our view unlikely, even under circumstances of accelerating human-induced climate change.[4]

The Water Stress Paradox

Let us begin by considering what the standard measures of water stress and scarcity suggest about our divided environments. The most widely used indicator of these, often termed the 'Water Stress Index', measures total renewable water resource availability per person. Using this and the Food and Agricultural Organization's (FAO) Aquastat database, the established reference point for global water data, Table 2.1 presents water stress levels in Cyprus, Israel–Palestine, Sudan–South Sudan, and Syria (we opt not to include Lake Chad here, given that it is a basin with no equivalent data available).

By this data, and using the FAO's threshold levels for water stress – developed from work by Malin Falkenmark and colleagues in the late 1980s – Israel, the West Bank and Gaza are in a condition of 'absolute water scarcity' (defined as water availability of less than 500 cubic metres

---

[4] R. D. Kaplan, 'The revenge of geography', *Foreign Policy*, 172 (2009), 96–105; R. D. Kaplan, *The Revenge of Geography: What the Map Tells Us about the Coming Conflicts and the Battle against Fate* (Random House, 2012).

Table 2.1 *Falkenmark's Water Stress Index, for selected countries, 2017.*

	Cyprus	Israel	Palestine	South Sudan	Sudan	Syria
Renewable water resources (bcm/y)[1]	0.78	1.78	0.84	49.5	37.8	16.8
Population (millions)	1.18	8.24	4.75	10.9	40.81	17.1
Renewable water resources per person (cm/cap/y)	661.2	215.9	176.3	4,537	926.2	982.8

[1] billion cubic metres per year
Source: FAO Aquastat core database.

per capita per year (cm/cap/y)); Cyprus, Sudan and Syria face 'chronic water scarcity' (less than 1,000 cm/cap/y); while South Sudan experiences just 'occasional or local water stress' (water availability of between 1,700 and 5,000 cm/cap/y).[5]

The same index also suggests that water stress levels are worsening near continuously in the face of population growth (see Figure 2.1). Since the 1960s, this data suggests, per capita water availability has declined by more than 400 per cent in Syria and the West Bank and Gaza, and by 350 per cent in Israel. The only registered reduction in water stress levels is for Syria since 2012 – a result of civil war–induced population displacement.

An alternative index of 'water stress', which has now become Sustainable Development Goal (SDG) indicator 6.4.2, measures 'freshwater withdrawal as a proportion of available freshwater resources', after accounting for environmental water requirements (see Table 2.2). By this measure, Israel, Sudan and Syria are in the highest water stress category (defined as withdrawing more than 70 per cent of total renewable water resources), and Cyprus and Palestine in an intermediate position (withdrawing more than 25 per cent of their resources); South Sudan, meanwhile, is in the lowest water stress category (withdrawing less than 10 per cent of its water resources, indeed far less than this).[6]

Both of these indicators represent 'water stress' as essentially shaped by geography and demography – by water availability in battle with populations and use. Both also suggest a pattern of generally rising water

---

[5] FAO, 'Total renewable water resources per inhabitant in 2014 (m³/year)', Aquastat core database; M. Falkenmark, 'The massive water scarcity threatening Africa: why isn't it being addressed?', *Ambio*, 18:2 (1989), 112–18.
[6] FAO, 'SDG indicator 6.4.2 – water stress', Aquastat core database.

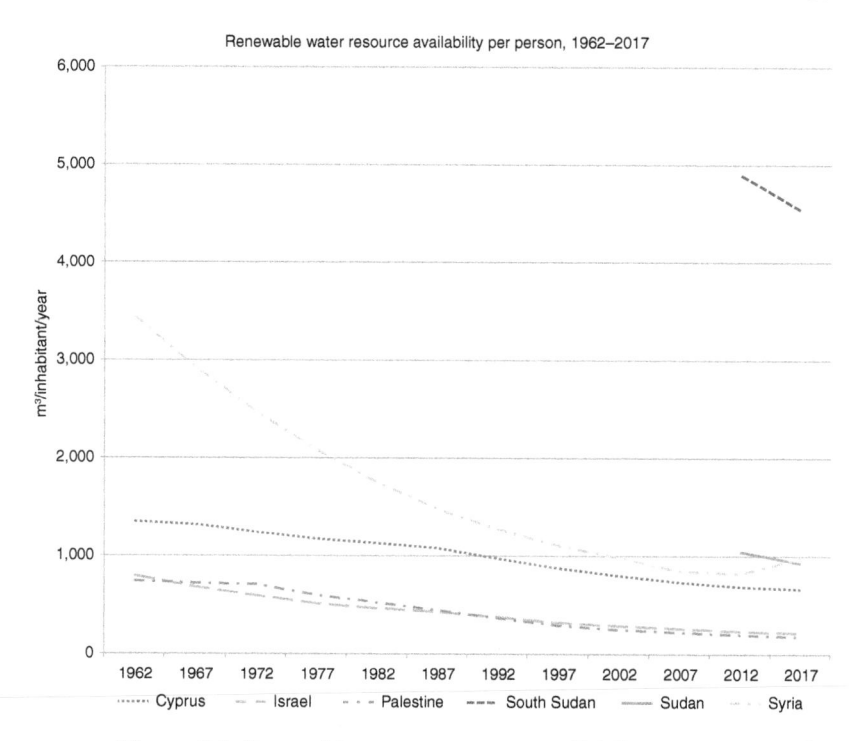

Figure 2.1 Renewable water resource availability per person, for selected countries, 1962–2017. *Source*: FAO Aquastat database.

Table 2.2 *SDG Indicator 6.4.2: water stress, for selected countries, 2017.*

	Cyprus	Israel	Palestine	South Sudan	Sudan	Syria
Renewable water resources (bcm/y)[1]	0.78	1.78	0.84	49.5	37.8	16.8
Withdrawal (bcm/y)	0.31	2.30	0.38	0.66	26.93	16.76
Environmental flow requirements (bcm/y)	0.05	0.62	0.14	33.93	15.1	5.57
Water stress (%)	29.52	103.4	41.08	4.23	118.7	124.4

[1] billion cubic metres per year
Source: FAO Aquastat database.

scarcities, as static and limited resources are overwhelmed by near-constant increases in demand. But is this accurate? Do such 'geography versus demography' indices accurately reflect patterns of water scarcity and insecurity, as experienced on the ground? Are scarcities ever-increasing with population and demand growth? And, more broadly, are geography and demography the foundational determinants of water scarcity, insecurity and vulnerability that they are so often taken to be?

That the answer to these questions is 'no' is clearly illustrated by the two constants in Tables 2.1 and 2.2 and in Figure 2.1: the positions held by Israel and South Sudan. Israel, these figures suggest, faces the acutest levels of water stress. Yet this is manifestly not so. Israel has in recent years had a large water surplus, as has been widely reported in its national press and acknowledged by the Israeli Water Authority (IWA). Per capita domestic consumption in Israel is higher than in most European countries. The boulevards and gardens of Tel Aviv are verdant green. According to UN measures, 100 per cent of Israelis have access to safe drinking water and all but a tiny proportion have assured supply. Moreover, in recent years Israel has possessed so much water that it has been able to significantly expand its allocation of water to agriculture. It has sought to develop new markets for its surplus water, selling it to the West Bank, Gaza, Jordan and elsewhere. It has officially recognised 'nature' as a water user, and at least promised a rehabilitation of the Jordan River. In 2020 the country's main natural reservoir, Lake Kinneret (Lake Tiberias in Arabic), was completely full for the first time in decades. And just a year before that, a new artificial lake – at 22.5 acres, the second largest body of water in the country after Kinneret – was opened in Beer Sheva on the edge of the Negev Desert; '[i]t is truly visionary to see boats sailing, fish swimming, and swans resting in the heart of the desert', boasted the city's mayor.[7]

In short, Israel is not currently a land of absolute water scarcity and high water stress, as the most widely used international measures suggest – but

---

[7] 'Water authority: Israel emerging from water crisis', *Globes* (24/09/2012); Z. Rinat, 'Water, water everywhere: desalination trumps drought in Israel', *Ha'aretz* (07/06/2015); H. Cohen, 'Israel Water Auth head: we've solved water crisis', *Globes* (17/05/2016); IWA, *Long-Term Master Plan for the National Water Sector, Part A – Policy Document, Version 4* (08/2012), 14; UK Environment Agency, 'International comparisons of per capita domestic consumption' (2008); FAO Aquastat core database; Y. Elizur, 'Over and drought: why the end of Israel's water shortage is a secret', *Ha'aretz* (24/01/2014); Israel Ministry of the Environment, 'The right of nature to water in Israel' (02/2005); Z. Rinat, 'For first time, Israel's water authority to pump Kinneret water into Jordan River', *Ha'aretz* (17/05/2013); M. Gilad, 'The Sea of Galilee is completely full – and empty', *Ha'aretz* (23/04/2020); 'Beer Sheva to build biggest artificial lake in Israel', *i24 News* (29/08/2015); D. Israel, 'Israel's 2nd largest lake inaugurated in ... Beersheba', *Jewish Press* (15/10/2019).

instead a land of water abundance. Except for among its minority Palestinian population, there is little water insecurity or vulnerability. And water availability per person is currently higher than it has ever been, despite the country's relatively high population growth. There is little sign here of livelihoods, economic development or expected modern standards of water provision foundering under the combined pressures of geography and demography.

What of South Sudan? The indices above suggest a land of water plenty, which has exceptionally high per capita availability and withdraws only a tiny fraction of available resources. And yet water-related insecurities in South Sudan are acute. Average domestic consumption is as low as 20 litres per capita per day (l/c/d), just a fraction of that in Israel. Less than 60 per cent of the population has access to safe drinking water. It is estimated that only 7.4 per cent of households have piped water, with the vast majority instead sourcing water from public taps, wells, tankers, and other sources, and at least one-third of the population travelling thirty minutes or more to reach water sources. Even in the capital, Juba, only a quarter of the population has access to piped water, and per capita consumption is just 30–36 l/c/d. Moreover, access to sanitation facilities is no better: it is estimated that, in 2015, only 19.5 per cent of the population had access to improved sanitation facilities.[8]

We are faced, then, with a paradox: while standard metrics show Israel as suffering from acute water stress, and South Sudan not, the reality of water insecurity in these two countries is the exact reverse – South Sudan being a land of multiple local water insecurities and Israel a land of abundance and of high and growing demand. The relationship between geography and demography on the one hand and water insecurity on the other is evidently neither necessary nor particularly close.

Furthermore, the extent to which the above data actually reflects physical geographical and demographic pressures is debatable. The figures for total renewable water resources shown in Tables 2.1 and 2.2 all include, except for Cyprus, surface and groundwater flows originating from other countries as well as flows required downstream – and thus reflect upstream and downstream patterns of water use and the political and treaty arrangements between riparians, as much as they do any 'natural' availability. For example, over 96 per cent of the 37.8 billion

---

[8] AfDB, *South Sudan: An Infrastructure Action Plan – A Program for Sustained Strong Economic Growth* (2013), 262; FAO Aquastat core database; WHO/UNICEF Joint Monitoring Programme for Water Supply and Sanitation data; AfDB, *Resilient Water and Sanitation Project for Improved Lives and Health in Juba: Environmental and Social Management Summary* (2016); JICA, Juba Urban Water Supply and Capacity Development Study in the Southern Sudan: Final Report (Summary) (2009), 2, 3–8.

cubic metres per year (bcm/y) of water that is characterised as 'Sudan's total water resources' is external in origin, with this figure being essentially a function of the low level of upstream Nile withdrawals, especially in Ethiopia and South Sudan, combined with the terms of Sudan's 1959 Nile Waters Agreement with Egypt, which requires an average 55 bcm/y to be allowed to reach Egypt each year.[9] The figures for water withdrawal likewise reflect complex socially and historically determined patterns of use, not natural or inevitable demands, as will be discussed further below. Population numbers are also, of course, the contingent products of history. And the data given for 'environmental flow requirements' is arbitrary: how else is it possible to account, for instance, for the fact that South Sudan is given an environmental flow requirement more than fifty times higher than its total withdrawal for human purposes, while Cyprus has less than a sixth of the same; how else can we make sense of the fact that Israel's environmental needs are depicted as 620 million cubic metres per year (mcm/y), despite its actual governmental allocation of water to nature being nowhere near this figure?

The point here is not so much that all of the above data is flawed (though some of it evidently is; Palestine, for instance, has 'available freshwater resources' of 840 mcm/y and a far lower water stress level than Israel only if one supposes that it has full access to West Bank groundwater, which, as will be discussed in Chapters 6, 7 and 8, it patently does not). And neither is it to suggest that the above data is politicised (though it undoubtedly is). The point, rather, is that each and every data point in Tables 2.1, 2.2 and Figure 2.1 bears the marks of diverse, and contingent, political, economic and social histories, and does not just reflect physical geography, the dictates of demography or, in short, nature. At best, water stress indicators afford us snapshot – and 'methodologically nationalist' – pictures of a particular dimension of aggregate water availability and scarcity.[10] But they are not the place to begin if one wants to understand the nature of contemporary water scarcities and insecurities, their causes, or their implications for conflict and for climate change. To understand these a return to first principles is required.

Water Is Not Life

Discussions of water security routinely start out by invoking biology. The report of the recent Global High-Level Panel on Water and Peace, for

---

[9] UAR and Sudan, *Agreement for the Full Utilization of the Nile Waters* (1959).
[10] H. Martins, 'Time and theory in sociology', in J. Rex (ed.), *Approaches to Sociology* (Routledge and Kegan Paul, 1974), 276.

instance, begins by making the superficially incontrovertible point that: 'Water is life. It is a fundamental condition of human survival and dignity, and is the basis for the resilience of societies and of the natural environment. Unlike other natural resources, water has no substitute: the only substitute for water is water.' And the report immediately continues: 'Water is scarce: about two million people still lack access to safe drinking water.'[11] Such formulations recur regularly across policy, NGO and academic treatises on urgent water issues. Yet they are misleading on at least four levels.

To start with, far from being scarce, water is our planet's most abundant natural resource. The total volume of water on Earth is in the order of 1,386 million km^3; known global oil reserves, by contrast, are estimated at just 270 km^3, a tiny fraction of this amount. Admittedly only 2.5 per cent of this water is freshwater, and the majority of this is held in glaciers, ice caps and permafrost. But non-ice surface and underground freshwater still amounts to over 10 million km^3 – an availability of well over 1 million cubic metres per person on a global basis.[12] In addition, given that in liquid and gaseous form water is a mass of flows, and not just a stock, it follows that even this figure overstates the limits to human water consumption: instead, water's flows can be reused and recycled ad infinitum. There exist local scarcities, no doubt: this most abundant of resources is very unevenly distributed, and even where abundant often carries so many other substances – salts, chemical pollutants, faecal bacteria and so on – as to drastically limit its utility. But the idea that water is scarce per se, and the implication that this scarcity somehow explains or is evidenced by poor access to safe drinking water, are both deeply misleading.

Second, water is not non-substitutable. Wind, railways, deodorant and ploughing, among many other things, have all at one point or another served as substitutes for water. Wind historically provided an alternative way of generating mechanical energy in regions where fast-flowing surface water for powering water mills was particularly scarce. The development of railways established an alternative means of long-distance freight and passenger transport from seas, rivers and canals. Deodorant provides an alternative detergent of smell. And ploughing, by helping to reintegrate salts back into the soil, can function as an alternative to desalting through irrigation. Even the bodily cooling function of water can be partly

---

[11] Geneva Water Hub, *A Matter of Survival: Report of the Global High-Level Panel on Water and Peace* (2017), 11.

[12] I. A. Shiklomanov, 'Appraisal and assessment of world water resources', *Water International*, 25:1 (2000), 11–32; British Petroleum, *Statistical Review of World Energy 2018* (2018), 12.

performed by shade or air conditioning. For other organic vegetative and animal processes, there are, of course, no alternatives to water. But water has many other uses besides these essential biological ones. Direct biological needs account for only a fraction of domestic water consumption: according to one widely used basic needs index, only 5 l/c/d – or 2 cubic metres per year – are required for drinking, and only 50 l/c/d for all basic needs.[13] Indeed, water is used for so many different purposes that it may be considered the acme of multi-tasking, the ultimate multi-purpose resource: solvent, reactant, habitat, excavator, transporter, mirror, cleanser, barrier, temperature regulator, energy provider, thirst quencher, firefighter, play inducer and vehicle for meaning, metaphor, meditation and myth.

The language of 'non-substitutability' – with drinking water as its basic reference point – suggests that human demand for and use of water has inexorably deepened alongside population and economic growth. But this too is only partially correct. Historically, different uses of water have ebbed and flowed in importance. The relative importance of inland waterways for transport has declined over the last century or so, with rail lines, roads and air travel progressively taking their place. The predominant use of water to drive Britain's cotton mills during the early industrial revolution had given way, by the late 1830s, to primary reliance on coal. There have also been sudden surges in water demand: in many areas of pre-twentieth-century England and France, half the population would die without ever having taken a bath whereas today, on the back of newly manufactured expectations and anxieties, especially historically unprecedented insecurities about smell, a daily shower is closer to the norm.[14] Without doubt, global demands on water for crop irrigation, industrial production and domestic uses are currently rising rapidly. But the above suggests that there is nothing linear or trans-historical about this. Certain uses are no longer what they once were; others have risen rapidly; and some of these could well go into reverse in the future.

The language of 'non-substitutability' can also be misleading in implying that a specific type of water resource – namely, locally available 'blue water' – has no substitutes. In reality, water is a highly fungible resource. Treated wastewater can be used for most purposes, technologies and standards permitting. Brackish and saline water can be desalinated. Water can, at some expense, be imported in bulk. And water can also be imported in 'embedded' or 'virtual water' form – that is, as agricultural

---

[13] P. H. Gleick, 'Basic human requirements for human activities: meet basic needs', *Water International*, 21:2 (1996), 83–92.

[14] A. Malm, *Fossil Capital: The Rise of Steam Power and the Roots of Global Warming* (Verso, 2016), 77–81; I. Illich, *H₂O and the Waters of Forgetfulness* (Marion Boyars, 1986), 59.

or industrial commodities which have been produced using rain, rivers or groundwater in distant locales.[15] Moreover, for many purposes – mining and transport, for example – the freshness or otherwise of water is of little bearing, and saline water is more than adequate. None of this is to suggest that the use of 'non-conventional waters' is always easy or cost-free. It is, however, to insist that a limited availability of surface – or underground – freshwater resources does not necessarily place any limits on human use.

Third, as accurate as the slogan 'water is life' undoubtedly is in biological terms, if human experience – whether historical or present-day – provides our reference point, then it must also be acknowledged that the opposite also holds true: that water is death. For, more than any other element or feature of the natural environment, water is simultaneously a 'good' and a 'bad', a blessing and a curse, a resource as well a source of danger, destruction and disaster. In 1931, flooding from the Yangtze, Yellow and Huai Rivers killed an estimated 3.7 million people. In early nineteenth-century Europe, increasing water use among the upper and middle classes, and the consequent development of mains sewers, was a major cause of deadly cholera epidemics in areas downstream. And historically, across Europe and elsewhere, many lowland plains were barely farmed or populated, out of deference to their waterlogged soils, stagnant pools and malaria. Still today, many more people die directly from water than from its absence. Floods, for example, caused a reported 4,731 deaths worldwide in 2016 – while in the same year no deaths were directly attributable to drought. Even more strikingly, while the overall impact of poor drinking water on mortality is unknown (since global statistics are not collected for this), its impact via diarrhoea alone is estimated to account for over 500,000 deaths annually. Yes, water is a giver of life. But all too often it also takes it away.[16]

Last, only in a very limited sense is water 'the basis for the resilience of societies', as the Global High-Level Panel suppose. A dearth of clean or usable water to meet basic needs can no doubt have adverse humanitarian, livelihood and developmental consequences; and the destruction or

---

[15] T. Allan, *Virtual Water: Tackling the Threat to the Planet's Most Precious Resource* (IB Tauris, 2011).
[16] CRED, EM-DAT data, https://public.emdat.be/data; Illich, *H_2O and the Waters of Forgetfulness*, 69–70; F. Braudel, *The Mediterranean and the Mediterranean World in the Age of Philip II*, Vol. I, trans. S. Reynolds (University of California Press, 1995), 60–75; D. Guha-Sapir et al., Annual Disaster Statistical Review 2016: The Numbers and Trends (CRED, 2017), 33, 36; A. Prüss-Ustün et al., 'Burden of disease from inadequate water, sanitation and hygiene in low- and middle-income settings: a retrospective analysis of data from 145 countries', *Tropical Medicine and International Health*, 19:8 (2014), 894–905.

capture of natural water resources can be socially devastating, especially for excluded and Indigenous peoples for whom water is indeed a fundamental source of life. However, as Nick Estes observes, the idea that water is life 'exists outside the logic of capitalism'.[17] Nowhere under contemporary global economic and political conditions is limited local water availability the fundamental cause of low development, resilience or human security. If water availability were causally fundamental to these, then South Sudan and other central African countries would be near the apex of the Human Development Index and other such indices, while Israel – along with Saudi Arabia, the United Arab Emirates and Kuwait, with water stress levels of 1,242, 2,346 and 2,603, respectively – would be languishing at their foot. That they are not suggests, at least at present, that natural blue water availability is not a significant determinant of resilience or development.

In truth, the reasons for this are not difficult to grasp, for under capitalism neither water specifically nor nature in general are the keys to secure life. More than this, capitalism is arguably one giant apparatus of superficially paradoxical transformations, inversions and dislocations. Under capitalism, machines, commodities and money are suddenly everywhere, revolutionising everything in their wake. Workers are dislocated en masse to burgeoning cities where they become appended to machines, subjected to new techniques of discipline and surveillance, and made hostage to the compulsions of the market. Life chances become determined by impersonal worldwide relations of production and exchange. Space is 'annihilated', as Marx put it, as capital searches voraciously for new markets, resources, armies of labour, comparative advantages, and more efficient supply chains. All manner of 'fictitious commodities' are created – from land and time, to water and carbon credits. The virtual becomes real; 'all that is solid melts into air'. The 'dead labour' of previous generations lives on as never before, in the form an ever-growing weight of machines, products and infrastructures. And state wealth and power cease to be defined by the size of territories and populations – by geography and demography – and are instead calculated as functions of the circulation of digits, notes and coins.[18]

---

[17] N. Estes, *Our History Is the Future: Standing Rock versus the Dakota Access Pipeline, and the Long Tradition of Indigenous Resistance* (Verso, 2019), 15–16, 257.

[18] K. Marx, *Grundrisse: Foundations of the Critique of Political Economy*, trans. M. Nicolaus (Penguin, 1993), 539; K. Polanyi, *The Great Transformation: The Political and Economic Origins of Our Time* (Beacon Press, 2014 [1944]); K. Marx and F. Engels, *The Communist Manifesto*, trans. S. Moore (Verso, 1998 [1848]), 38; Marx, *Capital*, 342; A. Bashford and J. E. Chaplin, *The New Worlds of Thomas Robert Malthus: Rereading the Principle of Population* (Princeton University Press, 2016), ch. 1; T. Mitchell, 'Fixing the economy',

As an inevitable corollary of this, the material transformation of nature is also taken to new heights. All of the biosphere is now humanised; nowhere within it does a pristine external nature any longer exist. The circulation of water in the biosphere is so transformed that its flows are no longer essentially hydrological – as represented by the 'hydrological cycle' – but are instead 'hydro-social', weaving through multiple human systems and uses and human-affected natural processes. Landscapes and waterscapes are not just reconfigured, but in extremis completely upended: in such situations, peninsulas are turned into islands while islands are born out of the sea. Saltwater can suddenly be turned into freshwater. Pollution can suddenly be remade and reimagined as a tradable resource. And, as Marc Reisner observed in *Cadillac Desert*, his famous study of the waters of the American West, water can suddenly 'flow uphill to money' – or perhaps, more accurately, 'to money and power':

In a hundred years, actually less, God's riverine handiwork in the West has been stood on its head. A number of rivers have been nearly dried up. One now flows backward. Some flow through mountains into other rivers' beds. There are huge reservoirs where there was once desert; there is desert, or cropland, where there were once huge shallow swamps and lakes.

As Reisner comments of the Central Arizona Project, which diverts water from the Colorado River up to Phoenix and beyond, this is 'as incongruous a spectacle as any on Earth: a man-made river flowing uphill in a place of almost no rain'. Such is what Marx called the 'magic and necromancy' of capitalism.[19]

Today, as a result, the key to water security is not local natural availability, but access to those means – those socio-technical relations and resources – through which water can be produced, contained, captured, repelled, distributed, reused and if necessary substituted. The cast of characters will be familiar. Technologies and infrastructures, without which a hydrological state of nature would still prevail. Reliable energy sources, without which these material means of provision could not be powered or produced. Investment and operating capital, without which neither major infrastructure projects nor most everyday maintenance

---

Cultural Studies, 12:1 (1998), 82–101; Wood, *Origin of Capitalism*; R. Marsden, *The Nature of Capital: Marx After Foucault* (Routledge, 1999).

[19] K. Marx, *Early Writings*, ed. Tom B. Bottomore (Watts and Co., 1963), 161; K. Marx and F. Engels, *The German Ideology* (Lawrence and Wishart, 1965), 59; B. McKibben, *The End of Nature* (Viking, 1990); Linton and Budds, 'The hydrosocial cycle'; M. Davis, 'Fear and money in Dubai', *New Left Review*, 41 (2006), 47–68; J. Gabbatiss, 'Saudi official hints at plan to turn Qatar into island by building canal', *Independent* (01/09/2018); Reisner, *Cadillac Desert*, 12, 293; Allan, *Virtual Water*, 19; Marx, *Capital*, 169.

would proceed. Knowledge and data, relating to a whole gamut of natural and social domains and processes. The modern state which, as Foucault emphasised, defines its mission as one of 'making live', of orchestrating and maximising the productive life of populations.[20] And, not least, power and authority – whether based on coercion, compulsion, co-optation or consent – over how water should be managed and used. The bulk of this book is effectively an extended discussion of how these various instruments and agents of transformation and substitution underpin patterns of water security and vulnerability – and of the highly uneven ways in which they do so. Our guiding premise is that, under conditions of contemporary capitalism, it is these powers and the relations between them which are the ultimate determinants of water security and vulnerability. Under capitalism, water is only in the most limited sense life.

Where Are the Water Wars?

For all this, many nonetheless argue that water is likely to become an increasingly important source of political power and cause of violent conflict as the twenty-first century progresses. 'As we approach the twenty-first century', wrote Peter Gleick in 1993 in one of the most influential early post–Cold War discussions of the subject, 'the probability of water-related violence is increasing', including over the Nile, Jordan and Euphrates Rivers. '[D]uring the next ten years, water problems will contribute to instability in states important to US security interests', concluded a 2012 US Intelligence Community Assessment, 'and distract them from working with the United States on important US policy objectives'. The wars of the twenty-first century 'will be over water', claimed a World Bank Vice President in 1995 – just as those of the twentieth century 'were about oil'. The 'plateau peoples of Turkey will dominate the Arabs in the 21st century because the Turks have water and the Arabs don't', proclaims Robert Kaplan, in his 2009 paean to the 'wisdom of geographical determinism.' And the World Economic Forum's landmark *Water Security* report asks:

[C]ould the 21st century see an emergence of economic and political power among water-endowed countries, akin to the 20th century geopolitics dominated by the rise of oil-rich states? Brazil, Canada, Iceland, northern Europe, and Russia are hydro-blessed in ways that India, Mexico, China and the Middle East are not. By 2030, might they form an Organization of Water-Endowed Countries?[21]

---

[20] M. Foucault, *'Society Must Be Defended': Lectures at the Collège de France, 1975–76*, trans. D. Macey (Picador, 2003), 241.
[21] P. H. Gleick, 'Water and conflict: fresh water resources and international security', *International Security*, 18: 1 (1993), 79, 80, 111; US ODNI, *Global Water Security*, Intelligence Community Assessment 2012-08 (2012), iii; B. Crossette, 'Severe water

Such claims are not easy to evaluate let alone refute, partly because they are often unclear – what is 'water-related violence'?; what does it mean to say that water problems 'might contribute to instability'? – and partly because the future is, as the saying goes, another country. Yet there are good reasons to reject them nonetheless.

First, a litany of such forecasts have already been proven wrong. In 1988, Joyce Starr and Daniel Stoll, among the leading global purveyors of the 'water wars' thesis, predicted that water scarcity could lead to 'unprecedented upheaval' across the Middle East '[b]efore the twenty-first century'. Around the same time, then Egyptian Foreign Minister Boutros Boutros Ghali offered the even more daring forecast that 'the next war in our region will be over the waters of the Nile, not politics'. In her 1994 book, Israeli geographer Nurit Kliot claimed that, 'by the year 2000' water would replace oil as 'the dominant subject of conflict for the Middle East'. Beyond water, the late twentieth century's most famous Malthusian, Paul Ehrlich, predicted that '[i]n the 1970s and 1980s hundreds of millions of people will starve to death', and that 'nothing can be done to prevent a substantial increase in the world death rate'. Malthus himself argued in his 1798 Essay on the *Principle of Population* that future agricultural production would inevitably fall short of natural population growth, and that this would inevitably lead, as historically, to recurrent waves of malnutrition, disease, famine and war.[22] None of this exactly inspires confidence in the claims that water wars, or eco-collapse more broadly, really are looming on the horizon.

Evidence from particular cases also provides reason for scepticism. It has often been said, for example, that the 1967 Arab–Israeli War was partly – even mainly – caused by conflict over the Jordan River.[23] The Six Day War was indeed preceded by strong tensions over rival Israeli and Arab League schemes for the diversion of water from the upper Jordan River; these included military clashes between Israeli and Syrian forces as

---

crisis ahead for poorest nations in next 2 decades', *New York Times* (10/08/1995); Kaplan, 'The revenge of geography', 105; WEF, *Water Security: The Water–Food–Energy–Climate Nexus* (2011), 97.

[22] J. Starr and D. Stoll (eds.), *The Politics of Scarcity: Water Scarcity in the Middle East* (Westview, 1988), ix; T. Walker, 'Arab and Israeli blood may flow if more water doesn't', *Sydney Morning Herald* (26/11/1988); N. Kliot, *Water Resources and Conflict in the Middle East* (Routledge, 1994), 12; P. Ehrlich and A. Ehrlich, *The Population Bomb* (Buccaneer Books, 1968), xi (Anne Ehrlich was co-author of this book but not initially credited); T. Malthus, *An Essay on the Principle of Population*, 1st ed. (Oxford University Press, 2008 [1798]), esp. 12–22, 61.

[23] See, e.g., J. Cooley, 'The war over water', *Foreign Policy*, 54 (1984), 3; T. Naff and R. Matson (eds.), *Water in the Middle East: Conflict or Cooperation?* (Westview, 1984), 44; J. Bulloch and A. Darwish, *Water Wars: Coming Conflicts in the Middle East* (Victor Gollancz, 1993), 34.

well as raids by Palestinian national groups, which often targeted water installations. Yet these water-related tensions during 1964–5 were not a proximate cause of the Six Day War itself: if they had been, the war would have broken out in 1965 and not suddenly, taking the world by surprise, after a gap of two years. Indeed, the central causes of and motivations behind the 1967 war are well known: the existence of a European settler–colonial state within a revolutionary, post-colonial Middle East; political rivalries between radical Arab states; the re-emergence of organised Palestinian nationalism; Israeli counter-insurgency strikes into the (then Jordanian-controlled) West Bank; global Cold War bipolarity; and the political insecurities of Israel's Eshkol government, combined with its desire to shatter Nasser's standing in the Arab world to increase the country's 'strategic depth' and to fulfil longstanding territorial ambitions. As Walter Laqueur commented in his classic account, the Jordan waters dispute 'was not among the major causes' of the Six Day War.[24]

The Nile is also regularly identified as a potential theatre for water wars; indeed, the prospect of war over the Nile is arguably more discussed than any other case.[25] There are some good reasons for this: the large number of Nile riparians, now numbering eleven, including South Sudan; the extreme hydro-social asymmetries between these riparians, with Ethiopia and other upstream states providing the bulk of the Nile's waters but withdrawing very little from them, and Egypt, conversely, providing negligible flow while being the river's major consumer; Egypt's consequent extreme Nile dependency, with the Nile accounting for 97 per cent of its total freshwater resources; pressure from 'late developing' riparians, most importantly Ethiopia, for increased access to and use of the river for irrigation and electricity generation purposes; and, not least, the lack of any agreed, basin-wide water allocation or management regime. Moreover, the Nile has repeatedly been an object of fierce political rhetoric, as discussed further below.

Yet caution is warranted. Nile politics has historically been marked more by active cooperation than violent conflict: Egyptian–Sudanese relations remain governed, sixty years on, by the terms of the 1959 Nile Waters Agreement and the Permanent Joint Technical Commission which it established, while the renegotiation of Egyptian 'hydro-hegemony' being demanded by Ethiopia and its neighbours has not led, despite all the rhetoric, to any military mobilisations or clashes.

---

[24] W. Laqueur, *The Road to War: The Origin and Aftermath of the Arab–Israeli Conflict 1967–8* (Penguin, 1968), 63.

[25] See e.g., J. Starr, 'Water wars', *Foreign Policy*, 82 (1991), 17–36; Gleick, 'Water and conflict'; US ODNI, *Global Water Security*.

Moreover, the Nile's waters are not nearly as stretched and contested as is often assumed. Since the 1960s, here has been no decline in average Nile flow through Sudan, no decline in the water level of Lake Nasser, and no decline in the volume of water released into Egypt through the Aswan High Dam. During the ten years 2000–2010, an average 60.8 bcm/y were released through the Aswan High Dam, well in excess of the 55.5 bcm/y allocated to Egypt under the Nile Waters Agreement. Sudan's withdrawals remain well below its 18.5 bcm/y Nile Waters Agreement allocation (its annual abstraction from the Nile varies between 10 and 16 bcm/y). Egypt continues to expand its irrigated agriculture, notably in the Toshka, East Owainat, Darb al-Araba'in and As-Salam Canal projects. And the Grand Ethiopian Renaissance Dam (GERD), which at the time of writing has just completed its second year of filling, need not have particularly negative long-term implications for downstream riparians. Future population and agricultural expansion in Egypt, and any equivalent development in Ethiopia and other upstream riparians, may of course place greater pressure on Nile waters. But there is no evidence as yet of any natural limits to Nile utilisation being reached, let alone of such limits generating, or contributing to, trans-boundary conflicts.[26]

The Tigris–Euphrates presents a similar story – but we defer discussion of this basin until Chapter 5. For, beyond such individual cases it is now widely accepted that there has not been a single modern inter-state 'water war' (i.e. a war fought primarily over water); that the only recorded water war in history occurred 4,500 years ago between two Mesopotamian city-states; and that active cooperation over trans-boundary water resources is much more common today than violent water-related conflict.[27] This is often accepted even within otherwise gloomy policy prognoses on water conflict; moreover, there is little disagreement on these questions among international water politics researchers. Some quantitative research does admittedly claim to identify connections between geographical water scarcities and violent conflict. However, the evidence to this effect is incredibly weak. A widely cited study by Paul Hensel and colleagues,

---

[26] A. E. Cascão, 'Changing power relations in the Nile River basin: unilateralism vs. cooperation?', *Water Alternatives*, 2:2 (2009), 245–68; M. M. Abdel-Latif and M. Yacoub, 'Effect of change of discharges at Dongola station due to sedimentation on the water losses from Nasser Lake', *Nile Basin Water Science and Engineering Journal*, 4:1 (2011), 89–90; A. M. Omer, 'Water resources and freshwater ecosystems in Sudan', *Renewable and Sustainable Energy Reviews*, 12:8 (2008), 2070; J. Barnes, 'Pumping possibility: agricultural expansion through desert reclamation in Egypt', *Social Studies of Science*, 42:4 (2012), 518.

[27] A. Wolf, 'Conflict and cooperation along international waterways', *Water Policy*, 1:2 (1998), 251–65; S. Yoffe et al., 'Conflict and cooperation over international freshwater resources: indicators of basins at risk', *Journal of the American Water Resources Association*, 39:5 (2003), 1109–26.

for example, concludes that '[a]reas with high levels of water scarcity will experience more frequent conflicts' to the extent that 'militarized disputes will be more likely in water scarce regions'. Yet Hensel et al.'s supposedly 'large-N' dataset includes just seventeen cases of militarised inter-state disputes over water resources, of which twelve are disputes about navigational issues – which, as the authors themselves admit, 'would not be expected if scarcity was the primary motivating factor behind militarization'.[28] Evidence of modern water wars is strikingly thin.

While many reasons could be adduced to explain the dearth of modern water wars and the absence of solid evidence of geographical water scarcities contributing to conflict, the most fundamental follow directly from the premises outlined above. Stocks of water are often not as scarce as is typically assumed, as in the case of the Nile. Individual flows of water can also be reused multiple times, effectively flowing through a succession of uses – first, say, through hydro-electricity production, then through homes and industries, then being reused in agriculture, perhaps more than once, before being used further downstream by container ships and river cruisers (it is often assumed that the water used in irrigation is only used once, but even this only applies to highly 'efficient' irrigation processes).[29] Some water uses have long been in decline (the world's trans-boundary rivers would surely be a graver focus of conflict than they are today if asphalt, railways or aeroplanes had not been invented). Local water scarcities can also be readily adapted, whether through more efficient delivery (for instance, using drip irrigation or low-flow showers); through reduced use of water for non-essential purposes; through the industrial production of new freshwater (e.g. through wastewater treatment or desalination); or by conveying water, or the benefits derived therefrom, from distant locales. The last of these 'adaptation mechanisms' is especially important. For, according to some analysts, if one includes 'virtual water' – that is, the water which has been used to produce imported agricultural and industrial commodities – in calculations of national water budgets, it becomes clear that the water needs of many Middle Eastern and North African countries are today being met more by international trade, than by local rivers and aquifers. In all of these respects, the foundational premise of water wars reasoning that ever-increasing local geographical and demographic pressures inevitably

---

[28] P. Hensel et al., 'Conflict management of riparian disputes', *Political Geography*, 25:4 (2006), 385, 407–8.

[29] J. Barnes, 'Water, water everywhere but not a drop to drink: the false promise of virtual water', *Critique of Anthropology*, 33:4 (2013), 371–89; J. Trottier and J. Perrier, 'Challenging the coproduction of virtual water and Palestinian agriculture', *Geoforum*, 87 (2017), 85–94.

presage increased water conflict – that, in Falkenmark's words, 'water resources are finite; future increases in population therefore imply increased water competition' – is inherently flawed.[30]

This aside, water wars discourse also rests upon flawed assumptions about water's economic and political value. Its proponents often assert or assume that, owing to deepening scarcities, water is becoming 'increasingly important for economic and agricultural development' and 'evolving into an issue of "high politics"', to quote Peter Gleick. It is these premises that underpin the fantasies of global water cartels and Anatolian domination of the Middle East, mentioned above, as well as claims such as that in a leading popular text that 'whoever controls water or its distribution can dominate the Middle East and all its riches'.[31] But such premises and claims are demonstrably false. For, just as water, under capitalism, is no longer the basis of secure life, so good natural availability of water resources is no longer fundamental to economic development or political power. If it were, then it would not be Israel but South Sudan in possession of a leading high-tech economy and nuclear arsenal.

Furthermore, the relative economic and political importance of water resources, far from increasing, is evidently in long-term decline. This can be seen with respect to both the consumption of water and its production. With regard to the former, the economic sector which accounts for the majority of water withdrawals worldwide and over 75 per cent of withdrawals in many arid and semi-arid countries – agriculture – is of declining structural economic and political importance, and hence the water used within this sector is of declining economic and political importance too. This is so both worldwide and within most individual countries, and is a straightforward consequence of economic development and diversification. Whether one considers agriculture's contribution to gross domestic product (GDP), to employment or to exports, the pattern is broadly similar (see Figure 2.2). Thus agriculture's contribution to global GDP has declined from an estimated 7.6 per cent in 1995 to just 3.5 per cent in 2016, with this pattern reproduced across each of our divided environments. Its contribution to global employment has also declined steadily, from around 43 per cent in 1991 to just 26 per cent in 2018. And agriculture's worldwide contribution to exports has been in long-term decline since

---

[30] M. Falkenmark, 'Fresh water: time for a modified approach', *Ambio*, 15:4 (1986), 192; on virtual water, see e.g., T. Allan, 'Water in the Middle East and in Israel–Palestine: some local and global resource issues', in M. Haddad and E. Feitelson (eds.), *Joint Management of Shared Aquifers: The Second Workshop* (Palestine Consultancy Group and Harry S. Truman Institute for the Advancement of Peace, 1997), 31–44.
[31] Gleick, 'Water and conflict', 80; Bulloch and Darwish, *Water Wars*, 161.

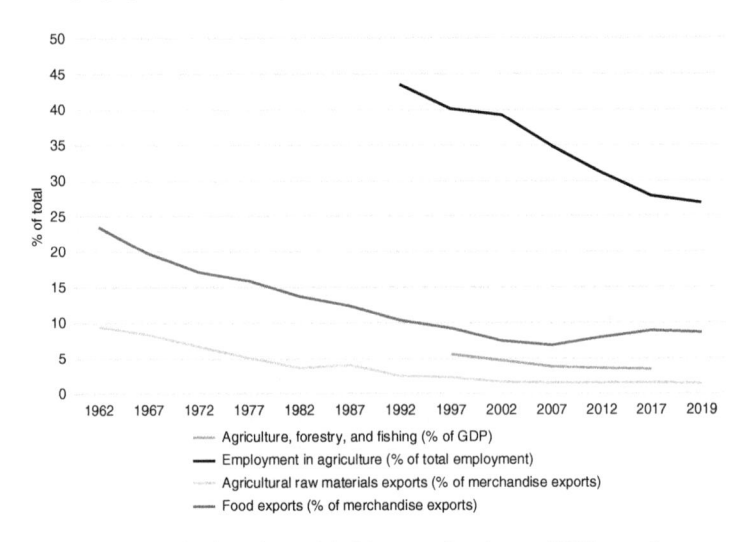

Figure 2.2 Agriculture's worldwide contribution to GDP, employment, and exports, 1962–2019. *Source*: World Bank database.

at least the 1960s – albeit with the qualification that this long-term trend has stalled, and by some measures gone into reverse, since 2000.[32]

As a corollary of this, the political importance of agriculture and agricultural interests and agendas are also in secular decline. The key economic bases of regime and elite power are now much more likely to lie in minerals, industry or services, than in agricultural production. Elites are less and less of the landed variety. Moreover, as ever-more people migrate to expanding cities, so the mass social bases of regimes, parties and opposition movements tend to shift away from the peasantry towards urban slums (contrast, for instance, the peasant social bases which brought Arab nationalist post-colonial regimes to power with the largely urban roots of contemporary Islamist movements).[33] There are of course important differences between different countries; and not everything is determined by economic rationality alone. Yet relations of production and exchange are foundational to patterns of political power and conflict.

---

[32] World Bank, 'Agriculture, forestry, and fishing, value added (% of GDP)', 'Employment in agriculture (% of total employment) (modeled ILO estimate)', 'Agricultural raw materials exports (% of merchandise exports)' and 'Food exports (% of merchandise exports)', World Development Indicators DataBank, https://databank.worldbank.org/source/world-development-indicators.

[33] R. Hinnebusch, 'Rural politics in Ba'athist Syria: a case study in the role of the countryside in the political development of Arab societies', *Review of Politics*, 44:1 (1982), 110–30; G. Kepel, *Jihad: The Trail of Political Islam* (I.B. Tauris, 2002), ch. 3.

And it is not difficult to see that if agriculture were as economically significant today as it was in the early or mid-twentieth century, let alone earlier, then the risk of local water scarcities being translated into political animosities and in turn water wars would be considerably higher. Stated differently, it should be no surprise that there have been no modern water wars given that the sector where most water is used, agriculture, is far from a priority for most of today's urban-based, and globally and financially oriented, regimes and elites.

The same conclusion is suggested by the economics of freshwater production and supply. At least on a global scale, the costs of producing and supplying freshwater for agricultural, industrial and domestic consumption are both low, and declining in absolute as well as relative terms. Desalination costs have dropped from US$1.25–US$1.50 per cubic metre (cm) in the 1990s to about US$0.50/cm today. Admittedly, some riders are necessary here. The capital costs associated with major water infrastructure projects can be huge: the construction of Ethiopia's GERD, for instance, is projected to cost around US$5 billion, equivalent to around 7 per cent of the country's GDP. Real costs of production are often obscured by government subsidies. Conversely, prices have often been driven up in recent years in line with neo-liberal 'full cost pricing' doctrines. And local scarcities or restrictions can result in exorbitantly high local prices, especially for poorer classes and communities: many Palestinians living in Area C of the West Bank, for instance, have no option but to rely on unregulated tankers for their water supplies – which are, on average, five times more expensive than equivalent piped water, and can account for a quarter, or even a third, of total household income.[34] Yet for all these riders, water is generally not a significant expense to national economies, regimes or elites, and is becoming ever less so. Consider the following observation, made in 1998 by a leading Israeli water expert, on the subject of the Israeli–Palestinian water conflict: 'The whole issue is about 100 mcm in the foreseeable future, and 100 mcm desalinated from the sea is $100 million, $100 million when Israel's GDP is already $100 billion. That makes it 0.1 per cent of GDP. So from an economic or financial point of view, it's irrelevant, water is irrelevant'.[35] Twenty years on, Israel's GDP has trebled while, as noted above, the cost of desalination has more than halved. It should be little wonder, given such trends, that there have been no recent water wars.

---

[34] S. Loutatidou et al., 'What is sustainable desalination?', in H. A. Arafat (ed.), *Desalination Sustainability: A Technical, Socioeconomic, and Environmental Approach* (Elsevier, 2017), 19; W. Abtew and S. B. Dessu, *The Grand Ethiopian Renaissance Dam on the Blue Nile* (Springer, 2019), 161; Amnesty International, *Troubled Waters: Palestinians Denied Fair Access to Water* (2009), 28, 42–3, 50, 59, 61.

[35] Selby interview with Saul Arlosoroff (05/04/1998).

Eco-Determinism: The Zombie Theory of Politics

All these substantive shortcomings aside – the history of wayward forecasts, the paucity of evidence from cases, the flawed understanding of the political economy of water, and the contrary evidence of widespread water cooperation – water wars discourse also rests on the most unstable of theoretical foundations. These foundations are provided principally by the work of the English parson and political economist Thomas Malthus, assorted environmental and geographical determinists, and 1990s scholarship on environmental security – work which, collectively, advances what may be thought of as a fundamentally 'eco-determinist' reading of humankind's relationship with nature. Given that such eco-determinist thought also provides the main theoretical underpinning for scholarly and policy discourse on climate security, it is well worth reflecting on precisely where it goes so wrong.

In a recent essay on the history and future of famine, Alex de Waal observed that Malthusianism is a 'zombie theory', a shallow theoretical edifice which somehow keeps rising from the dead even in the face of overwhelming evidence to the contrary. But Malthusianism – and indeed the eco-determinist tradition as a whole – is the ultimate zombie theory of politics in a more substantive sense, too.[36] Eco-determinist thought rests upon an essentially zoological understanding of the human condition, wherein the human species is understood as fundamentally akin to animals or indeed zombies, a mass of mindless, multiplying bodies whose interactions with the external environment are basically defined by their constant, voracious, and in aggregate terms ever-rising appetites. Eco-determinism also depicts the environment and its resources as essentially external forces, that repeatedly stand in the way of humanity's restless consumption. It views conflict and crisis as occurring when the demands of the flesh hit against inviolable environmental scarcities and limits. It has, in consequence, a localist and often internalist bias, viewing these conflicts and crises as occurring in those locations, often countries, where bodily demands and environmental constraints meet. And it is both apocalyptic and often fantastically imaginative about the future. Very much like the zombie genre – and no doubt owing something to it – the eco-determinist tradition is a pedagogy of civilisational collapse.

Moreover, in each of these respects eco-determinism is an essentially fictional or faith-based mode of thought, its zoological conception of the human condition being supported by neither evidence nor reason. Thus

---

[36] A. de Waal, 'The end of famine? Prospects for the elimination of mass starvation by political action', *Political Geography*, 62 (2018), 188, 191; D. W. Drezner, *Theories of International Politics and Zombies*, rev. ed. (Princeton University Press, 2014).

Malthus' belief in a universal 'principle of population' – according to which populations naturally expand geometrically but their means of subsistence at most arithmetically – and his conviction that the races of plants, animals and man all equally 'shrink under this general restrictive law' were just that: ungrounded a priori suppositions. The key source used by Malthus in developing this principle, an essay by American colonist Benjamin Franklin, provided no evidential basis for it and was either misunderstood or misrepresented by him; as Alison Bashford and Joyce Chaplin observe, Malthus' political arithmetic 'was an item of faith, quite literally so'. Equally, the likes of Garrett Hardin's 'tragedy of the commons' thesis, which argued that the degradation of common land is inevitable where growing biological and population needs combine with an absence of private property rights, and Jared Diamond's *Collapse*, which interprets a range of ancient and modern societal 'collapses' as centrally caused by population growth, rest on a remarkable absence of evidence, or even consideration, of precisely how population pressures translate into social and environmental change.[37]

Ceteris paribus, larger populations of course have bigger consumption needs than smaller ones. Yet, at the risk of protesting the obvious, this truism also needs to be considered alongside several others. First, that humankind's material relationship with the environment involves not just the latter's use and consumption, but also its transformation through human labour, and thus the continuous creation, re-creation and production of new resources and a newly humanised nature; as Erich Zimmermann famously observed, 'resources are not, they become'. Second, that neither the consumptive nor the productive dimensions of the nature–society relationship are unthinking and animalistic in form, but instead are always rooted in purposive human actions informed by learning, innovation and adaptation – in capacities which make humans the 'ultimate resource', as Julian Simon famously declared – and which at the same time are always shaped by social structures and processes, whether norms, technologies or relations of power. Third, that human population growth does not follow the dynamics of the animal kingdom either, but is instead always configured by social forces, including by local and global inequalities, hierarchies and patterns of conflict. Fourth, that population growth not only, *ceteris paribus*, increases demands on the environment, but can simultaneously spur productivity and innovation, enhancing the capacity of societies to adapt to, and even transcend, their resource problems. And fifth, that the unprecedented pressures which

---

[37] Malthus, *Essay*, 1st ed., 14; Bashford and Chaplin, *New Worlds*, 70-2, 83-4; G. Hardin, 'The tragedy of the commons', *Science*, 162:3859 (1968), 1243-8; Diamond, *Collapse*.

humankind is today placing on the Earth's resources derive less from trans-historical, let alone trans-species, dynamics of population growth than from qualitatively unprecedented consumption patterns underpinned by a historically specific social form: in three words, mass consumer capitalism. Lest there be any doubt about this, consider not just that average global economic growth currently far exceeds global population growth (around 3 per cent compared to just 1 per cent), but that the latter rate is declining while the former is not, and that population growth is also predominantly concentrated among the poorest and least resource-consumptive segments of the global South. While population undoubtedly matters – such that the question of population should not be taboo, as in some quarters it has become – its causes and consequences, and its value and politics, are much more complex than imagined within the eco-determinist canon.[38]

Eco-determinism is also mistaken in privileging scarcity and natural limits as the central motors of resource conflicts and crises. 'Scarcity', as we have seen, is the central preoccupation of environmental security discourse, as illustrated by Homer-Dixon's dictum that 'environmental scarcity causes violent conflict', by the countless variations on this theme within academic studies – claims that '[r]esource poor areas create environments that are highly competitive', that '[m]ost conflicts are over some type of resource perceived as scarce' and so on – and by the omnipresence of scarcity talk within policy discourse on the subject.[39] However, the same is also true of eco-determinism more broadly: from Malthus right through to Diamond and Kaplan, it is demographically and geographically defined scarcities that are held to be the central factors in societal crisis and collapse.

Yet this axiomatic recourse to scarcity is problematic on many levels. First, just as environmental resources are produced and made available through technologies and human labour – in the sense that neither the seabed, seawater nor sewage are resources until technologies and economies allow – so by the same token 'scarcities' are social constructs too: as David Harvey has put it: '[t]o declare a state of ecoscarcity is in effect to say that we have not the will, wit or capacity to change our state of knowledge, our social goals, cultural modes, and technological mixes,

---

[38] E. Zimmermann, *World Resources and Industries*, rev. ed. (Harper and Row, 1951), 15; J. Simon, *The Ultimate Resource* (Princeton University Press, 1981); E. Boserup, *The Conditions of Agricultural Growth: The Economics of Agrarian Change under Population Pressure* (Transaction Publishers, 2006 [1965]); D. Harvey, *Justice, Nature and the Geography of Difference* (Blackwell, 1996), 120–49.

[39] Homer-Dixon, 'Environmental scarcities', 39; Hensel et al., 'Conflict management', 385; N. P. Gleditsch et al., 'Conflicts over shared rivers: resource scarcity or fuzzy boundaries?', *Political Geography*, 25:4 (2006), 362.

or our form of economy, and that we are powerless to modify either our material practices or "nature" according to human requirements'.[40] Second, outside of environmental security debates, it is the 'abundance' of resources rather than their scarcity which is usually viewed as key, with work on the political economy of civil wars, for instance, often identifying locally abundant diamonds, coltan, oil and other resources as 'curses' that are correlated with civil war onset and intensity.[41] That there exist such diametrically opposed lenses on resource conflict is at the very least puzzling and provides reason to doubt whether 'scarcity' really is the most appropriate conceptual vehicle for analysing environment–conflict linkages. More fundamentally, however, the main theoretical problem with scarcity talk is that 'scarcity' and 'abundance' are relational concepts which, like the terms 'master' and 'slave', only make sense because of the simultaneous existence of the other; given this, it is illogical to associate certain conflicts with abundance but not simultaneously with scarcity – and no more logical to claim, as Homer-Dixon does, that it is 'environmental scarcity', but not abundance, that 'causes violent conflict'.

Furthermore, there is no good reason to suppose that environment- or resource-related conflicts are fundamentally caused by deepening scarcities – that is, by degradation, over-exploitation or rising demand. Homer-Dixon portrays environmental 'resource capture' as only occurring when 'a fall in the quality and quantity of a renewable resource interacts with population growth to encourage powerful groups within a society to shift resource distribution in their favor'.[42] But *pace* this view, projects of resource capture often occur without being preceded by a worsening of, or any other change in, resource–demand equations. 'Blood diamond' conflicts in Angola, Sierra Leone and elsewhere have not been sparked by changes in diamond availability or demand, but rather by changing local political forces and strategies, within the context of the consistently high monetary valuation of diamonds on global markets. Likewise, the 2003 US-led invasion of Iraq was not caused by any sudden or secular changes in the availability of oil, but rather by specific ideological and political configurations within the broader context of oil's centrality to capitalist reproduction. Just as in these examples, there is no good reason why projects of environmental resource capture, including the capture of water resources,

---

[40] Harvey, *Justice, Nature*, 147.
[41] P. Collier and A. Hoeffler, 'On economic causes of civil war', *Oxford Economic Papers*, 50:4 (1998), 563–73; J. D. Fearon, 'Primary commodity exports and civil war', *Journal of Conflict Resolution*, 49:4 (2005), 483–507; and for a useful review, V. Koubi et al., 'Do natural resources matter for interstate and intrastate armed conflict?', *Journal of Peace Research*, 51:2 (2014), 227–43; Selby and Hoffmann, 'Beyond scarcity', 361–3.
[42] Homer-Dixon, *Environment, Scarcity and Violence*, 73.

need be rooted in worsening resource availability; as Philippe Le Billon puts it, 'the availability *in nature* of any resource' is 'not in itself a predictive indicator of conflict'. Indeed, as political ecologists have long argued, 'resource conflicts' are never in-the-moment reactions to resource disequilibria but rather long-term and strategic contestations with complex sociopolitical and moral dimensions.[43] Moreover, what the above examples suggest is that it is the economic and political value of environmental resources, rather than their relative scarcity or abundance, that is the crucial determinant of their conflict potential (and as discussed in the previous section, water's value is low and declining). The eco-determinist assumption that resource conflicts and insecurities are sparked by sudden or deepening scarcities is just that: an eco-centric and deterministic assumption without logical basis.

Eco-determinism's traditionally localist and internalist biases are similarly problematic, licensing a persistent inattention to the international, global and imperial dimensions of environment-related insecurities. For not only do eco-determinist analyses generally interpret scarcity as locally caused and constituted, as arising from locationally specific geography–demography imbalances; in addition, they typically view it and its sociopolitical consequences as conterminous, that is, as occurring within the very same locales and from there spilling over to affect other countries or regions – for example, through out-migration. Admittedly, climate security discourse does depart somewhat from this localism, in that it insists on the impacts of greenhouse gas emissions, most of which have emanated from the global North, on patterns of scarcity and insecurity in the global South. Yet on most other non-local dynamics, eco-determinism is either woefully silent, or else relegates them to the realm of ad hoc or contextual factors. The first and only sustained decline in Easter Island's population, which Diamond pins on local 'ecocide', actually took place following the arrival of Dutch explorers and their germs on Easter Sunday, 1722; it was this catastrophic encounter with empire that wiped out Easter Island's civilisation, not just internal causes. Equally, the Irish potato famine, which Malthus and followers interpreted as a 'most dreadful resource of nature', actually took place within the context of a food surplus and was in large measure an outcome of British colonial policy.[44] Fast forward to our contemporary world of globally

---

[43] Le Billon, 'The political ecology of war', 563; M. D. Turner, 'Political ecology and the moral dimensions of "resource conflicts": the case of farmer-herder conflicts in the Sahel', *Political Geography*, 23:7 (2004), 863–89.

[44] T. Hunt and C. Lipo, 'Ecological catastrophe, collapse, and the myth of "ecocide" on Rapa Nui (Easter Island)', in P. A. McAnany and N. Yoffee (eds.), *Questioning Collapse: Human Resilience, Ecological Vulnerability, and the Aftermath of Empire* (Cambridge

integrated commodity, communications, financial and migration networks, and it is hard to imagine that any social, political or environment-related conflict could be essentially local in its causes. Yet one searches in vain through environmental security studies for systematic discussion of the impacts of international trade, multinational corporations, liberalising reforms, new financial instruments or transnational political forces on local environmental vulnerabilities and insecurities. That eco-determinism is so blind to such dynamics is not for want of evidence, but because at a fundamental level its zoological model of the human condition precludes any sustained analysis of extra-local social relations.

Last, the spectres of civilisational collapse which are so central to eco-determinist thought – from Malthus' grim prediction of the world's population being levelled 'with one mighty blow', to Diamond's fear of mass ecocide and Kaplan's dream of geography enacting its 'revenge' – are also deserving of scepticism. Viewed charitably, such fears and prognoses arise from, and are illustrative of, the conservative tradition in the philosophy of history, according to which human history oscillates between periods of progress and others of tragedy, regress and decline. Viewed less charitably, however, the idea of ecologically-induced collapse is essentially a religiously-inspired myth. This is evident in the closing theodicy of *Principle of Population*, where Malthus argues that scarcity-induced misery and death is ultimately for the best, ordained by God to 'soften and humanise the heart' and propel humanity into productive labour. It is evident in how collapse discourse invokes and updates symbols and storylines from Christian theology. It is indicated too by the fact that apocalypticism has been a recurring feature of American public life – which is from where contemporary collapse narratives derive, above all – ever since the arrival of the first Puritans in 1620.[45] And it is revealed too by the poverty of the evidence relating to historical collapses. Collapse discourse typically takes its cues from visual spectacles of abandoned ruins: Easter Island, the Maya ruins of the Yucatán and, as noted in Chapter 1, Shelley's 'Ozymandias'. Yet understood as a social rather than just an archaeological phenomena, ecologically-induced collapse has seldom, and perhaps never, occurred. The outstanding modern 'collapse' – the decimation of the population of the Americas between 1492 and 1650, when 70–90 per cent of the Indigenous population died and

University Press, 2010), 38–9; Malthus, *Essay*, 1st ed., 61; C. Woodham-Smith, *The Great Hunger: Ireland 1845–9* (Harper and Row, 1962).

[45] Malthus, *Essay*, 1st ed., 141–58; D. Rothe, 'Governing the end times? Planet politics and the secular eschatology of the Anthropocene', *Millennium*, 48:2 (2020), 143–64; B. Hartmann, *The America Syndrome: Apocalypse, War, and Our Call to Greatness* (Seven Stories Press, 2017).

the resulting reforestation led to a conspicuously sharp decline in global atmospheric carbon dioxide levels – was caused by European settlement and strategies of genocide, not local geography or demography. Moreover, most claimed historical 'collapses' bear witness to adaptation, migration, resilience and social transformation, not just decline or extinction. Classical Mayan civilisation may have 'collapsed', but Maya society continued to thrive long afterwards, such that seven million Maya descendants are alive today. Like the rhetoric of modern 'state failure', the idea of civilisational 'collapse' is premised on a misleadingly binary understanding of social change – of there being essentially two conditions, the successful status quo and its fall – which naturalises the status quo, obscures internal differences in power and experience, and is insensitive to the structural historical transformations that all societies undergo.[46] Neither collapse discourse specifically, nor eco-determinism more broadly, offer much of scientific value.

The Politics of Scarcity Narratives

Before exploring the implications of all this for a world with a changing climate, the political and normative shortcomings of eco-determinist thought also deserve brief consideration. For, with a constancy which is surely beyond accident, eco-determinism has repeatedly revealed itself to be a hyper-conservative – and indeed regressive – tradition of political thought. Malthus argued, in his *Principle of Population*, not only against advocates of the 'perfectibility of mankind', but also against the possibility of 'any very marked and striking change for the better in the form and structure of general society'. He called for the complete abolition of poor relief on the grounds that it enabled that part of society 'that cannot in general be considered as the most valuable' to have larger families. And he accompanied this, in his *Principles of Political Economy*, with a call for the propertied classes to increase their consumption, in order to forestall economic stagnation. In Malthus, as Harvey remarks, 'the Enlightenment project [is] reserved for a small elite while everyone else is condemned to live by natural law. This is an appalling instance of that

---

[46] S. N. Eisenstadt, 'Beyond collapse', in N. Yoffee and G. L. Cowgill (eds.), *The Collapse of Ancient States and Civilizations* (University of Arizona Press, 1988), 236–43; J. A. Tainter, 'Archaeology of overshoot and collapse', *Annual Review of Anthropology*, 35 (2006), 71–2; S. L. Lewis and M. A. Maslin, 'Defining the Anthropocene', *Nature*, 519 (2015), 171–80; S. L. Lewis and M. A. Maslin, *The Human Planet: How We Created the Anthropocene* (Penguin, 2018), ch. 5; P. A. McAnany and T. G. Negrón, 'Bellicose rulers and climatological peril: retrofitting twenty-first-century woes on eighth-century Maya society', in McAnany and Yoffee, *Questioning Collapse*, 163–5; J. Hill, 'Beyond the other? A postcolonial critique of the failed state thesis', *African Identities*, 3:2 (2005), 139–54.

awful habit of denying one section of our species the right to be considered human'.[47]

Malthus is no outlier here. Ellsworth Huntington's views blurred similarly into elitist ultra-conservatism – into claims that the 'whole lesson of biology is that America is seriously endangering her future by making fetishes of equality, democracy, and universal education'; into condemnation of modern societies for permitting 'the people with the greatest mental power to have the fewest children, while the stupid breed like rabbits'; into proposals for strict US immigration control; and ultimately into eugenics, his 1935 book *Tomorrow's Children* being 'probably the simplest and clearest exposition of eugenic principles then available'. William and Paul Paddock, and later Paul Ehrlich in *The Population Bomb*, advocated the abolition of international food aid to those countries in the 'can't be saved' and 'will die regardless' category (which included Haiti, Egypt and India, among others); 'to send food to them is to throw sand in the ocean', the Paddocks observed. And Garrett Hardin famously argued in his 'lifeboat ethics' essay that nations should be viewed as separate 'lifeboats', each with their limited 'carrying capacity', and – on the strength of this extreme methodological nationalism – that humanitarian aid and North–South immigration should be disallowed.[48]

Indeed, eco-determinism has been an inveterate ally to imperialism, authoritarianism and militarism. During the heyday of European empire, eco-determinism repeatedly licensed attempts to 'civilise' supposedly 'natural' subjects, whether through population control, forced relocation or the destruction of traditional livelihoods (we explore this issue further in Chapter 4). During the Cold War, it became a major component of US foreign policy, as the United States pursued a war on female fertility with the aim of reducing Third World population growth and containing Communism. And in post-colonial Africa, eco-determinist premises have often been invoked by authoritarian states to excuse or obscure their own violent state-building endeavours. Also noteworthy is that eco-determinist thought includes no environmentalists: it includes no environmental movements, grassroots organisations or writer–activists in

---

[47] Malthus, *Essay*, 1st ed., 113, 39; T. Malthus, *Essay*, variorum ed., ed. P. James (Cambridge University Press, 1989), II, 137–47; T. Malthus, *The Principles of Political Economy*, variorum ed. (Cambridge University Press, 1989); Harvey, *Justice, Nature*, 145.

[48] E. Huntington, *The Yale Review*, 6:3 (1917), 670, in G. Martin, *Ellsworth Huntington: His Life and Thought* (Archon Books, 1973), 177; E. Huntington, *World Power and Evolution* (Yale University Press, 1919), 18; Martin, *Ellsworth Huntington*, 186; W. Paddock and P. Paddock, *Famine – 1975! America's Decision: Who Will Survive?* (Little, Brown and Company, 1967), 207; P. Ehrlich and A. Ehrlich, *Population Bomb*, rev. ed. (Ballantine Books, 1978), 146–8; G. Hardin, 'Living on a lifeboat', *BioScience*, 24:10 (1974), 561–8.

possession of political agency, and definitely no 'environmentalism of the poor' (the only environmentalists within the eco-determinist imaginary are its proponents; their objects of analysis are essentially assumed to be bodies without agency). Moreover, as illustrated by Diamond's account of Easter Island and Malthusian accounts of the great Irish famine, eco-determinism routinely misrepresents history's victims as unwitting causes of their own marginalisation and demise. As Piers Blaikie observed in his classic political ecology of soil erosion, eco-determinist reasoning is in this sense inherently colonial. Or as another critique has put it: '[e]nvironmental determinism is not critical environmental politics; it is a bourgeois social theory'.[49]

Something very similar applies to water scarcity and water wars discourse specifically. Thus narratives of 'absolute water scarcity' are most typically invoked by, and fit the interests of, hydro-hegemonic powers such as Egypt and most clearly Israel, where the language of natural scarcity serves simultaneously as a celebration of national adaptive genius, and as an argument against redistribution to, or cooperation with, other riparians. The rhetoric of water scarcity also complements and reflects status quo interests: for example, as Samer Alatout has shown, the Zionist movement initially portrayed Palestine's waters as abundant, only embracing the diametrically opposed language of scarcity after its dominant position over regional water resources had been successfully achieved. As for 'water wars', such framings have repeatedly been invoked by authoritarian and military elites to project themselves as defensive representatives of the life and blood of the nation: 'we depend upon the Nile 100 per cent in our life ... if you do not go to war for a matter of life or death, and water is life for your people, then for what reason do you go to war?' was the rhetorical question posed by Egyptian President Anwar Sadat in 1978; if the Nile 'diminishes by one drop, then our blood is the alternative', observed his successor Mohammed Morsi in 2013; the Six Day War really began as an existential defence of Israel's water resources, claimed Ariel Sharon.[50] In all such rhetorical framings the securitisation

---

[49] B. Hartmann, *Reproductive Rights and Wrongs: The Global Politics of Population Control* (South End Press, 1995), 93–112; H. Verhoeven, 'Gardens of Eden or Hearts of Darkness? The genealogy of discourses on environmental security and climate wars in Africa', *Geopolitics*, 19:4 (2014), 793–8; R. Guha and J. Martinez-Alier, *Varieties of Environmentalism: Essays North and South* (Earthscan, 1997); Nixon, *Slow Violence*; Blaikie, *The Political Economy*, 53, 56; D. Correia, 'F**k Jared Diamond', *Capitalism Nature Socialism*, 24:4 (2013), 5.

[50] S. Alatout, '"States" of scarcity: water, space, and identity politics in Israel, 1948–59', *Environment and Planning D: Society and Space*, 26:6 (2008), 959–82; 'Egypt will go to war if Ethiopia tampers with Nile water or attacks Sudan', *Jordan Times* (31/05/1978); 'Egypt escalates war of words over Ethiopia's Nile dam project', *Sudan Tribune* (11/06/

of water follows inexorably from eco-determinist premises, with local geographical 'scarcities' dictating a supposed need for actions to enforce 'security'. Not only is such rhetoric misleading, standing in stark contrast to the actual historical record of modern-day 'water wars'; it is also politically questionable. Whether in relation to water, climate or environmental security more broadly, arguments from geography and demography to security are essentially a weapon of the powerful.

A Revenge of Geography?

Before turning to climate change, let us recapitulate the argument thus far. Our main arguments have been fivefold. First, that physical geography and its battle with demography tell us little about actual patterns of water abundance, water shortage and water-related insecurity as they are experienced by populations and communities. Second, that the underlying reason for this is that, under capitalism, it is wealth, power, technologies and so on, and not local environmental endowments, which are the keys to secure life. Third, that there is negligible evidence of geographically or demographically defined water scarcities causing or contributing to recent violent conflicts, the essential reason for this being the low and declining economic and political importance of water to states, economies and their elites. Fourth, and at a more theoretical level, that the privileging of geography and demography as the central determinants of conflict and crisis – a privileging which we label 'eco-determinist' – is deeply misleading, misrepresenting not just the human condition, but also the causes and significance of scarcity under conditions of capitalism, the extra-local causes of environmental insecurities and the actual historical record of societal transformation and 'collapse'. And fifth, that in addition to its substantive flaws, eco-determinist and water wars reasoning is also politically questionable, routinely segueing into advocacy of highly regressive social policies and repeatedly legitimising or complementing imperial and hegemonic state agendas. Such are the multiple flaws of 'geography versus demography' reasoning on questions of water and environmental security.

What, though, of climate change; is climate change not of such gravity as to completely transform the human–environment relationship to the extent that climate security discourse is resistant, or even immune, to the above lines of critique? We have already seen, in Chapter 1 and earlier in this chapter, that standard climate security discourse shares much in common

2013); A. Sharon with D. Chanoff, *Warrior: The Autobiography of Ariel Sharon* (Simon and Schuster, 1989), 167.

both with water wars arguments and the eco-determinist tradition more broadly. Like water wars discourse and the entire eco-determinist tradition, climate security discourse identifies resource scarcity crises as a central cause – even the central motor – of conflict and disorder. Like water wars discourse, it foresees a twenty-first century of deepening scarcity-induced conflict. And it is very much a pedagogy of civilisational collapse. But does this thereby render climate security reasoning invalid?

Let us first consider the evidence on how climate change is affecting and will affect water resource availability, before then placing these changes in economic and social context. It is often said that as climate change deepens there will be a dual pattern of 'wet gets wetter' and 'dry gets drier' – indeed, there is a firm scientific consensus on this.[51] For dryland regions like those examined in this book, it thus seems highly likely that climate change will place extra pressures on water sectors and their resources. This much can be stated with some confidence. Yet a number of qualifications are also in order.

First, there is only limited evidence that climate change is already reducing water resource availability; indeed, on certain scores the evidence suggests the contrary. Since 1900 there has been an increase, not a decrease, in globally averaged precipitation, alongside increased water loss from evapotranspiration thanks to higher temperatures – this much is clear. However, the relative global importance of these and other countervailing trends is disputed, so that some studies identify a global drying trend but others do not.[52] Complicating matters still further, in recent decades there has been an overall global greening trend, with a larger land area witnessing greening than that seeing browning: in relation to drylands specifically, between the 1980s and 2000s less than 10 per cent of the world's dryland area saw a decline in vegetation productivity; the large majority of dryland did not. At regional levels, West Asia has overall seen a weak but non-significant decline in average precipitation; the Sahel saw reduced rainfall over most of the twentieth century but recovery after 1980, alongside a clear pattern of greening; and most of the Mediterranean has, by most analyses, become drier.[53]

---

[51] IPCC, *Climate Change 2013*, 984.
[52] IPCC, *Climate Change 2013*, 201; A. Dai, 'Increasing drought under global warming in observations and models', *Nature Climate Change*, 3 (2013), 52–8; J. Sheffield et al., 'Little change in global drought over the past 60 years', *Nature*, 491 (2012), 435–8.
[53] IPCC, *Climate Change and Land*, 9, 251; IPCC, *Climate Change 2014: Impacts, Adaptation, and Vulnerability. Part B: Regional Aspects. Contribution of Working Group II to the Fifth Assessment Report of the Intergovernmental Panel on Climate Change* (Cambridge University Press, 2014), 1209, 1333; P. M. Sousa et al., 'Trends and extremes of drought indices throughout the 20th century in the Mediterranean', *Natural Hazards and Earth*

Projections of climate change impacts on precipitation and water resources present a similarly mixed picture. Globally, a gradual increase in global precipitation is considered 'virtually certain' over the course of the twenty-first century, alongside a 'general intensification of the global hydrological cycle'. Alongside this, many regions are expected to become considerably drier: drying is, for instance, considered likely in the Mediterranean region and West Asia. Yet several qualifications to this overall picture are required. It is not clear whether the world will witness greening or browning overall, nor whether there will be a positive or negative change in the extent of drylands. Precipitation projections vary greatly between models, much more so than in the case of temperature projections. For some transitional zones there are particularly large uncertainties, even over the sign of the projected change: individual Global Circulation Model projections for Nyala in Darfur, for example, suggest anywhere from a 21 per cent increase in precipitation to a 25 per cent decrease by the mid-twenty-first century. Moreover, for the near term, the magnitude of changes in precipitation is projected to be small compared to natural variability. And most of the long-term projections of significant precipitation or drying trends rely on one particular greenhouse gas emissions scenario – Representative Concentration Pathway (RCP) 8.5 – which, though often framed as a 'business as usual' scenario, is now widely considered highly unlikely or even implausible, including by the IPCC.[54]

Israel–Palestine illustrates this, in microcosm. It is regularly claimed that climate change will have severe negative impacts on precipitation and hence water resource availability there – for instance, that Israeli supplies may decrease by 60 per cent by the turn of the century, and that under certain scenarios Jordan River flow may drop by a staggering 98 per cent. It is also often claimed that climate change impacts on precipitation are 'already evident' in much of West Asia and North Africa, including Israel–Palestine; that climate change 'is no longer a gloomy projection ... but a new hydrological reality', having already led to a 10 per cent decline in annual rainfall in Israel; and in Prime Minister Binyamin Netanyahu's

---

System Sciences, 11:1 (2011), 33–51; M. Hoerling et al., 'On the increased frequency of Mediterranean drought', *Journal of Climate*, 25:6 (2012), 2146–61.

[54] IPCC, *Climate Change 2013*, 1032, 1055, 984, 45, 1079, 984, 986; IPCC, *Climate Change and Land*, 144, 254; A. Bruggeman et al., *Climate Outlooks for CLICO Case Study Sites* (Cyprus Institute, 2010); E. Ho et al., 'Not all carbon dioxide emission scenarios are equally likely: a subjective expert assessment', *Climatic Change*, 155 (2019), 545–61; Z. Hausfather and G. P. Peters, 'Emissions – the "business as usual" story is misleading', *Nature*, 577 (2020), 618–20; J. Ritchie and H. Dowlatabadi, 'The 1000 GtC coal question: are cases of vastly expanded future coal combustion still plausible?', *Energy Economics*, 65 (2017), 16–31; IPCC, *Climate Change 2021: The Physical Science Basis. Working Group I Contribution to the Sixth Assessment Report of the Intergovernmental Panel on Climate Change* (2021), 304.

extreme formulation, that rainfall levels in Israel have more than halved since 1948.[55] In turn, many maintain that these environmental changes will generate crisis and conflict. Thus International Alert has identified Israel–Palestine as 'facing a high risk of armed conflict as a knock-on consequence of climate change'. Israeli scientist Arie Issar argues, on the strength of an eco-determinist reading of Israeli history, that a climate change-induced 'water shortage catastrophe ... seems imminent'. Leon Fuerth, onetime national security advisor to Al Gore, has speculated that by 2037 efforts to resolve the Israeli–Palestinian conflict might be indefinitely abandoned because of climate change, following a 'collective conclusion that the problem of sharing water supplies must be regarded as permanently intractable'.[56] Many others have argued or speculated similarly.

Such claims rest on the flimsiest of foundations, however. There is no clear existing trend of declining rainfall within Israel: rainfall levels may be represented as either increasing or decreasing, depending on the time period selected. Moreover, while projections of climate change impacts on Israeli–Palestinian water resources vary in their findings, most suggest relatively small changes in rainfall and river flows: Jonathan Chenoweth and colleagues' projections using a mid-range IPCC scenario, for example, suggest a mid-century precipitation decline in Israel of just 13 per cent (15 per cent in Palestine), a decline by 2070–99 of 19 per cent (23 per cent in Palestine) and a decrease in Jordan River discharge by 2070–99 of 30 per cent (other studies have found changes of a similar order, or in some cases much less).[57] To put these figures in perspective,

---

[55] Israeli Ministry of Environment, *Israel National Report on Climate Change: First National Communication to the Conference of the Parties to the United Nations Framework Convention on Climate Change* (2000), 23; A. Kitoh et al., 'First super-high-resolution model projection that the ancient "Fertile Crescent" will disappear in this century', *Hydrological Research Letters*, 2 (2008), 3; J. Sowers et al., 'Climate change, water resources, and the politics of adaptation in the Middle East and North Africa', *Climatic Change*, 104: 3–4 (2011), 599–627; A. Tal, 'Building a hydrological future', *Ha'aretz* (22/05/2009); D. Pine, 'Netanyahu visits Silicon Valley, signs Israel–California pro-business pact', *J* (07/03/2014).

[56] D. Smith and J. Vivekananda, *A Climate of Conflict: The Links between Climate Change, Peace and War* (International Alert, 2007), 44; A. S. Issar, 'The forecasted negative impact of global warming on the water resources of the Middle East and how to mitigate it', in H. Shuval and H. Dweik (eds.), *Water for Life in the Middle East* (Israeli/Palestine Center for Research and Information, 2006), 388; L. Fuerth, 'Security implications of climate change 2: severe climate change over next 30 years', in K. Campbell et al., *The Age of Consequences: The Foreign Policy and National Security Implications of Global Climate Change* (Center for Strategic and International Studies/Center for a New American Security 2007), 76.

[57] C. Messerschmid, 'Nothing new in the Middle East: reality and discourses of climate change in the Israeli-Palestinian conflict', in J. Scheffran et al. (eds.), *Climate Change, Human Security and Violent Conflict* (Springer-Verlag, 2012), 435–6; J. Chenoweth et al.,

at current official growth rates it will take just 7.5 years for the Palestinian population, and just 6.5 years for the Palestinian economy, to expand by 23 per cent; by these figures (and putting to one side, for a moment, the fact that people are producers as well as consumers of water), population and economic growth will have much greater impacts on aggregate levels of water stress than climate change-induced changes in rainfall. Moreover, as hydro-geologist Clemens Messerschmid has convincingly demonstrated, in Palestine specifically, where access to water is shaped above all by Israeli occupation policies (as will be demonstrated in Chapters 6, 7 and 8), the projected impacts of climate change on Palestinian water supplies 'pale into complete insignificance' relative to the political questions of water access and rights. By Messerschmid's calculations, under a 'full water rights + climate change' scenario Palestine would have access to 6.5 times as much freshwater as under a scenario of 'no water rights + no climate change'. Climate change impacts on rainfall are also likely to be of marginal economic significance: a 19 per cent decline in rainfall over Israel is equivalent to a 338 mcm/y decrease in the country's renewable water resource base (assuming a direct correspondence between rainfall and resource availability and using the FAO figures cited above) which, if compensated for through desalination, would at current rates amount to less than 0.05 per cent of Israel's GDP. Moreover, contrary to the suggestion that Jordan River flow might decrease by 98 per cent by the end of the century, the reality is that the lower Jordan *already* has an annual flow of just 2 per cent of pre-twentieth-century levels as a consequence of upstream abstractions by Israel and, to a lesser extent, Syria.[58]

This suggests, in turn, that climate change-induced changes in water availability will probably have only marginal impacts on patterns of conflict and violence in Israel–Palestine as well as beyond. Climate change will, no doubt, have additional effects on water beyond those considered above: warming will increase evapotranspiration rates while rising sea levels will, in due course, contribute to the salinisation of many low-lying aquifers. But even the addition of such factors does not alter the basic fact that climate change is unlikely to affect levels of water stress and availability nearly as much as population growth, economic growth, levels of technological innovation and investment in water

---

'Impact of climate change on the water resources of the Eastern Mediterranean and Middle East region: modelled 21st century changes and implications', *Water Resources Research*, 47:6 (2011), 9.

[58] Messerschmid, 'Nothing new', 432–3; Friends of the Earth Middle East, *The First Palestinian Symposium on Water Rights and Access to the Jordan River and Development Needs in the Jordan Valley* (2011).

infrastructure or political restrictions on access. Even quite mainstream scholars have reached similar verdicts, noting, for instance, that 'climate change effects on water resources are unlikely to have a discernible effect on Israeli–Palestinian relations' and that 'the likelihood that climate change will lead to a conflagration and violence beyond what might occur without climate change is low'. It is only when climate is abstracted from these other processes and likely developments – when the future is 'reduced to climate', as Mike Hulme puts it – and when stationary 'natural' patterns are assumed, anachronistically, to constitute the appropriate baseline for analysing this future, that it becomes plausible to conclude the opposite.[59]

Stated differently, while anthropogenic climate change will undoubtedly have far-reaching consequences for weather patterns, food production, rural livelihoods and more, such environmental changes will not, by themselves, upend the relationship between local geography and demography on the one hand, and patterns of politics and conflict under conditions of global capitalism on the other; within the context of the constant revolutionising of socio-environmental relations that characterises capitalism, global climate change is not so exceptional as to justify bringing eco-determinist thinking back from the dead. This means, in turn, that the multiple shortcomings of water wars and eco-determinist discourse discussed above will almost certainly continue to apply, even under conditions of accelerating global warming. Natural local geographical scarcities and population growth will not suddenly become key drivers of conflict, crisis and instability, simply because of additional pressures emanating from global climate change. There will be no 'revenge of geography'.

This does not mean, of course, that climate change has only marginal or negligible conflict and security implications; rather, what it means is that the presumed battle between local physical geographical constraints and demographic pressures is the wrong place to start if one wants to properly understand them. Where, then, should we begin; what should our starting point be? And before that – since our positive reconstruction of the relations between climate change, water, security and insecurity will begin only in Chapter 4 – are there not other types of evidence in the mainstream climate security and eco-determinist armoury which first require consideration?

---

[59] E. Feitelson et al., 'Climate change and security in the Israeli–Palestinian context', *Journal of Peace Research*, 49:1 (2012), 253; Hulme, 'Reducing the future'; P. C. D. Milly et al., 'Stationarity is dead: whither water management?', *Science*, 319:5863 (2008), 573–4.

3 Drought

More Drought, More Conflict?

Drought features more prominently than any other meteorological phenomenon in predictions of future climate chaos and climate-induced conflict. 'Climate change leads to drought, drought accelerates rural exodus, rural exodus leads to overpopulation in the cities, overpopulation fuels conflict and causes health and environmental problems. The chain of causality is never ending', then German Foreign Minister Heiko Maas claimed in 2019. 'There is little scientific dispute that if we do nothing, we will face more drought, more famine, more mass displacement – all of which will fuel more conflict for decades', Barack Obama pronounced on receipt of his Nobel Prize. 'Look at the Sahel where we see conflict spreading, terrorism spreading and one of the key factors ... is climate change accelerating drought and accelerating the competition between farmers and herders for scarce resources and facilitating the work of terrorists', UN Secretary-General António Guterres claimed on World Peace Day 2019.[1] As in these examples, drought is routinely portrayed as the surest route through which climate change will deepen and exacerbate conflict.

This focus on droughts is in many respects quite understandable, even reasonable. For most of human history and in most parts of the world, drought has been a perennial cause of hardship, malnutrition, destitution, famine, migration and mortality; and this history inevitably stalks fears about possible climate change futures, given that climate change is widely expected to usher in more and more extreme weather events, droughts included. There is an intuitively commonsensical – and also frightening – quality to claims that climate change will bring 'more droughts, more famine, more conflict'. Moreover, droughts are also compellingly visualisable and spectacular phenomena, readily providing images of cracked

---

[1] H. Maas, Speech at Partner Towns Conference on Sustainability, Bremen (04/03/2019); B. Obama, 'A just and lasting peace', Nobel Peace Prize lecture, Oslo (10/12/2009); A. Guterres, 'Remarks at Peace Bell Ceremony' (20/09/2019).

earth, parched fields and caravans of migrants that are perfect for communicating the threat of climate change, and are much less abstract seeming than '1.5°C average global warming'.

Yet there are also good reasons to question the supposition that increased drought – understood here as 'a period of abnormally dry weather long enough to cause a serious hydrological imbalance', that is, defined in 'meteorological' terms – will generate widespread conflict and instability under conditions of accelerating global heating.[2] Its intuitive appeal aside, the case for thinking that this will be the case really rests on four things: evidence from the first supposed 'climate wars', evidence from history, projections of future climate change and quantitative analyses. However, on all four fronts the evidence, and thus the grounds for concluding that climate change–induced droughts will trigger ever-increasing conflict and insecurity in future, is remarkably thin. We have already seen, in Chapter 1, that quantitative studies do not provide much clear guidance on climate–conflict relationships. Hence, in this chapter we turn our attention elsewhere, focusing mainly on evidence from cases – specifically, the major civil conflicts in Darfur from 2003 to 2005, the Lake Chad region from 2010 and Syria from 2011 – and complementing this with reflections on the global history of drought, and on the likely future of drought as global temperatures continue to rise. Through this we show that there is little solid evidence that climate change and drought are already sparking large-scale armed conflicts – and that the little evidence that does exist has consistently been overstated. We argue, against widespread assumptions to the contrary, that there has been a long-term historical trend of decreasing drought vulnerability, and we identify reasons for this in the development of structures, systems and practices of drought adaptation. We suggest on the basis of this that, while it is entirely possible that the recent trend of decreasing drought vulnerability may go into reverse, it is unlikely that such a reversal will be driven primarily by anthropogenic climate change. And we contend that it is not the weight of evidence but the questionable politics of drought narratives – the range of interests and agendas that 'drought' can be mobilised to serve – that accounts, above all, for their prominence within discussions of climate change and security. If the previous chapter showed that eco-determinism is misleading when applied to supposed long-term imbalances between resource availability and population demands, here we make a parallel case in relation to drought. Whether their focus is long-term imbalances or short-term environmental shocks, scarcity–conflict narratives are, in our assessment, misleading tout court.

---

[2] IPCC, *Climate Change 2014: Synthesis Report. Contribution of Working Groups I, II and III to the Fifth Assessment Report of the Intergovernmental Panel on Climate Change* (2014), 122.

Darfur 2003–5: The 'First Climate Change War'?

Let us begin with the 2003–5 war in Darfur, western Sudan, which became a poster child of climate security thinking in 2007 after UN Secretary-General Ban Ki-moon penned an op-ed in *The Washington Post* proclaiming that the crisis there 'began as an ecological crisis, arising at least in part from climate change'. From this statement, this latest of Darfur's many wars became widely represented as the world's 'first climate change war', with a multitude of actors – democratic and authoritarian political leaders, climate change activists, aid and development NGOs, defence planning organisations, international security think tanks, public intellectuals of various hues, and more – all converging on the claim that carbon emissions were a factor in the region's deep troubles. Darfur became represented not just as a localised African conflict, or even as a local genocide to which the world's humanitarians must urgently respond, but as a paradigmatic illustration of the dangers of planetary warming. 'Let Darfur stand as a starkest of warnings about what the future could bring', pronounced one leading development NGO.[3]

Unfortunately, such claims are not just overstated but fundamentally flawed – as has been widely, indeed consistently, demonstrated in scholarly analyses. They receive, in the first place, only the thinnest support from, and in some respects are contradicted by, the actual record of recent environmental change in Sudan. The Darfur war did not begin 'during a drought' as Ban Ki-moon claimed: to the contrary, during the three year period 2003–5 most areas of Darfur experienced above average rainfall. *Pace* the United Nations Environment Programme (UNEP) and others, rainfall levels in Darfur were also not gradually declining prior to 2003: during the 1990s and early 2000s they were generally above the thirty-year average, with no major droughts after 1990. This rainfall evidence is corroborated by evidence of a greening of Darfur and the African Sahel more widely from the early 1990s, which is visible through satellite imagery. Hence, if poor rains and resulting ecological and socio-economic changes were a causal factor in the Darfur war, the drought or droughts in question must have been those from the 1970s and 1980s, or earlier, with a multi-decadal time lag to the war itself. This possibility cannot be discounted, as discussed below. Yet there is little scientific

---

[3] B. Ki-moon, 'A climate culprit in Darfur', *Washington Post* (16/06/2007); Christian Aid, *Human Tide: The Real Migration Crisis* (2007), 2; and see, e.g., S. Faris, 'The real roots of Darfur', *Atlantic Monthly* (04/2007); T. Homer-Dixon, 'Terror in the weather forecast', *New York Times* (24/04/2007); J. D. Sachs, 'Ecology and political upheaval', *Scientific American* (26/06/2006); N. Stern et al., *The Economics of Climate Change* (Cambridge University Press, 2006), 112; A. Gore, *An Inconvenient Truth*, dir. D. Guggenheim (Paramount, 2006); Mazo, *Climate Conflict*, 73–86; Welzer, *Climate Wars*, 61–5.

evidence that global anthropogenic climate change was a factor in the Sahel droughts of the 1970s and 1980s: if anything, the evidence suggests that climate change has been a factor in the Sahel's recent greening, rather than in the preceding droughts.[4]

This environmental evidence aside, climate–conflict accounts of the war in Darfur are also consistently wanting in their characterisations of local livelihoods and the region's political ecology. The only sensible causal narrative of drought–conflict linkages in Darfur runs as follows: that the multiple, severe droughts and associated famines that the region experienced during the 1970s and 1980s undermined traditional subsistence and especially pastoralist livelihoods, fracturing long-established customary relations and boundaries between tribes, prompting migration from areas of particularly acute environmental stress and scarcity, and in turn generating conflicts among pastoralist groups and between them and sedentary farmers – with all this being a significant and maybe even primary factor in the onset of war. Informed qualitative accounts of the drought–conflict nexus in Darfur make just such claims. And they are not, it must be acknowledged, completely without merit. Large-scale famine in Darfur during 1984–5 was both sparked by a severe two-year drought and an element in sweeping late-twentieth-century social transformations. The three decades prior to the war witnessed both unprecedented southward migrations, and severe and growing conflicts over access and rights to land. And the Khartoum-backed militias, the Janjaweed, which were at the forefront of the death, displacement and destruction during 2003–5, were mainly recruited from northern Darfuri pastoralist tribes, most prominently the camel-herding Northern Rizeigat.[5] At first glance, such facts may seem to corroborate drought and climate–conflict

---

[4] UNEP, *Water Resource Management in Humanitarian Programming in Darfur: The Case for Drought Preparedness* (2008), 8; UNEP, *Sudan: Post-Conflict Environmental Assessment* (2007), 62; M. Kevane and L. Gray, 'Darfur: rainfall and conflict', *Environmental Research Letters*, 3:3 (2008), 3–5; L. Olsson et al., 'A recent greening of the Sahel: trends, patterns and potential causes', *Journal of Arid Environments*, 63:3 (2005), 556–66; I. A. Brown, 'Assessing eco-scarcity as a cause of the outbreak of conflict in Darfur: a remote sensing approach', *International Journal of Remote Sensing*, 31:10 (2010), 2513–20; B. Dong and R. Sutton, 'Dominant role of greenhouse-gas forcing in the recovery of Sahel rainfall', *Nature Climate Change*, 5:8 (2015), 757–60.

[5] A. de Waal, *Famine That Kills: Darfur, Sudan, 1984–1985* (Oxford University Press, 1989); A. de Waal, 'Famine mortality: a case study of Darfur, Sudan 1984-5', *Population Studies*, 43: 1 (1989), 5–24; A. de Waal, 'Is climate change the culprit for Darfur?', *African Arguments* (25/06/2007); M. Mamdani, *Saviors and Survivors: Darfur, Politics and the War on Terror* (Verso, 2009); J. Tubiana, 'Le Darfour, un conflit pour la terre?', *Politique Africaine*, 101 (2006), 111–31; M. Abdul-Jalil and J. D. Unruh, 'Land rights under stress in Darfur: a volatile dynamic of the conflict', *War and Society*, 32:2 (2013), 156–81.

narratives. Yet when properly contextualised it is evident that drought's contribution to the 2003–5 war could not have been more than secondary.

To start with, climate-centred interpretations of Darfur's war typically – and wrongly – assume that Darfur had a settled and traditional socio-economic structure until this was torn asunder by extreme drought. Indeed, this assumption is no accident since, without it, the idea that drought could have suddenly caused migration and conflict would be impossible to sustain. Yet Darfuri livelihoods have long failed to conform to traditionalist stereotype. The subsistence tribal peasant has not existed in Darfur for decades. The distinction between 'nomadic pastoralists' and 'settled farmers' which is invariably invoked in accounts of Darfur's war as a climate conflict is in truth extremely blurred: as long ago as the early 1980s, most 'farmers' were raising livestock and the large majority of 'pastoralists' were cultivating crops. Both livestock and crop production – of cattle, sheep, goats, camels; groundnuts, sesame, gum Arabic, *tombac*, oranges – are heavily commercialised and export-oriented. And rural livelihoods are not just dependent on local production, but also on remittances (especially from Libya, Saudi Arabia and other oil producer states), international aid, and public, especially security sector, employment. Darfur's economy and society are, in sum, distinctly modern, as well as asymmetrically integrated into assorted national and international economic structures and relations.[6]

During the decades prior to 2003–5, Darfur had gone through profound economic changes – most of them completely independent of drought. Improved transportation, especially following the opening of the railway to Nyala in 1959, had spurred the development of the local cash crop economy, within South Darfur in particular. Groundwater exploitation had expanded rapidly from the 1960s, enabling both multi-season cropping and the horizontal expansion of irrigated agriculture beyond the *wadis* dropping down from Darfur's central highland area, the Jebel Marra. Both the intensity of, and the area under, irrigated horticulture had increased dramatically. The rate of bush clearance had accelerated to enable this increased cultivation. Together, these developments had transformed patterns of land control and ownership, with formerly open rangelands being demarcated, enclosed and subsequently

---

[6] de Waal, *Famine That Kills*, 55; J. F. Morton, Agricultural Development in Darfur Region, Sudan (PhD thesis, London University, 1993), 44; H. Young et al., *Darfur: Livelihoods under Siege* (Feinstein International Famine Center, Tufts University, 2005), 10, 51–108; M. Buchanan-Smith et al., *Taking Root: The Cash Crop Trade in Darfur* (UNEP, 2013); H. Young et al., 'Darfurian livelihoods and Libya: trade, migration, and remittance flows in times of conflict and crisis', *International Migration Review*, 41:4 (2007), 826–49.

commodified. And paralleling this, Darfur's livestock sector had expanded significantly, with meat and live animal exports to Arab Gulf countries driving increases in livestock numbers among both pastoralists and sedentary farmers; Darfur's cattle population jumped from under five million in 1987 to almost forty million in 2002, while its sheep population rose from around three to nearly fifty million during the same period.[7]

The socio-political consequences of these developments were many and varied. On the one hand, the horizontal expansion of Darfur's agricultural sector constituted a significant challenge to the expanding livestock sector and to pastoral livelihoods, simultaneously reducing the overall area of grazing land, placing new obstacles before seasonal livestock migrations and generating new conflicts over access and rights to key watercourses. And yet these developments also generated opportunities for pastoralists to move into the cash crop economy, including by migrating to and claiming land in areas of agricultural expansion, especially in South Darfur. That the consequent north–south migration began prior to the major droughts of the 1970s and was led by some relatively wealthy tribal groups – most notably the Zaghawa, who had formerly dominated northern Darfur trade routes but quickly developed important trading communities in southern towns – clearly suggests that this migration was not driven, above all, by drought-induced immiseration. Overall, Darfur was a region of net in-migration and thus high population growth during the pre-war decades, with substantial numbers of Chadian and west African pastoralists settling there, especially in the agricultural frontier zone to the north of the Jebel Marra. As poor as it was, Darfur was neither in economic decline nor teetering on the edge of drought-induced collapse. Rather, it was a relatively attractive and developing region, home to 'among the richest agricultural lands in Sudan' – and it was in large part these characteristics that underpinned the increase in local conflict over land rights and access.[8]

A critical additional factor here was Khartoum-led changes in Darfur's – and indeed the whole of Sudan's – legal and political regimes. In 1970 the government of Jaafar al-Nimeiri brought all unregistered land in Sudan under government ownership, and in 1971 it dissolved the Native Administrations which had prevailed since colonial times. These

---

[7] Abdul-Jalil and Unruh, 'Land rights', 166–70; Buchanan-Smith et al., *Taking Root*, 19; Morton, *Agricultural Development*, 34–5, 80, 104; Y. S. S. Takana, *Darfur: Struggle of Power and Resources, 1650–2002: An Institutional Perspective*, trans. A. Al-Awad (Chr. Michelsen Institute, 2016), 88; Young et al., *Darfur*, 41.

[8] Abdul-Jalil and Unruh, 'Land rights', 166–77; Morton, *Agricultural Development*, 30–32; de Waal, 'Is climate change the culprit?'; Mamdani, *Saviors and Survivors*, 10.

actions simultaneously challenged Darfur's established land tenure system, providing a new basis for migrants from northern Darfur to claim rights to land, and undermined extant conflict resolution mechanisms. Under the Native Administration system, most tribes had been granted *'dars'* – tribal homelands – but a minority had not. Following al-Nimeiri's reforms, the land rights claims of these 'dar-less tribes' suddenly acquired newfound legitimacy. It is no accident that the Northern Rizeigat, who came to form the core of the Janjaweed during the 2003–5 war, were one such 'dar-less tribe'.[9]

None of this means that drought played no role in Darfur's many prewar migrations and land rights disputes: such a claim could not be substantiated. In Darfur, as elsewhere, drought is one of the very many contingent factors – alongside floods, pests, increased production costs, decreased crop prices, local violence or the expenses of things such as family weddings – which, alone or in conjunction, is capable of pushing vulnerable rural households 'below the water'. In Darfur specifically, the 1984–5 drought undoubtedly caused both mass mortality and extreme social dislocation. Yet there is little solid reason to think that either this or earlier droughts were the principal long-term determinants of the region's pre-war socio-economic transformations and ensuing tensions. Suggestions that the region was in the midst of a Malthusian crisis caused by overpopulation and over-grazing, and then exacerbated by droughts, are contradicted by the aforementioned evidence of greening and in-migration (and by the absence of evidence of over-grazing). Moreover, though some analysts have identified significant correlations between areas of high rainfall and the geography of violence within Darfur during 2003–5, the evidence to this effect could equally be explained by the pattern of economic development discussed above, and so is not robust.[10] More broadly, what is perhaps most striking about drought-conflict narratives of the war in Darfur is how little they even consider, let alone attribute any causal importance to, economic development and its contradictions. Ignoring these, and interpreting the war instead as essentially a consequence of environmental stresses on traditional tribal relations, these narratives are – just like the eco-determinist tradition of which they form a part – deeply and misleadingly anachronistic.

---

[9] Abdul-Jalil and Unruh, 'Land rights', 162–5, 169–71; Mamdani, *Saviors and Survivors*.
[10] T. M. Li, *Land's End: Capitalist Relations on an Indigenous Frontier* (Duke University Press, 2014), 135–7; Morton, *Agricultural Development*, 87–110; A. De Juan, 'Long-term environmental change and geographical patterns of violence in Darfur, 2003–2005', *Political Geography*, 45 (2015), 22–33; O. Olsson and E. Siba, 'Ethnic cleansing or resource struggle in Darfur? An empirical analysis', *Journal of Development Economics*, 103 (2013), 299–312.

Last, 'climate war' readings of the Darfur conflict are also often problematic in their representations of its politics. At the extreme, such readings ignore politics altogether, as in the case of an influential UNEP report that chose 'not to investigate in detail the social and political aspects of conflicts in Sudan, focusing instead on their environmental dimensions', but nonetheless felt able to conclude that there was 'a very strong link between land degradation, desertification and conflict in Darfur'.[11] More common, however, is the recurring eco-determinist problem of misplaced localism: that in focusing primarily on internal and local causes of crisis – specifically on the claimed linkages between local drought and local violence – such readings can end up downplaying national and geopolitical factors, including the role and agency of the state. Of course, not all accounts do this and there is no reason in principle why a sensitivity to the impacts of local socio-ecological changes cannot be juxtaposed with analysis of the non-local causes of conflict, as the most sensitive accounts of the Darfur crisis illustrate. In practice, however, politics is often peripheral to, if not completely written out of, climate war accounts of the Darfur crisis.

The importance of this lies in the fact that the 2003–5 war was above all a brutal counter-insurgency operation, launched and financed by the government in Khartoum and conducted by a combination of Sudanese army, intelligence and air force units and paramilitary Janjaweed militias. The latter were in many respects arms of the Sudanese state, often wearing army uniforms, operating together with regular army units and attacking villages immediately after Sudanese Air Force bombing raids. The initial rebellion to which this state violence was a response was not merely local, either: politically, it was a result of the collapse of the political accommodations that had held together the Sudanese state's clientelistic networks; while militarily, Darfur's rebel movements received significant political and military support from the principal national opposition, the Sudan People's Liberation Movement/Army (SPLM/A). The violence in and militarisation of Darfur were also profoundly shaped by Cold War–related regional legacies, most notably political crisis in Chad and Libyan President Muammar al-Gaddafi's pan-Africanist ambitions.[12] The Darfur war, in sum, had national and

---

[11] UNEP, *Sudan*, 8, 73.
[12] A. de Waal, 'Counter-insurgency on the cheap', *London Review of Books*, 26:15 (05/08/2004); G. Prunier, *Darfur: The Ambiguous Genocide* (Hurst, 2005), 98–100; A. A. Gallab, *The First Islamist Republic: Development and Disintegration of Islamism in the Sudan* (Ashgate, 2008), 149–64; P. Roessler, *Ethnic Politics and State Power in Africa: The Logic of the Coup–Civil War Trap* (Cambridge University Press, 2016), 103–204; Mamdani, *Saviors and Survivors*, 212–27.

geopolitical as much as regional causes, the most significant of which lay on the banks of the Nile, 1,000 km away from the burning villages of the Jebel Marra. And these extra-regional causes clearly had little to do with drought, human-induced or otherwise.

The Syrian Civil War: Drought, Migration and Conflict Revisited

The evidence that Darfur was home to the 'first climate change war' is, in sum, exceedingly thin. But what of Syria? The standard narrative here is in key respects identical: that human-induced climate change was a factor in a severe pre–civil war drought; that this drought caused large-scale internal migration, specifically from the north-east of the country; and that this migration exacerbated the socio-economic stresses that underpinned Syria's descent into war. Yet the Syria case presents us with a weightier test. Where the climate–conflict narrative about Darfur rose to prominence on the back of journalistic reports and op-eds, the claim that the Syrian civil war was partially caused by climate change was underpinned by a series of detailed studies, including by Peter Gleick – whose views on looming water conflicts have already been discussed (Chapter 2) – and, most influentially, by earth scientists Colin Kelley and colleagues in an article published in *Proceedings of the National Academy of Sciences*. In turn, the Syria climate–conflict thesis has been widely reported and repeated, including by political leaders, international organisations, environmental NGOs, defence think tanks, numerous academics, activists and media commentators, and in outlets ranging from *Nature* to the *Daily Mail*.[13] The language used in relation to Syria has also generally been more nuanced than it was in relation to Darfur, with few claiming that climate change and drought were the sole, or even primary, causes of Syria's 2011 unrest. The question facing us, therefore, is not whether the Syrian civil war should be considered a 'climate conflict' – since such an essentialist formulation would risk erecting a straw man for critique – but whether, on the balance of evidence, the contribution of climatic factors to civil war onset was as significant as has been claimed. As in the case of Darfur, there are basically three issues to address. Here

---

[13] P. Gleick, 'Water, drought, climate change, and conflict in Syria', *Weather, Climate and Society*, 6:3 (2014), 331–40; C. Kelley et al., 'Climate change in the fertile crescent and implications of the recent Syrian drought', *Proceedings of the National Academy of Sciences*, 112:11 (2015), 3241–6; M. Zastrow, 'Climate change implicated in current Syrian conflict', *Nature* (02/03/2015); R. Gray, 'Did climate change trigger the war in Syria? Severe drought may have contributed to uprising, study reveals', *Daily Mail* (02/03/2015).

we consider each of these briefly; fuller substantiation of our arguments can be found elsewhere.[14]

The first issue concerns the depth of Syria's pre–civil war drought and the role of global anthropogenic climate change therein. It is clear that, during 2006/7–2008/9, the eastern Mediterranean region experienced 'the most severe drought in the instrumental record': proponents and critics of the Syria climate–conflict thesis can, at least on this, agree. Within north-east Syria the same applied: for example, the city of Qamishli, on the border with Turkey, received only 25 per cent of its average rainfall during 2007-8 and Deir ez-Zor, further south, only 12 per cent. This pattern was not countrywide: most of western Syria including its three largest cities, Aleppo, Damascus and Homs, received above average rainfall during 2006/7–2008/9. Moreover, the drought was not unique to Syria: during 2007/8–2008/9 northern Iraq experienced even larger precipitation declines.[15] That said, there is no doubt that north-east Syria experienced unprecedentedly severe one- and three-year droughts during the pre–civil war period.

According to Kelley and colleagues, as well as others, there are clear signs of anthropogenic impacts here: 'a drought of the severity and duration of the recent Syrian drought', say Kelley et al., 'became more than twice as likely as a consequence of human interference in the climate system'. It is important to explain the underpinnings of this claim. Effectively, Kelley et al reach this conclusion first by identifying what they view as a long-term drying trend in the Fertile Crescent during the period 1931–2009, then by attributing this trend to human influence, before finally estimating the increased likelihood of the 2006/7–2008/9 drought occurring because of it. In principle this is a reasonable enough 'attribution method'. The problem, though, is that there is no clear evidence of progressive long-term drying either in Syria or the Fertile Crescent more widely: very few rainfall stations in the region display significant long-term negative rainfall trends, and there is no good reason why regional rainfall should be modelled as a linear trend either. Hence, whatever its intuitive plausibility, Kelley et al.'s claim about the increased likelihood of severe drought in Syria is a statistical artefact. Moreover, the most that can be claimed, even accepting Kelley et al.'s analysis, is that human influences increased the *likelihood* of the 2006/7–2008/9 drought

---

[14] Selby et al., 'Climate change and the Syrian civil war revisited'; Selby et al., 'Climate change and the Syrian civil war revisited: a rejoinder'.

[15] Kelley et al., 'Climate change', 3241; Selby et al., 'Climate change and the Syrian civil war revisited', 234–7; R. Trigo et al., 'The intense 2007–2009 drought in the Fertile Crescent: impacts and associated atmospheric circulation', *Agricultural and Forest Meteorology*, 150:9 (2010), 1245–57.

(a probabilistic formulation which is very different from maintaining that climate change is causally 'implicated' in it, as Kelley et al. claim).[16] But if our comments above are valid, even this may be overstated.

The second issue for us to consider is whether this drought did indeed spark large-scale internal migration within Syria. According to Kelley et al., Gleick and others, the region-wide 2006/7–2008/9 drought resulted in wholesale agricultural 'collapse' in the north-east of the country, which in turn pushed 2–3 million people into extreme poverty and precipitated the displacement of around 1.5 million from rural areas of the north-east to Damascus, Aleppo, the southern town of Dara'a, and elsewhere. There is, to be sure, abundant evidence both of a significant agricultural production downturn and of large-scale out-migration from north-east Syria during the 2006/7–2008/9 period.[17] Yet there are major problems here.

To start with, and most straightforwardly, the key statistical claims underpinning this narrative are without foundation. The claim that 2–3 million were driven into extreme poverty by drought was based, extraordinarily, on United Nations Development Programme (UNDP) analyses of *pre-drought* poverty levels. Moreover, there is no sound evidence for the widely reproduced claim that around 1.5 million were displaced by the drought: this claim derives from a single humanitarian news report, was probably meant to refer to the number of people *affected* not *displaced* by it (the UN estimated that 1.3 million were affected) and, most important, was completely out of line with Syrian government, UN and other estimates, most of which suggested figures in the order of 40,000–60,000 families.[18] There is no good evidence that the rural economic, humanitarian and migration impacts of Syria's pre–civil war drought were on anywhere near the scale that has typically been claimed.

---

[16] Kelley et al., 'Climate change', 3241; M. Hulme, 'Attributing weather extremes to "climate change": a review', *Progress in Physical Geography*, 38:4 (2014), 499–511; M. Hulme et al., 'African climate change: 1900–2100', *Climate Research*, 17: 2 (2001), 145–68.

[17] Kelley et al., 'Climate change', 3241–2; Gleick, 'Water, drought, climate change', 334; UN-OCHA, *Syria Drought Appeal* (2008); UN-OCHA, *Syria Drought Response Plan* (2009); UN-OCHA, *Joint United Nations Drought Assessment Mission: The Syrian Arab Republic 2008/09* (2009); UN-OCHA, *Syria Drought Response Plan 2009–10: Mid-Term Review* (2010); O. De Schutter, *Report of the Special Rapporteur on the Right to Food, Mission to the Syrian Arab Republic*, report to UNHRC (27/01/2011).

[18] H. El Laithy and K. Abu-Ismael, *Poverty in Syria, 1996–2004: Diagnosis and Pro-Poor Policy Considerations* (UNDP, 2005); H. Laithy and K. Abu-Ismael, *Poverty and Distribution in Syria* (UNDP, 2009); 'Drought driving farmers to the cities', *Integrated Regional Information Networks* (02/09/2009); UN-OCHA, *Syria Drought Appeal*, 1; UN-OCHA, *Syria Drought Response Plan*, 1, 5; UN-OCHA, *Joint United Nations Drought Assessment*, 18–19; UN-OCHA, *Syria Drought Response Plan 2009–10*, 6.

Conversely, there is evidence that there existed deep agrarian problems in Syria even prior to the 2006/7–2008/9 drought. Both in the northeastern province of Hasakah, and nationwide, agricultural production was in decline from the early 2000s, including the production of the key government-designated 'strategic crops', wheat and cotton. According to recent remote sensing work, in central Hasakah land abandonments occurred at a higher rate during 2002–7 than during the drought years that followed. Settlements were also being abandoned well before the drought, while annual net out-migration from Hasakah was higher than from any other region. The particular district within Hasakah identified by the Syrian government and UN as especially affected by the drought also, or to the contrary, experienced particularly high levels of pre-drought abandonment and out-migration. Moreover, there was high rural–urban migration from across rural Syria from around 2000 onwards, as evidenced, most conspicuously, by the rapid expansion of peri-urban informal housing areas around Syria's cities. There was, in sum, a structural agrarian crisis in rural north-east Syria and nationwide from the early 2000s which, independently of drought, was already resulting in production declines and out-migration. At the very least, this structural crisis must have heightened drought vulnerabilities (this perhaps explaining why Syria was more deeply affected by the 2006/7–2008/9 drought than Iraq, despite the latter having faced the deeper climatic shock). But it is also evident that some of the so-called 'drought migration' was nothing of the sort.[19]

At the national level, at least (we consider Hasakah-specific dynamics in Chapter 6), the central cause of this agrarian crisis was state-led economic change. Starting in the 1990s and intensifying following Bashar al-Assad's accession in 2000, Syria made a decisive turn towards the market economy, including through the privatisation of state farms, the removal of smallholder and sharecropping rights, trade liberalisation and reduced input subsidies and food price controls. These policy changes, combined with increased mechanisation and rising price

---

[19] J. Selby, 'Climate change and the Syrian civil war, part II: the Jazira's agrarian crisis', *Geoforum*, 101 (2019), 262–3; B. Mohr, *Modelling Human Migration Using Remote Sensing and Proxy Indicators – a Case Study: Syria* (MA thesis, Lund University, 2020); F. Hole and B. F. Zaitchik, 'Policies, plans, practice, and prospects: irrigation in northeastern Syria', *Land Degradation and Development*, 18:2 (2007), 147; F. Hole, 'Drivers of unsustainable land use in the semi-arid Khabour River Basin, Syria', *Geographical Research*, 47:1 (2009), 6–7; M. Khawaja, Internal Migration in Syria: Findings From a National Survey (Fafo, 2002), 27; 'Persecution and discrimination against Kurdish citizens in Syria', report to UNHRC, Working Group on the Universal Periodic Review (10/2011), 3; . R Goulden, 'Housing, inequality, and economic change in Syria', *British Journal of Middle Eastern Studies*, 38:2 (2011), 187–202.

inflation, sparked a steep decline in farming employment and underpinned the mass exodus to the cities discussed above. Between 2001 and 2007, an estimated 104,000 jobs were created each year outside of the agricultural sector, but 67,000 jobs per annum were lost within it. As Syrian economist Samir Aita put it, this steep decline was 'clearly linked to major structural transformations of the sector, more than to drought and climatic conditions'. Yet Kelley et al., Gleick and other proponents of climate–conflict reasoning ignore this causality altogether and as a result repeatedly misattribute important socio-economic changes to the drought, when they were in fact long-term corollaries of liberalisation.[20]

Last, drought was not even the only spark and catalyst of migration during the 2007–9 period. The year 2008 saw the enactment of a Presidential Decree which tightened restrictions on land sales across the whole of Hasakah province, resulted in some losing their land rights altogether and has been credited as leading to increased out-migration. Moreover, during 2008–9 rural Syria was hit by triple-digit increases in the prices of crucial agricultural inputs: in May 2008 fuel subsidies were halved, leading to an overnight 342 per cent spike in the price of diesel, and then, in May 2009, fertiliser subsidies were removed, causing price rises ranging from 200 to 450 per cent. The fuel subsidy cuts had particularly devastating economic consequences, for reasons which are not difficult to understand. Until 2008, government subsidies in Syria had covered around 80 per cent of the local diesel purchase price. Syria's agricultural sector, in turn, had become highly dependent on cheap fuel. Herding livelihoods had become more reliant on subsidised inputs – including cheap fuel for transporting flocks from one region to another and for trucking in food and water – than on the availability of natural forage. Wheat and cotton farmers, equally, had come to rely heavily on cheap diesel for groundwater irrigation, with one study calculating that cotton and wheat production in Syria's three driest agro-climatic zones would cease being profitable altogether at fuel cost increases of 100 per cent (let alone over 300 per cent). In consequence, as one commentator noted, for many farmers in Syria the cutting of input subsidies in 2008–9 'formed a greater burden than the successive years of drought and spurred their decision to abandon their land'. Indeed, while the relative importance of all these factors cannot be known for certain – and while they would not, of course, have operated entirely independently – the above suggests that it was probably the shock

[20] S. Aita, *Labour Markets Performance and Migration Flows in Syria* (Robert Schuman Centre for Advanced Studies, 2009), 24; S. Aita, *Labour Market Policies and Institutions, with a Focus on Inclusion, Equal Opportunity and the Informal Economy: The Case of Syria* (ILO, 2009), 76.

therapy-style increase in input prices, more than drought, which was the principal trigger of out-migration from north-east Syria prior to the civil war.[21]

Be that as it may, how did this migration from Syria's north-east figure in the country's descent into war? According to Kelley et al., Gleick and others, drought migration produced a 'population shock' within Syria's urban peripheries, that 'contributed to urban unemployment and economic dislocations and social unrest', and became one 'driver' of unrest. Yet the evidence in support of these claims is remarkably thin. Syria's cities grew rapidly not only during the drought years of 2008–9 but throughout the 2000s, with the 'excess migration' from the north-east during 2008–9 probably accounting for a mere 4–12 per cent of Syria's 2003–10 urban growth; hence drought can hardly have been a major cause of the pressures on Syria's cities. Moreover, there is no evidence that migrants from the north-east were significantly involved – whether as mobilisers, participants or targets – in the early civil war protests. None of the political demands made by Syria's early 2011 protest movements related directly to either drought or migration: the thirteen demands presented to the President's emissary in the city of Dara'a, site of the first major clashes, in March 2011, for instance, called for the removal of the head of the local Political Security branch and town Governor; apologies to be made to families offended by the Governor; political reform and greater political freedoms; a meeting with the President; the return of women teachers expelled following a new law banning them from wearing the niqab; the expulsion from the town of all companies owned by Rami Makhlouf (the President's cousin and the country's leading business magnate); the removal of new laws regarding land sales; and the lowering of the price of fuel – but nothing on either migration or drought. And interview evidence suggests that, in Dara'a at least, migrants from the north-east did not participate to any significant degree in the spring 2011 protests, were not targets of the demonstrations or subsequent repression, and left as soon as the protests started: '[t]hey had nothing to do with politics. They went to work and back home', observed one interviewee; '[t]hey left at the beginning of the demonstrations', observed another; '[t]hey had no opinion. Their life revolved solely

---

[21] Austrian Centre for Country of Origin and Asylum Research and Documentation and Danish Immigration Service, *Human Rights Issues Concerning Kurds in Syria* (2010); IMF, *Syrian Arab Republic: IMF Country Report No. 10/86* (2010), 10; De Schutter, *Report of the Special Rapporteur*, 16; A. Gül et al., 'Economic analysis of energy use in groundwater irrigation of dry areas: a case study in Syria', *Applied Energy*, 82:4 (2005), 291, 296–7; F. De Châtel, 'The role of drought and climate change in the Syrian uprising: untangling the triggers of the revolution', *Middle Eastern Studies*, 50:4 (2014), 526.

around their work.' Hence we concur with Marwa Daoudy that there is 'little evidence' that 'climate change in Syria sparked popular revolt in 2011' – but 'a lot of evidence' that 'suggests it did not.'[22]

Lake Chad: Who Needs Evidence?

Alongside Syria, the other case to have been most often invoked in recent climate security discourse is the ongoing crisis around Lake Chad.[23] Since 2009–10, after attempts by the Nigerian military to suppress what would become known as 'Boko Haram', this region has been home to a deep security, governance and humanitarian crisis which has embroiled all four states neighbouring the Lake – Cameroon, Chad, Niger as well as Nigeria – and involved widespread civilian casualties, the internal and trans-border displacement of over 2.5 million people and acute economic and food insecurity. Climate change has been widely identified as a significant contributory factor to this crisis, including by heads of all four Lake states, UN Secretary-Generals Ban Ki-moon and António Guterres, by Barack Obama, John Kerry, Emmanuel Macron and Margot Wallström and across the mainstream European, American and African media. In turn, the case of Lake Chad has repeatedly been invoked in policy-oriented forums and studies on climate security, including UN Security Council meetings on the subject. Lake Chad was the focus of the first ever Security Council resolution to explicitly refer to the security impacts of climate change, and was the most important case study focus of the G7 initiative, *A New Climate for Peace*.[24] Across these

---

[22] Kelley et al., 'Climate change', 3242; Gleick, 'Water, drought, climate change', 333; F. Femia and C. Werrell, *Syria: Climate Change, Drought and Social Unrest* (CCS, 2012); Selby et al., 'Climate change and the Syrian civil war revisited', 242; M. J. Barout, *The Past Decade in Syria: The Dialectic of Stagnation and Reform* (Arab Center for Policy and Research Studies, 2011), 186–8 (in Arabic); C. Fröhlich, 'Climate migrants as protestors? Dispelling misconceptions about global environmental change in pre-revolutionary Syria', *Contemporary Levant*, 1:1 (2016), 38–50; M. Daoudy, *The Origins of the Syrian Conflict: Climate Change and Human Security* (Cambridge University Press, 2020), 203.

[23] This section draws upon G. Daoust and J. Selby, 'Understanding the politics of climate security policy discourse: the case of the Lake Chad basin', *Geopolitics* (2022). https://doi.org/10.1080/14650045.2021.2014821.

[24] M. Buhari, Speech at the 73rd Session of the UN General Assembly, New York (25/09/2018); M. Issoufou, Speech at the First Summit of the Sahel Climate Commission, Niamey (25/02/2019); UNCCD, 'World leaders renew commitment to strengthen climate resilience through Africa's Great Green Wall' (12/02/2015); P. Biya, Speech at the Mini-Summit on Africa on Climate Challenges and African Solutions, Paris (01/12/2015); International Institute for Sustainable Development, 'UN Secretary-General visits Nigeria, discusses 2030 Agenda' (27/08/2015); UN Secretary-General, 'Secretary-General's message to International Conference on Lake Chad' (28/02/2018); P. Kasperowicz, 'Obama: climate change contributed to rise of Boko Haram, Syrian civil war', *Washington Examiner* (20/05/2015); 'Climate change intensifies conflicts, John

diverse policy terrains, global anthropogenic climate change has been widely identified as a significant factor in the basin's deep security and humanitarian crises. The basic thesis here – which bears an uncanny resemblance to the Darfur and Syria narratives discussed above – is that Lake Chad has been shrinking markedly; that this trend is attributable to climate change–induced drought; and that this particular local environmental scarcity shock is impacting livelihoods, feeding insecurity and contributed to the rise of Boko Haram. Thus is the widely reproduced policy narrative. Unfortunately, though, it has hardly any basis in fact – much less, indeed, than in the parallel cases of Syria and Darfur.

Lake Chad itself is a large, shallow endorheic lake which drains an area from the Central African Republic to Algeria, and creates one of the few islands of green along the entire southern rim of the Sahara. During the Sahelian droughts of the 1970s and early 1980s Lake Chad declined significantly, from an area of about 25,000 km^2 in 1963 to 2,000 km^2 in 1985. In 1973, the Lake split into two pools, and since 1985 its northern part has frequently been dry. Since around 1995, however, Lake Chad's total surface area has been relatively stable. Indeed, if anything the Lake has in recent years been expanding: in 2018 it had a total surface area of approximately 12,000 km^2 and there is evidence that the total volume of water stored within the Lake, including the aquifers beneath it, has been steadily increasing. These trends are extensively documented in recent peer-reviewed articles and other expert studies. As the main output of the aforementioned G7 initiative on Lake Chad put it, in recent years the Lake has not been 'a shrinking lake, but a fluctuating one'.[25]

One would not know this, however, from the vast bulk of policy and media commentary on the subject. Thus Lake Chad is routinely described as not just 'shrinking' and 'drying' but even 'disappearing' and 'vanishing', and regularly characterised as having shrunk by '90 per cent since the 1960s' – with the implication that these are all ongoing trends. Particularly influential in this regard has been a set of NASA satellite images dating back to 2001, which were taken up by

---

Kerry says', *Al Jazeera* (11/11/2015); E. Macron, Speech at the World Economic Forum, Davos (24/01/2018); M. Wallström, Speech at the UN Security Council, New York (11/07/2018); UN Security Council, *Resolution 2349* (S/RES/2349) (31/03/2017), 7; adelphi, 'G7 Working Group: what will come next for G7 action on climate and fragility' (24/10/2017).

[25] GIZ and LCBC, *Report on the State of the Lake Chad Basin Ecosystem* (2013); J. Lemoalle and G. Magrin (eds.), *Le Développement du Lac Tchad: Situation Actuelle et Futurs Possibles* (IRD Éditions, 2014); B. Pham-Duc et al., 'The Lake Chad hydrology under current climate change', *Scientific Reports*, 10 (2020), 5498; Lake Chad Information System, 'Water: Lake Chad', https://lis.cblt.org/lis/water/surface/sub/lake-chad; J. Vivekananda et al., *Shoring Up Stability: Addressing Climate and Fragility Risks in the Lake Chad Region* (adelphi, 2019), 43.

UNEP in the late 2000s and have since been reproduced in everything from Al Gore's *An Inconvenient Truth* to media reporting, to major policy events like the 2018 'Saving the Lake Chad' conference in Abuja. Juxtaposing images from 1973, 1987, 1997 and 2001 – as well as 1963 and 2007 in the UNEP illustrations – these provide a stark visual representation of the Lake's demise: 'If you're trying to tell the story ... these satellite images, aerial images [provide] a very powerful message', one think tank researcher has observed. The NASA and UNEP images are thoroughly misleading, however. Lake Chad's surface area fluctuates to such a degree that, as one leading French scholar of the Lake's geography observes, 'choosing different dates could have shown exactly the opposite'. The images and accompanying 'decline statistics' focus solely on the Lake's open water, excluding from consideration the Lake's vegetation-covered area which extends to thousands of square kilometres and by some accounts represents up to three-quarters of its total area. In addition, 1963 is misleading as a baseline year for judging lake shrinkage, since the 1950s and early 1960s saw well above average rainfall such that at this point Lake Chad was abnormally large. And, most obviously, the images are now rather dated. There exists, to the best of our knowledge, not a single scientific study that finds that Lake Chad has been drying or shrinking since 2000.[26]

Nonetheless, Lake Chad's 'decline' is consistently attributed within climate security discourse to increased and more extreme drought associated with global anthropogenic climate change. This is, to put it bluntly, doubly misleading. As yet there have been no analyses of the impacts of global anthropogenic climate change per se on lake levels, the basis for the linkage between the two being instead evidence of the effects of 'climate variability' – which is, of course, a natural phenomenon that can occur independently of greenhouse forcing. Moreover, rainfall levels have been increasing across the Lake Chad basin since around 1990, mirroring patterns across the Sahel and in line with climate change attribution and modelling studies, which consistently link climate change with rainfall increases rather than drought. There has also been an increase in extreme rainfall events over the eastern Sahel region, including Lake Chad, since

---

[26] W. Ross, 'Lake Chad: can the vanishing lake be saved?', *BBC* (31/03/2018); UNEP, 'The tale of a disappearing lake' (28/02/2018); World Bank, 'Restoring a disappearing giant: Lake Chad' (27/03/2014); NASA Earth Observatory, 'Africa's disappearing Lake Chad' (12/03/2001); UNEP, 'Lake Chad: almost gone' (2009); A. Gore, *An Inconvenient Truth*; I. K. Musa et al., *Saving Lake Chad* (LCBC, 2008); G. Daoust interview with think tank researcher (28/06/2019); G. Magrin, 'The disappearance of Lake Chad: history of a myth', *Journal of Political Ecology*, 23:1 (2016), 211; M. Leblanc et al., 'Thermal remote sensing of water under flooded vegetation: new observations of inundation patterns for the "Small" Lake Chad', *Journal of Hydrology*, 404:1–2 (2011), 87–98.

the 1990s, which should have translated into increased run-off into the Lake. In actual fact, however, river flows into Lake Chad have not recovered since the 1990s: while rainfall levels have significantly increased, river flow has not. The reasons for this trend divergence are clear, with all studies to have explored the issue identifying local human activities, especially irrigation withdrawals and dam construction, as the major cause of low river flow in the basin. It is these local activities rather than global carbon dioxide levels which account for why Lake Chad has only minimally recovered since the droughts of the 1970s and 1980s.[27]

In case this is not damning enough, the evidence of links between Lake Chad's ostensible decline and the region's security and humanitarian crises is even weaker. Policy-oriented and media reports repeatedly claim that climate change is creating or exacerbating local resource scarcities around the Lake, increasing migration, intensifying competition over land and water resources and, in turn, exacerbating livelihood insecurities and tensions between communities and supporting militant recruitment. Political leaders and diplomats have claimed similarly: thus former Chadian President Idriss Déby asserted 'the disappearance of Lake Chad is a security crisis that is fuelling terrorist groups like Boko Haram', while German Foreign Affairs Minister Heiko Maas observed that '[a]s Lake Chad shrinks, the livelihoods of entire populations are disappearing – the perfect breeding ground for extremism and terrorism'.[28] Climate change, in such accounts, is usually portrayed not as the sole or primary cause of the regional Lake Chad crisis, but instead as one crucial factor among others. Yet even with this qualifier the case for linking environmental change with the region's ongoing crisis is remarkably thin.

In the first place, there exist no peer-reviewed studies of which we are aware that show how environmental changes have fed into the ongoing armed conflict around Lake Chad. The one peer-reviewed study to have

---

[27] O. E. Adeyeri et al., 'Analysis of climate extreme indices over the Komadugu-Yobe basin, Lake Chad region: past and future occurrences', *Weather and Climate Extremes*, 23 (2019), 100194; C. Okonkwo et al., 'Characteristics of Lake Chad level variability and links to ENSO, precipitation, and river discharge', *Scientific World Journal* (2014), 145893; G. Panthou et al., 'Rainfall intensification in tropical semi-arid regions: the Sahelian case', *Environmental Research Letters*, 13:6 (2018), 064013; C. M. Taylor et al., 'Frequency of extreme Sahelian storms tripled since 1982 in satellite observations', *Nature*, 544 (2017), 475–8; M. Hoerling et al., 'Detection and attribution of twentieth-century northern and southern African rainfall change', *Journal of Climate*, 19:16 (2006), 3989–4008; Dong and Sutton, 'Dominant role of greenhouse-gas forcing'; R. Mahmood and S. Jia, 'Assessment of hydro-climatic trends and causes of dramatically declining stream flow to Lake Chad, Africa, using a hydrological approach', *Science of the Total Environment*, 675 (2019), 122–40; W. Zhu et al., 'Relative contribution of climate variability and human activities on the water loss of the Chari/Logone River discharge into Lake Chad', *Journal of Hydrology*, 569 (2019), 519–31.

[28] UNCCD, 'World leaders renew commitment'; UN Security Council, *8451st Meeting*, 12.

examined climate–water–conflict connections in the region since 2009 found no evidence of causal links. Moreover, it is not even clear what the relevant causal connections are claimed to be. While in the Syrian case, as discussed above, a particularly extreme three-year drought is claimed to have sparked mass migration from the country's north-east, in turn contributing to unrest in host regions, no equivalent causal account is found within reports on Lake Chad. Lake and rainfall levels were increasing when Boko Haram activity intensified around 2009 and cross-border violence expanded in 2013–14. Out-migration was not, as far as we are aware, increasing prior to the Boko Haram crisis. And most of the existing literature on the rise of Boko Haram does not even mention the role of local environmental changes or climate change therein. Moreover, while environmental changes have certainly affected resources, livelihoods and population movements around Lake Chad, studies describe both positive and negative consequences, pointing not just to tensions but also to shared resource management systems, and to shifts in livelihood strategies and patterns of mobility and settlement as forms of adaptation rather than sources of violent conflict.[29] While the possibility of important linkages between environmental change and the ongoing security crisis in the Lake Chad basin cannot be discounted, at present there is no meaningful evidence to this effect.

From Lake Chad to Syria and Darfur the 'climate war' storyline has essentially been the same: that regional droughts worsened by climate change have sparked socio-economic including migration crises, leading eventually to large-scale civil conflicts. Yet the parallels between the three cases do not end here. As argued above, neither the Lake Chad crisis nor the Syrian civil war, nor even the war in Darfur, should be considered 'climate conflicts' – and for broadly similar reasons. In each case, there exists only limited or questionable evidence of climate change having caused or amplified drought. In each case, the contribution of drought to socio-economic crisis, including migration, has been wildly overstated. And in each case, especially in the more recent cases of Syria and Lake Chad, there is a striking paucity of evidence that drought-induced

---

[29] U. T. Okpara et al., 'Using a novel climate–water conflict vulnerability index to capture double exposures in Lake Chad', *Regional Environmental Change*, 17:2 (2017), 351–66; E. E. Anugwom, *The Boko Haram Insurgence in Nigeria: Perspectives from Within* (Palgrave Macmillan, 2019); V. Comolli, *Boko Haram: Nigeria's Islamist Insurgency* (Hurst and Company, 2015); M. A. Pérouse de Montclos (ed.), *Boko Haram: Islamism, Politics, Security and the State in Nigeria* (African Studies Centre, 2014); H. Kiari Fougou and M. Abdourahamani, 'Une oasis à la porte du Sahara: le lac Tchad et ses systèmes', *Geo-Eco-Trop*, 42:2 (2018), 275–83; C. Rangé and M. Abdourahamani, 'Le lac Tchad, un agrosystème cosmopolite centré sur l'innovation', *Les Cahiers d'Outre-Mer*, 67:265 (2014), 43–66.

socio-economic crisis was a significant cause of conflict. The pattern here is striking, suggesting that the problems with drought–conflict narratives go well beyond the three cases considered here. Why, we need to ask, have commentators, analysts and policymakers got it so wrong about the role of drought in these and other conflicts? How, more fundamentally, should we understand the politics and political ecology of drought? And what, if anything, might this tell us about the future? It is to these questions that we now turn.

A Brief History of Drought

The power and appeal of drought-crisis narratives is evidently in large part rooted in the fact that most of human history has been plagued by variable and uncertain rains. From the annual dry seasons of monsoon and Mediterranean climes through to those extreme multi-year and multi-decadal dry spells which, it is now clear, are often driven by global climatic oscillations, drought has historically had far-reaching implications for livelihoods, economies, polities and environments alike. 'Upper Egypt has become an empty waste', recounts the 'Lament of Ipuwer' from Egypt's Middle Kingdom period; '[t]he desert is throughout the land ... there is no food ... cattle moan because of the state of the land ... the storehouse is empty ... the laws of the council-chamber are thrown out ... men have fallen into rebellion'. 'The whole land has perished', observed the Prophecy of Neferty; 'not the back of the [finger]nail survives on what should be there ... no one speaks, and no eye weeps ... men take up weapons of war ... They beg for bread with blood'. Moreover, fast forward 4,000 years to the sixteenth-century Mediterranean chronicled by Fernand Braudel, and one finds a situation not altogether different; so at the mercy of the weather were Mediterranean societies of this period that they were 'always on the verge of famine', with 'everything affected accordingly, even politics'. The only details of daily life which regularly found their way into diplomatic correspondence concerned the weather and state of the harvests. Poor barley or wheat crops would lead to the cancellation of military campaigns and a corresponding increase in brigandage and piracy. At extreme, people would die in the streets. Islands like Cyprus were 'constantly threatened by famine'. Drought, Braudel comments, 'was no mere economic problem, but a matter of life and death'.[30]

[30] B. Bell, 'The dark ages in ancient history. I. The first dark age in Egypt', *American Journal of Archaeology*, 75:1 (1971), 12–13, 17–18; R. O. Faulkner, 'The admonitions of an Egyptian sage', *Journal of Egyptian Archaeology*, 51:1 (1965), 53–62; Braudel, *Mediterranean*, 243–5, 152.

During the late eighteenth and nineteenth centuries, the consequences of drought were in many places no less acute. The 1769–73 Bengal famine killed an estimated ten million, one-third of Bengal's population, while the 'great El Niño' of 1789–93 led to the death of an estimated eleven million people across south Asia and contributed to crop failures, epidemics and social unrest from the Caribbean to northern Europe – perhaps also contributing to revolution in France. The subsequent El Niño droughts of 1876–9, 1889–91 and 1896–1902 resulted in the death of somewhere between thirty and sixty million in India and China alone. Drought conditions in Ethiopia in 1889–92 were so catastrophic that grain prices increased a hundredfold or more, leaving an estimated third of the population to perish. Downstream in Sudan, meanwhile, famine conditions were so severe during 1889 that by the end of their long-planned march on Egypt, the Mahdist forces of General al-Nujumi had already been decimated – most of the *jihadiya* had deserted, nearly all their horses and baggage animals were dead, many of them eaten, and all but one gun had been abandoned for lack of transport – whereupon the Mahdists were promptly annihilated by Anglo-Egyptian forces at the Battle of Tushkah.[31] Numbers and details are often uncertain, but there can be little doubt about the historical importance and gravity of droughts within pre-capitalist and early capitalist societies: a recurring cause of major catastrophes and a near-constant threat to precarious rural livelihoods.

In the face of all this, the temptation may be to think of droughts and their impacts as timeless and recurring, natural and preordained features of human history. But this would be a mistake. The consequences of poor rains have always been socially constituted, shaped both by relations of production and exchange and by systems of drought management and response. These consequences have also always been highly uneven, visiting hunger and in extremis death on certain populations and groups (above all peasant producers) while placing relatively less stress on others (most notably urban classes protected by their greater mobility and purchasing power). And, in turn, this means that drought has a secular history that is inextricably tied to processes of political and economic change.

---

[31] M. Alamgir, *Famine in South Asia: Political Economy of Mass Starvation* (Oelgeschlager, Gunn and Hain, 1980), 59; R. H. Grove, 'The great El Niño of 1789–93 and its global consequences: reconstructing an extreme climate event in world environmental history', *Medieval History Journal*, 10: 1–2 (2007), 75–98; M. Davis, *Late Victorian Holocausts: El Niño Famines and the Making of the Third World* (Verso, 2002), 7, 129, 133–4; P. M. Holt, *The Mahdist State in the Sudan 1881–1898: A Study of Its Origins, Development and Overthrow*, 2nd ed. (Clarendon, 1970), 178–83.

The worldwide transformation in the geography of drought-related famine during the late nineteenth century provides perhaps the most telling demonstration of this. The late nineteenth century, as Mike Davis observes, was 'the very half-century when peacetime famine permanently disappeared from Western Europe'. Yet across large swathes of Asia, Africa and South America, it was also the era of 'late Victorian Holocausts', as Davis calls them, in which colonial rule and 'the imperialism of free trade' conspired to turn severe El Niño droughts into famines of unprecedented consequence; never before, it seems, had India and China in particular experienced such extreme famine mortality. Little over a century earlier, high Qing China had possessed powerful functioning drought mitigation and adaptation systems – including a national system of grain price stabilisation and large, well-managed granaries and hydraulic infrastructures – while Europeans were dying in their millions from famine and associated disease. In 'Europe's Age of Reason', Davis notes, 'the "starving masses" were French, Irish and Calabrian, not Chinese'. But by the Belle Époque, this pattern had changed utterly, mirroring the broader 'great divergence' of the European and non-European worlds after 1750. Both in terms of drought impacts and more broadly, this was 'a radical point of division in the experience of humanity'.[32]

This great divergence in drought vulnerability occurred for a range of reasons. The deepening incorporation of non-European land and labour into the London-centred capitalist world economy, first, meant that the peasantries of Asia, Africa and Latin America were increasingly compelled to turn from local subsistence to cash crops and exports. As a result they simultaneously reduced their own subsistence production and became increasingly embedded in and dependent on national and global circuits of capitalist commodity production, thus becoming vulnerable to market fluctuations (including those arising from good harvests and over-production). In Kautsky's memorable description:

> The peasant was now dependent on the market, which proved to be even more moody and unpredictable than the weather. At least the weather's perfidiousness could be prepared for. Ditches could mitigate the effects of a particularly wet summer, and irrigation could counter the consequences of a drought; smoking fires could protect vines from spring frosts and so on. But the peasant had no

---

[32] Davis, *Late Victorian Holocausts*, 8–9, 280–5, 139; J. Gallagher and R. Robinson, 'The imperialism of free trade', *Economic History Review*, 6:1 (1953), 1–15; K. Pomeranz, *The Great Divergence: China, Europe, and the Making of the Modern World Economy* (Princeton University Press, 2000); P.-E. Will and R. B. Wong, *Nourish the People: The State Civilian Granary System in China, 1650–1850* (University of Michigan, 1991).

means of arresting a collapse in prices, or of selling unsaleable corn. What was once a blessing – a good harvest – became a curse.[33]

Second, across Asia, Africa and Latin America this growing market dependence and vulnerability was exacerbated by colonial state predation, especially the Tolstoyan levels of taxation that were routinely imposed by colonial states in order to make their rule self-financing and to ensure a net flow of capital to European metropoles. From India to Nigeria, the typical consequence of this was progressively increasing rural impoverishment and indebtedness, which in turn precipitated still further commodity production and yet deeper market penetration and dependence. Third, this combination of predation and immiseration paved the way for the erosion, indeed collapse, of indigenous drought adaptation capacities, especially through the depletion of granaries and the decline of established water management and irrigation systems. Moreover, at least under British rule, colonial administrators were so enamoured of Malthusian and Smithian doctrines – famine is 'the last, most dreadful resource of nature', Malthus had asserted; 'a famine has never arisen from any other cause but the violence of government attempting, by improper means, to remedy the inconvenience of dearth', Smith claimed in *The Wealth of Nations* on the subject of the 1769–73 Bengal famine – that they refused to stockpile grain, control prices or limit food exports when drought and famine hit, or otherwise interfere with market forces. Not only Davis' *Late Victorian Holocausts* but also that other great study of the political ecology of drought and famine, Michael Watts' *Silent Violence*, repeatedly point to these factors as key to the transformation of drought vulnerability during the colonial era. And while some of these factors applied to Europe too, the crucial difference here was that the great European peasantry possessed a means and outlet for escaping from drought and famine: New World emigration.[34]

As for the post-colonial era – or more precisely the period since 1900 – two trends are especially noteworthy. One is that famine and starvation have become much less tied to drought specifically; indeed, if there is one factor that stands out as the primary cause of twentieth- and twenty-first-century famines, it is not drought but the state, its policies and its wars. Thus the two major peaks in famine incidence during the twentieth century coincided not with drought but the First and Second World

---

[33] K. Kautsky, *The Agrarian Question*, Vol. 1, trans. P. Burgess (Zwan, 1988 [1899]), 16.
[34] Malthus, *Essay*, variorum ed., I, 303; A. Smith, *An Inquiry into the Nature and Causes of the Wealth of Nations*, ed. E. Cannan (University of Chicago Press, 1976 [1776]), 2, 33; Davis, *Late Victorian Holocausts*, chs. 1, 5, 10; M. J. Watts, *Silent Violence: Food, Famine and Peasantry in Northern Nigeria*, new ed. (University of Georgia Press, 2013 [1983]), ch. 6.

Wars; they were also concentrated in Europe, as the atrocities which had formerly been visited upon the colonised world returned, boomerang-style, to their imperial core. The 1943 famines in Bengal and, on a smaller scale, northern Nigeria were precipitated by wartime grain requisition, price inflation and behind them British imperial indifference, and had little if anything to do with poor rains. The Great Leap Forward famine of 1958–62 in China, the single worst famine in the modern historical record, was a product of totalitarian social engineering. The 'world food crisis' of 1972 was caused less by droughts in the Sahel, Soviet Union and elsewhere, extensive as these undoubtedly were, than by the collapse of the Bretton Woods system and the Nixon administration's new economic policies, which together led to sharp global food commodity price rises. The 1984 Ethiopian famine, though partly a consequence of poor rains, was more directly caused and sustained as an instrument of counter-insurgency, specifically through government offensives into food-producing areas under opposition control, the intentional destruction of crops and food supplies, the blocking of food aid to areas under opposition control, and forced displacement and resettlement. South Sudan's 2017 famine was even more clearly caused by war than by drought, specifically the destruction of crops and food stores by armed groups, restrictions on planting and harvesting, the blocking of food aid delivery, war-induced mass displacement and hyperinflation. The same is true of Yemen, which became home to the world's worst humanitarian crisis in 2018 and 2019 thanks not to the weather but infrastructure destruction, non-payment of salaries within Houthi-controlled areas of the country and an economic embargo imposed by Saudi Arabia and the United Arab Emirates, with US and British warships in support. Reflecting all this, contemporary theorisations of famine consistently emphasise, in one way or another, the centrality of politics – of power, governance, violence or inequalities – in its aetiology.[35] Whereas in pre-capitalist societies like Braudel's Mediterranean, famine was principally caused by local

---

[35] de Waal, 'The end of famine?', 186–7, 194; M. Mukerjee, *Churchill's Secret War: The British Empire and the Ravaging of India during World War II* (Basic Books, 2010); Watts, *Silent Violence*, 329–36; F. Dikötter, *Mao's Great Famine: The History of China's Most Devastating Catastrophe, 1958–62* (Bloomsbury, 2010); R. V. Garcia, *Drought and Man, Vol. 1: Nature Pleads Not Guilty* (Pergamon, 1981); Africa Watch, *Evil Days: Thirty Years of War and Famine in Ethiopia* (HRW, 1991); T. D. Anei et al., *Accountability for Starvation Crimes: South Sudan* (Global Rights Compliance Group, 2019); D. Nikbakht and S. McKenzie, 'The Yemen war is the world's worst humanitarian crisis, UN says', *CNN* (03/04/2018); Sen, *Poverty and Famines*; S. Devereux (ed.), *The New Famines: Why Famines Persist in an Era of Globalization* (Routledge, 2007); A. de Waal, *Mass Starvation: The History and Future of Famine* (Polity, 2018).

environmental changes, especially drought, today the situation is altogether different.

Alongside this decoupling of famine from environmental shocks, drought and with it famine and hunger have all been in long-term decline as causes of human insecurity. Since 1945, the scale and lethality of famines have dropped precipitously (though as Alex de Waal observes there has recently been a war-related upturn in famine risks). Hunger and undernourishment have also been in long-term decline (though, paralleling famine, this trend has stopped since 2015, and there has been a slight absolute-terms increase in hunger). And, even more sharply, deaths from drought have plummeted. The widely used Emergency Events Database (EM-DAT), for instance, records not a single major drought-related fatality event since the mid-1980s (the 1984–5 famine in Darfur, which led to an estimated 100,000 deaths), in stark contrast to previous decades. While global exposure to drought has increased – if nothing else because of the rise in the global population – drought vulnerability has plummeted. In both absolute and relative terms, the overall global historical trends are beyond doubt.[36]

Moreover, it is not difficult to identify the reasons for this. Increased irrigation from surface and groundwater resources has reduced both dependence on regular rains and vulnerability to climatic anomalies and capricious weather. Economic development and diversification have, even in the poorest parts of the global South, reduced financial dependence on farming and pastoralism, as even isolated rural families and communities have come to depend on state employment, migrant remittances and links to informal urban economies as well as on the tending of crops and livestock. Urbanisation and the globalisation of trade in food staples have decoupled food security from local agricultural production. Modern transport and communications infrastructures can, when necessary – though also, of course, with significant hardship and suffering – usually facilitate escape from protracted drought conditions; migration in this sense is thus a critical means of adaptation to environmental shocks.[37] And, in a post-colonial international political system where mass starvation is not as easily ignored or hidden as previously, national and international systems of drought monitoring and response serve as powerful safety nets against drought impacts. Ultimately, in a world where the speeds of movement, exchange and communication have so radically increased, droughts are no longer short-term shock events but protracted socio-ecological

---

[36] de Waal, 'The end of famine?', 184–5; FAO et al., *The State of Food Security and Nutrition in the World 2019: Safeguarding against Economic Slowdowns and Downturns* (2019), 3–26; CRED, EM-DAT data; de Waal, 'Famine mortality'.

[37] R. Black et al., 'Migration as adaptation', *Nature*, 478 (2011), 447–9.

crises, such that adaptation and impact mitigation are, though painful, typically possible. Storm surges and flash floods typically strike with little warning, visiting immense destruction on all in their paths. The same, however, does not apply to droughts.

The point here, to avoid any doubt, is neither that the global trend of reduced drought vulnerability is universal, inevitable or even linear, nor that drought is rendered irrelevant by them. Some places and groups, especially politically excluded ethnic groups and women, are much more vulnerable to drought than others. Agriculture-dependent economies tend to be more vulnerable, while more developed and diversified economies are less so: hence north-east Syria was more affected by the 2006/7–2008/9 eastern Mediterranean drought than the rest of the country and Syria more affected than Cyprus (or than Israel would have been had the drought also struck there). Drought vulnerability can also ebb and flow, increasing, for instance, when groundwater reserves or international remittances run dry, or during times of commodity price crashes or war. Moreover, 'modernisation' does not just enhance adaptive capacities: it also typically involves the replacement, destruction or decay of traditional adaptive practices and infrastructures, thus in certain respects deepening local vulnerabilities.[38]

Yet for all this, at an aggregate, global level drought impacts have evidently been in historical decline as a result of those processes of capitalist technological, economic, infrastructural and political development which, by obliterating time and space, have made people and places less and less hostage to local environmental fluctuations. Clearly, these very developments have simultaneously generated forms of dependency – on technologies, infrastructures, markets, capital and the state – and, with this, new sources and dimensions of insecurity. Environmental risks have, in this sense, been progressively replaced by economic and political ones, with these new risks often themselves generating new patterns of vulnerability, marginalisation and violence. But we must not get ahead of ourselves, as this is a subject for later chapters. For now, the point is simply that the age-old concurrence of drought with death and chronic vulnerability has been decisively broken, that today droughts have nothing like the social consequences they once had – and that this needs to be borne in mind when considering what climate change is likely to bring in the future.

---

[38] T. Kakota et al., 'Gender vulnerability to climate variability and household food insecurity', *Climate and Development*, 3:4 (2011), 298–309; E. Feitelson and A. Tubi, 'A main driver or an intermediate variable? Climate change, water and security in the Middle East', *Global Environmental Change*, 44 (2017), 39–48; V. Lemire, 'The awakening of Palestinian hydropolitical consciousness: the Artas-Jerusalem water conflict of 1925', *Jerusalem Quarterly*, 48 (2011), 31–53.

Everybody Loves a Good Drought

To understand why analysts and policymakers are so inattentive to this – and more specifically why their claims about drought and conflict are so often wrong – it is worth engaging with Palagummi Sainath's powerful essays on poverty in rural India, *Everybody Loves a Good Drought*. In this work and especially in the essay of the same name, Sainath shows that India's countless severe water problems are typically framed as products of 'drought' even when they occur in the midst of, and in districts with, abundant rainfall. Motivated above all by pitches for funding, these drought framings are typically advanced first by local contractors, politicians and stringers, but then also receive support from district administrators, state governments, the urban press, the national government and even international donors – to the extent that the 'global aid community' may be 'mobilised into fighting drought in a district that gets 1500 mm of rainfall annually'. The money rolls in – drought relief is rural India's biggest growth industry, says Sainath – but given that water scarcity is typically caused more by political and economic inequalities than by poor rains, rarely ends up helping the rural poor. The drought industry is untroubled by this, however: as Sainath notes with characteristic understatement, '[i]t is so much nicer to just put the whole thing down to nature's vagaries'.[39]

Blaming the climate has a rich history. In 1852, a French colonial force laid siege to the town of Laghouat in what is now Algeria, killed a third of its population, and filled its wells with corpses (in one well, 256 corpses were discovered). The French romanticist Eugène Fromentin arrived in Laghouat just six months afterwards and, as recounted in his classic work *A Summer in the Sahara*, blamed the silence of the native population and the lack of welcome afforded him on the terrible desert sun – a sun, he said, 'which has fallen from the sky onto objects and from objects has transferred to their faces'. A half century later another French colonial expedition, intent this time on conquering the Lake Chad basin, left a trail of burnt ruins and corpses from Tombouctou eastwards before culminating in a violent stand-off between the French forces. The official French inquiry found only one conceivable explanation: the climate. Fast forward another half century, and the Great Leap Forward famine in China, discussed in the previous section, was explained away by the Communist Party leadership as a 'natural catastrophe'.[40] What is true of eco-determinist discourse in general also

---

[39] P. Sainath, *Everybody Loves a Good Drought: Stories from India's Poorest Districts* (Penguin, 1996), 317–24; L. Mehta, *The Politics and Poetics of Water: Naturalising Scarcity in Western India* (Orient Longman, 2005).
[40] S. Lindqvist, *Desert Divers*, trans. J. Tate (Granta, 1990), 49–63; S. Lindqvist, '*Exterminate All the Brutes*', trans. J. Tate (Granta, 1997), 161–70; Dikötter, *Mao's Great Famine*.

applies to drought narratives specifically: both have long been deployed, in particular by authoritarian and imperial powers, as a way of deflecting blame and responsibility.

In our era of heightened international concern with climate change, this discursive strategy has become, if anything, even more commonplace. Thus in pre–civil war Syria, as the International Crisis Group reported, the Assad regime would regularly take diplomats to the north-east and tell them 'it all has to do with global warming', blaming what was in essence a state-induced socio-ecological crisis (as will be detailed in Chapters 5 and 6) on climatic transformations beyond its control. The drought was 'beyond our powers', claimed President Bashar al-Assad; the drought was 'beyond our capacity as a country to deal with', claimed the regime's Minister of Agriculture; 'Syria could have achieved [its] goals pertaining to unemployment, poverty, and growth if it was not for the drought', proclaimed Deputy Prime Minister Abdullah al-Dardari. Ban Ki-moon's assertion that the Darfur war 'began as an ecological crisis' was an attempt to mollify Sudan at a time when he needed its cooperation over the establishment of a peacekeeping mission. His drought–conflict reasoning was subsequently embraced by Khartoum, with former President Omar al-Bashir claiming that what happened in Darfur was 'a traditional conflict' rooted in 'frictions between the shepherds and the farmers' that 'increased because of climate change and the dry weather', and the Sudanese government claiming at the UN Security Council that 'the main cause of the conflict in Darfur was desertification and drought'; as Harry Verhoeven has observed, the Sudanese regime 'loves the "climate war" rhetoric'. Meanwhile, the idea that climate change was a factor in the Boko Haram crisis was first advanced by Nigerian leaders, in keeping with their overall representational strategy of portraying the crisis as an external intrusion, rather than, as it primarily was, a consequence of internal marginalisation, repression and militarisation. As one European think tank researcher has observed, '[t]here's a very strong political interest in this narrative from the national governments, particularly the Nigerian government, because if this is how they portray the problem of terrorism, the instability is to do with a lack of water [and] nothing to do with weak governance, exclusion, marginalisation, [or] human rights violations by security forces'. Indeed in each of our three cases, the initial 'climatisation' of crisis – the discursive construction of climate change as a key driver of conflict – was either directly motivated by, or indirectly a function of deference to, regime interests in being represented as victims rather than instigators of violence.[41]

[41] ICG, *Popular Protest*, 23; 'President al-Assad delivers speech at people's assembly', *SANA* (30/03/2011), cited in De Châtel, 'The role of drought', 535; US Embassy Damascus, '2008 UN drought appeal for Syria', Diplomatic Cable 08DAMASCUS847_a (26/11/2008); US Embassy Damascus, 'SARG sheds light on its drought concerns', Diplomatic Cable

The subsequent reproduction and uptake of these narratives has also reflected various interests, agendas and power relations. International organisations and donors working in Syria repeatedly invoked climate change and drought as causes of the country's problems – partly, it seems, because of their dependence on the Syrian government for information and access, and to keep on acceptable working terms with the Assad regime; partly out of technical biases; partly because of their complicity in many of Syria's deep structural problems; and partly also, no doubt, out of performative concern with the challenge of climate change. Official UN reports on the drought, for instance, which were produced in collaboration with the Syrian regime, are predictably drought-centric, barely mentioning the role of factors other than drought in the country's pre-civil war rural crisis, failing to voice any criticisms of government policy and not touching on any other sensitive matters (e.g. not even mentioning the existence of a discriminated against Kurdish minority, which, as we shall see in Chapter 6, was at the heart of the crisis in Syria's north-east). At the conclusion of his 2010 mission to Syria, the UN Rapporteur on the Right to Food, Olivier De Schutter, likewise provided a highly drought-centric presentation of the country's poverty and migration crises – partly, we may surmise, to pressure foreign governments to respond to the UN's drought appeal, and partly out of deference to his Syrian government hosts and the international media. International media reports on the subject were similarly drought-centric – partly, no doubt, because of media preferences for simplified and striking Band Aid–style narratives, but also because they relied mostly on the UN sources discussed above and took these at their word. The same emphases were then reproduced and circulated by assorted European and US think tanks and researchers concerned with climate change, with even the academic researchers barely pausing to consider whether the UN reports, media stories and Syrian government statements they relied upon were trustworthy representations of Syria's problems. Thus did a flawed narrative that originated in the Assad regime's interests in deflecting responsibility for a crisis of its own making end up being regurgitated, in elaborated form, as a statement

10DAMASCUS97_a (01/02/2010); S. Tisdall, 'Omar al-Bashir: genocidal mastermind or bringer of peace?', *Guardian* (20/04/2011); UN Security Council, *6587'h Meeting* (S/PV.6587, Resumption 1) (20/07/2011), 34; Verhoeven, 'Climate change, conflict and development', 695; Bromwich, 'Power, contested institutions and land', 13; O. Nzeshi, 'Nigeria: NSA blames insecurity in climate change', *This Day* (24/04/2013); Daoust interview with think tank researcher (29/06/2019); A. Oels, 'From "securitization" of climate change to "climatization" of the security field: comparing three theoretical perspectives', in Scheffran et al. (eds.), *Climate Change, Human Security and Violent Conflict*, 185–205; M. L. Wine, 'Climatization of environmental degradation: a widespread challenge to the integrity of earth science', *Hydrological Sciences Journal*, 65:6 (2020), 867–83.

of fact in the leading American journal of political science. Friedrich Nietzsche's famous comment that 'truths are illusions about which one has forgotten that this is what they are' captures what transpired here perfectly.[42]

From such beginnings, international political interests in 'securitising' climate change typically take over – as all three of our cases attest. The Syria climate–conflict narrative rose to prominence in the lead-up to COP21, the 2015 Paris conference on climate change, when it was invoked by political leaders and commentators from a wide variety of political positions – from Barack Obama to Naomi Klein – to emphasise the gravity of the climate change challenge. Coinciding with the height of the European 'migration crisis', as Turkey opened is borders to the outflow of Syrian refugees, the Syria story was also widely mobilised, especially by western European environmental campaigners and liberal commentators, as evidence of the 'new normal' of mass 'climate refugee' flows.[43] The Lake Chad climate security narrative became prominent around the same time, mobilised by France – as host of COP21 and because of its extensive geopolitical interests in the region – to emphasise the need for decisive climate change mitigation; indeed, French Foreign Minister and COP President Laurent Fabius highlighted 'the spectacular drying of the Lake Chad and its disastrous consequences' in his opening remarks to the conference and France's then Environment Minister, Ségolène Royal, later warned researchers participating in the conference to 'stop saying the Lake Chad is not disappearing'. The Lake Chad narrative has also been instrumental to, and instrumentalised within, efforts to push climate change onto the UN Security Council's agenda, and to institutionalise 'climate security' within the UN system.[44] Darfur, similarly, became such a prominent focus of climate security discourse in 2007 not thanks to any new weight of evidence thereon, but because

---

[42] UN-OCHA, *Syria Drought Appeal*; UN-OOHA, *Syria Drought Response Plan*; UN-OCHA, *Joint United Nations Drought Assessment*; UN-OCHA, *Syria Drought Response Plan 2009–10*; UNCHR, 'Two to three million Syrians face food insecurity'; O. De Schutter, 'UN Special Rapporteur on the right to food: mission to Syria from 29 August to 7 September 2010', preliminary reflections' (07/09/2010); F. Nietzsche, 'On truth and lies in an extra-moral sense', in *The Portable Nietzsche*, trans. W. Kaufmann (Viking, 1976), 46–7.

[43] Obama, 'Remarks by the President at the United States Coast Guard Academy Commencement'; Box and Klein, 'Why a climate deal is the best hope'; E. M. O'Hagan, 'Mass migration is no crisis: it's the new normal as the climate changes', *Guardian* (18/08/2015); J. Wendle, 'Syria's climate refugees', *Scientific American* (12/2015), 98–103.

[44] L. Fabius, Speech to COP21, Paris (30/11/2015); Magrin, 'Disappearance of Lake Chad', 205; UN Security Council, *Resolution 2349*; UN Security Council, *8307th Meeting* (S/PV.8307) (11/07/2018); UN Security Council, *8451st Meeting*.

assorted commentators and policymakers, right the way up to Ban Ki-moon, wanted to strengthen the case for international and especially US action on climate change; and they chose to do so by invoking a war that was already a subject of extensive international debate, especially within the United States. More broadly, ever since the mid-2000s, when climate change started being widely represented as an international security issue within defence policies and diplomatic circles, the securitisation of climate change has been led neither by evidence nor by a concern with climate security specifically, but instead by an interest in 'dramatising' the seriousness of climate change and the need for deeper emissions reductions. The central underlying premise here has been – as Sharon Burke, one of the most influential figures in the climate security policy community, put it – that 'if you can't describe something as a security issue it's not important', at least in the United States.[45] Within policy discourse on climate security, the crises in Darfur, Syria and Lake Chad have essentially been servants to this higher political end.

Under contemporary globalised capitalism, as we have seen, poor rains are not nearly as locally devastating as they once were: just as water is no longer life, so extreme drought is no longer a clear harbinger of death either. Yet for all this, the spectre of drought-induced collapse lives on, much like the Malthusian tradition that so often informs it. In part this is no doubt a historical legacy, a function of just how embedded ideas of devastating droughts are within myth and culture. In part it also reflects the continuing pull of eco-determinist reasoning across many different sites and domains, as discussed in the previous chapter. Colonialism is a factor too: the myth of Lake Chad's ostensible decline, for instance, has been a constant since the scramble for Africa (witness Winston Churchill's turn-of-the-century observations on the worthlessness of the lands of the Sahel: 'Only one important river, the Shari [Chari], flows through them, and never reaches the sea: and even Lake Chad, into which the Shari flows, appears to be leaking through some subterranean exit, and is rapidly changing from a lake into a mighty swamp'). But above and beyond all this, drought narratives are so often invoked and reproduced because they are politically comfortable and convenient, helping to obscure and evade the fundamentally political causes of crises, and to blame failing rains instead. Among other things, they enable Pentagon

---

[45] N. Detraz and M. M. Betsill, 'Climate change and environmental security: for whom the discourse shifts', *International Studies Perspectives*, 10:3 (2009), 303–20; F. Sindico, 'Climate change: a security (council) issue?', *Carbon and Climate Law Review*, 1: 1 (2007), 29–34; Mamdani, *Saviors and Survivors*; Schwartz and Randall, *Abrupt Climate Change Scenario*, 7; S. Burke, Comments at panel on 'Advancing climate security: global views on US strategies', Planetary Security Conference, The Hague (19/02/2019).

military planners to imagine themselves as victims, called to do 'endless mop-up' by environmentally-induced chaos. Following James Ferguson, drought narratives are 'anti-politics' par excellence.[46]

Of course, not everybody 'loves a good drought' – as even Sainath's essay demonstrates. The rural poor often benefit little from relief programmes. Some climate change denialists would rather blame anything but poor rains. And drought narratives are often deployed in an inconsistent, even duplicitous, fashion. One irony of the Syria case, for example, is that those transatlantic actors who would later claim that drought played a crucial role in causing mass poverty and migration, sparking the country's civil war – and that the Syrian experience revealed a rapidly warming world its future – were largely uninterested in the drought at the time, opting for geopolitical reasons against supporting UN drought relief efforts.[47] Indeed, if there has been one constant in US and European engagement with Syria's 2006/7–2008/9 drought it has been this: whether ignored out of hostility to the Syrian regime and its Iranian allies or reified as a climate mitigation morality story, the actual experiences and suffering of the people of north-east Syria have been at most a secondary concern, and arguably much less than that. Whether mythologised, disregarded or actively inflamed, droughts in the modern world are always playthings in the service of power.

A Revenge of Uncertainty?

What, though, does all the above tell us about the future of drought, as global temperatures spiral ever upwards, towards and then exceeding 1.5°C and then 2°C? For, even if it is true that drought has not sparked civil conflicts in Darfur, Syria and Lake Chad; even if it is true that it is mostly coped with and adapted to; even if it is of declining overall importance as a cause of famine, migration and other insecurities; and even if it has become a subject of myth-making and politics – even if all this is true, so what? Is the past at all reliable as a guide to what is to come, if this future will be one of regular extreme environmental shocks, including droughts?

---

[46] W. S. Churchill, *The River War: An Historical Account of the Reconquest of the Soudan*, Vol. II (Longmans, Green, and Co., 1899), 319–20; Magrin, 'Disappearance of Lake Chad'; M. Funk, *Windfall: The Booming Business of Global Warming* (Penguin, 2014), 30; J. Ferguson, *The Anti-Politics Machine: Development, Depoliticization and Bureaucratic Power in Lesotho* (Cambridge University Press, 1990).

[47] 'Donald Trump tells California "there is no drought" as drought continues', *CBS Sacramento* (27/05/2016); De Schutter, *Report of the Special Rapporteur*, 12; De Châtel, 'Waiting for a response', *Syria Today* (02/2010); US Embassy Damascus, '2008 UN drought appeal'.

Paralleling our discussion in the previous chapter, there are a series of issues to consider here. To start with, the evidence on whether drought events have already become more frequent and more severe is not nearly as clear-cut as is often thought. Both the IPCC's Fifth and Sixth Assessment Reports, for instance, express only 'low confidence' that there has been a global-scale increase in the intensity or duration of droughts since 1950. Different models performed by the same authors even yield contradictory findings on how drought frequency, and the areas affected by it, have changed (with some models suggesting increases, others slight decreases). Certain regions are regularly identified as having experienced increasingly frequent and intense droughts, most notably the Mediterranean and western North America. Yet the evidence to this effect largely depends on drought models that place heavy emphasis on the contribution of (high) temperatures to drought; models focusing solely on (abnormally low) rainfall identify quite different regional drought trends.[48]

As for the future, the IPCC's Fifth Assessment Report expressed only 'low confidence' that the early twenty-first century will witness a global increase in the intensity and duration of droughts as well as 'low confidence' in long-term projections of the frequency and duration of megadroughts and acknowledged great uncertainties in regional drought projections – conclusions that are broadly reaffirmed in IPCC's more recent Sixth Assessment Report.[49] However, even these assessments rely on an end of century greenhouse gas emissions and concentration scenario, RCP 8.5, which is not 'business as usual' and is widely considered implausible, as noted in the previous chapter. While drought incidence and severity may well increase through the course of the twenty-first century, this is by no means a scientific certainty.

What does seem 'very likely' is that alongside the projected increase in global precipitation levels, there will be an increase in the incidence and intensity of extreme rainfall events. Assuming this is the case, flood risks will *ceteris paribus* increase (though with the qualifier that all else is not equal, since flooding and flood impacts are affected by many factors besides rainfall). The risk of soil erosion will do likewise.[50] If a higher proportion of rainfall falls as heavy rains, this may translate into more long

---

[48] IPCC, *Climate Change 2013*, 7, 211, 213; IPCC, *Climate Change 2021*, 117, 1910–12; IPCC, *Climate Change 2014: Synthesis Report*, 53; IPCC, *Climate Change and Land*, 8; J. Spinoni et al., 'A new global database of meteorological drought events from 1951 to 2016', *Journal of Hydrology: Regional Studies*, 22 (2019), 1–24.
[49] IPCC, *Climate Change 2013*, 7, 1115, 986; IPCC, *Climate Change 2014: Synthesis Report*, 69; IPCC, *Climate Change 2014. Part B*, 1279; IPCC, *Climate Change 2021*, 1946.
[50] IPCC, *Climate Change 2013*, 7; IPCC, *Climate Change and Land*, 53, 256.

dry periods. If rains increasingly fall at the wrong time of year, the consequences for agriculture could be profound. Moreover, more frequent heavy rains may also mean increased water losses – with flash floods ripping through river basins and communities and water being 'lost' by them and to the sea – in turn increasing drought vulnerabilities.

Overall, it seems sensible to assume and, more importantly, act on the basis that drought incidence and severity will very probably increase as the century unfolds, and that drought exposure will too (beyond that arising from population growth alone). Yet even allowing for this, a significant increase in drought vulnerability does not thereby follow. Over the past century, population growth has increased absolute levels of drought exposure, but this has been accompanied by a sharp decline in drought vulnerability. Moreover, as indicated above, this decoupling of drought exposure and vulnerability is the consequence of global integration and the development of multiple systems and mechanisms of drought adaptation in the areas of trade, human mobility, communications, emergency relief and more. This suggests, in turn, that there is no reason why drought exposure and vulnerability should be recoupled – unless these systems and mechanisms of adaptation themselves wither and decline. Although it is entirely possible that the long-term trend of decreasing vulnerability to drought may go into reverse, any such reversal is likely to be driven more by a reduction in adaptive capacities than by increased exposure arising from climate change.

The supposed climate crises in Darfur, Syria and Lake Chad are thus not close to being windows on the world's future, not only because of the poverty of evidence on them but because in an integrated capitalist world order local droughts do not neatly translate into local crisis. Just as we should not expect a 'revenge of geography' – as was argued in the previous chapter – so there is unlikely to be a 'revenge of uncertain rains' either. Of course, this does not mean that all is well, or even that drought risks can be dismissed entirely. The very structures and flows which have underpinned declining global drought vulnerability are themselves often sources of dependency and danger. Drought impacts, moreover, need not be defined by or confined within political boundaries – they are not subject to the 'territorial trap' – but may ripple outwards across the world, like many other such shock events. There is much more to be said on these issues. But before that we need to extend our critique of eco-determinism into a rather different terrain: the thorny question of identity.

4   Others

A Destroyer of Trees

In his otherwise unremarkable 1969 history of Cyprus, the English historian Hugh Dominic Purcell causally observes that 'one of the gravest charges which can be made against Islam is that, wherever it has penetrated, the trees have tended to disappear'. From the perspective of the early twenty-first century, when Islam and Muslims are routinely accused of all manner of crimes far graver than the disappearance of trees, this claim no doubt seems archaic if not downright bizarre. But this aside, two things about it are particularly noteworthy. One is that it has absolutely no evidential basis: neither in Cyprus specifically, nor beyond, is there any sound evidence that Islam is, or historically has been, particularly hostile to trees. And the second is that the claim should originate from Britain, which more than any other land *has* long been at the forefront of the global project of deforestation. By the end of the seventeenth century, England and Wales were more deforested than any other area of Europe, bar the Netherlands. It was the extent of this deforestation – driven by assorted social, economic and geopolitical factors – which precipitated Britain's uniquely early turn to fossil fuels. Britain then led the global charge against the world's forests during the nineteenth century, as its colonies were stripped for timber and commercial agriculture. And still today, upland 'conservation' in Britain is largely a matter of *preventing* natural forest regeneration. If there is one empire which merits being called a destroyer of trees, then it is surely the British one – not Islam.[1]

For all this, Purcell's passing comment is not that remarkable. Pejorative and hypocritical representations of Islam are not just common within but a defining and constitutive feature of high-modern European thought, as

---

[1] H. D. Purcell, *Cyprus* (Ernest Benn, 1969), 215; S. E. Harris, 'Cyprus as a degraded landscape or resilient environment in the wake of colonial intrusion', *Proceedings of the National Academy of Sciences*, 109:10 (2012), 3670–75; M. Williams, *Deforesting the Earth: From Prehistory to Global Crisis*, abridged version (University of Chicago Press, 2006), 150, 164–7; G. Monbiot, *Feral* (Allen Lane, 2013), ch. 12.

Edward Said most famously demonstrated. During the nineteenth and early twentieth centuries entire disciplines were invented and framed around supposed civilisational and racial hierarchies and their maintenance. The discipline of International Relations, for instance, began life in the United States as the study of race relations, devoted to maintaining white supremacy over what one of its formative figures, Reinhold Niebuhr, called 'the colored continents'. IR's first specialised journal was *The Journal of Race Development* (later relaunched and superficially de-racialised as *Foreign Affairs*), while the inventor of its discipline defining 'realist' tradition was Lothrop Stoddard, leading American interwar intellectual and author of *The Rising Tide of Color: The Threat against White World Supremacy*. Suffice to add that such concerns both informed colonial practice and were far from unique to International Relations.[2]

Pejorative representations of the environment and human–nature relations were, and remain, central to much of this Orientalist, racist and colonial thinking. This is not only because, informed by evolutionary biology and early anthropology, inferior races were generally assumed to be inherently closer to nature, along with women and the lower classes. In addition, colonial experiences of African, Asian and new world environments – of deserts, forests, jungles, swamps, disease and more, all wild and alien – often framed their interpretations of these colonised lands' peoples (as Roy Harvey Pearce put it, Englishmen 'found in America not only an uncivilized environment, but uncivilised men – natural men, as it was said, living in a natural world'). Within colonial discourse, uncivilised environments were routinely portrayed as objects for improvement and development, as well as exploitation, often on the premise of their prior degradation by local cultures and indigenous practices. Islam was widely portrayed as a destroyer of trees. 'Tropical environments', especially climates, were also important for the practical barriers, indeed dangers, they posed to colonisation. All this and more – including the continuation, and echoes, of these motifs in the post-colonial and contemporary eras – is documented in a now large body of work on imperial ecologies, 'environmental Orientalism', and the misreading of Southern environments.[3]

---

[2] E. Said, *Orientalism: Western Conceptions of the Orient* (Penguin, 1978); R. Vitalis, *White World Order, Black Power Politics: The Birth of American International Relations* (Cornell University Press, 2015); R. Niebuhr, 'The colored continents', in H. R. Davis and R. C. Good (eds.), *Reinhold Niebuhr on Politics* (Wipf and Stack, 2007), 308–12; L. Stoddard, *The Rising Tide of Color: The Threat against White World Supremacy* (Scribner's, 1920).

[3] R. H. Pearce, *Savagism and Civilization: A Study of the Indian and the American Mind*, rev. ed. (University of California Press, 1988), 3; M. Leech and R. Mearns (eds.), *The Lie of the Land: Challenging Received Wisdom on the African Environment* (Currey, 1996); S. Sawyer and A. Agrawal, 'Environmental orientalisms', *Cultural Critique*, 45 (2000), 71–108;

Some may wonder how this helps us to understand the relations between water, climate and security or the likelihood of future water scarcity- and climate change-driven chaos. Our response would be unequivocal: that racialised constructions of the Other deriving above all from European colonialism are foundational both to the eco-determinist tradition in general and water and climate security discourse specifically, as well as to the divided environments which are the particular focus of this book. This chapter seeks to make this case. In doing so, it simultaneously revisits and extends the critique of eco-determinism developed in Chapters 2 and 3, demonstrating that this mode of thought is not just substantively flawed, but also indebted to and reproductive of racialised colonial reasoning; and provides a starting point for our positive reconstruction of the political ecology of water and climate (in)security. Thus we start by interpreting eco-determinism and mainstream water and climate security discourse as premised on a racialised imaginative geography. Then, in the longer second half of the chapter, we turn to our divided environments – and specifically the three ex-British colonies among our cases, Sudan, Cyprus and Israel–Palestine – in each case showing and comparing how their politics and political ecologies have been decisively forged by racialised identity constructions, and continue to be through to today. This paves the way, in conclusion, for a set of reflections on the politics of identity and alterity under circumstances of accelerating climate change. No single factor or element, of course, can fully explain modern waterscapes and contemporary patterns of water- and climate-related vulnerability. But if there is one which underpins all the others, it is that considered here: the simultaneously divisive and transformative legacies of racism and colonialism, in thought as well as in practice.

The Global Environmental Colour Line

Eco-determinism was discussed in Chapter 2 as a simultaneously distinctive and universal approach to human–environment relations that offers a particular reading of these relations without tying this reading to specific times or places. But in truth eco-determinism is more complex and contradictory than this. For, far from offering a distinctively anti-liberal understanding of human relations with nature or standing on the margins of 'Western' thought, eco-determinism instead permeates right through

D. K. Davis and E. Burke III (eds.), *Environmental Imaginaries of the Middle East and North Africa* (Ohio University Press, 2011).

the latter's intellectual heartlands. And far from advancing a universal theory, it always involves a specific imaginative geography and imaginative history – and not by accident. James Blaut points to this in his characterisation of environmental determinism as a 'colonizer's model of the world'. Building on W. E. B. Du Bois, it may also be suggested, more sharply, that an imagined 'global environmental colour line' runs right through eco-determinist discourse.[4]

There are two variants of this colour-line thinking within eco-determinist thought. In its classic, colonial formulation, geography, ecology and climate are invoked to explain, and often justify, the rise of Europe and 'the West', as well as the assumed non-rise of the world beyond. On one level nature is understood here as a universal determinant, the ultimate cause of both European superiority and non-European inferiority, with propitious environments being viewed as the fundamental reason for the former, and less propitious ones the basic reason for the latter. More than this, though, European environments are also assumed to have been so advantageous as to allow escape from natural imperatives and constraints, and to facilitate the consequent flourishing of morality, law and science – or, in short, 'civilisation'. The global divide imagined within this strand of eco-determinism, then, is not merely environmental but between a zone of nature on the one hand and one of freedom on the other, between one zone where people are imagined as subject to ecological determinations and another where human subjectivity is thought to have transcended them.

The most important European Enlightenment philosophies of history all adhered to this basic logic. Thus Montesquieu identified naturally high soil fertility, from which livelihoods could too easily be extracted, as '[t]he cause of there being such a number of savage nations' in pre-conquest America; suggested, in an 'imaginary division of the earth', that 'great nations' arise 'where nature seems to refuse everything'; and held that the result was that '[n]ature and the climate rule almost alone over the savages', whereas 'morals' and 'maxims of government' rule in Europe. Immanuel Kant, arguably the father of both modern cosmopolitanism and scientific racism, claimed that environmental and especially climatic factors accounted for humanity attaining 'its highest degree of perfection in the white race'; averred that '[t]he inhabitant of the temperate zone . . . is more beautiful in body, harder working, more witty, more moderate in his passions, and more sensible than any other kind of people in the world.

---

[4] J. M. Blaut, *The Colonizer's Model of the World: Geographical Diffusionism and Eurocentric History* (Guilford Press, 1993); W. E. B. Du Bois, *The Souls of Black Folk* (Fawcett, 1961 [1903]), 23.

Consequently, these people have always taught the rest, and vanquished them by the use of weapons'; and elsewhere remarked, building on the same reasoning, that the 'Negroes of Africa have no feeling that rises above the ridiculous' and the 'savages' of North America 'no conception of the morally beautiful' or traces of 'finer sentiments'. And the pre-eminent Enlightenment philosopher of history, Georg Wilhelm Friedrich Hegel, contended in his exploration of 'the geographical basis of world history' that 'neither the *torrid* nor the *cold regions* can provide a basis for human freedom or for world-historical nations' and that 'it is therefore the *temperate zone* which must furnish the theatre of world history'. In Hegel's view, while in Africa man had 'not progressed beyond a merely sensuous existence', instead existing in a 'state of animality' – one where 'there is no subjectivity, but merely a series of subjects who destroy one another' – modern Europe was 'the land of spiritual unity', possessed of culture, freedom and history. Many have debated whether these famous dead white men should be considered 'environmental determinists', but this is rather beside the point. More salient is that in each of their writings, determinist reasoning is applied to some peoples but much less to others, in turn establishing some of them as fully human, but others as more akin to animals.[5]

If we jump forward to the twentieth century we find the very same pattern. Thus Karl Ritter, Henry Thomas Buckle, Friedrich Ratzel, Ellen Churchill Semple, Arnold Toynbee and Ellsworth Huntington, among other early twentieth-century thinkers, all articulated some version of the view, as expressed by Semple, that environmental factors 'give a certain zonal stamp to human ... development'. Numerous late twentieth-century works have done likewise: Eric Jones' *The European Miracle*, Jared Diamond's *Guns, Germs and Steel*, David Landes' *The Wealth and Poverty of Nations*, and more besides. In each case, ecology and geography – or what Diamond calls differences in 'real estate' – are invoked to explain, if not also extol, the specificity of 'the European miracle' and Europe's consequent global ascendancy.[6]

---

[5] Baron De Montesquieu, *The Spirit of the Laws*, trans. T. Nugent (Hafner, 1949), 272, 275, 294; I. Kant, 'Physical geography', in *Natural Science*, ed. K. Watkins, trans. L.W. Beck et al. (Cambridge University Press, 2012), 576–7; I. Kant, 'Observations on the feeling of the beautiful and sublime', in *Observations on the Feeling of the Beautiful and Sublime and Other Writings*, ed. and trans. P. Frierson and P. Guyer (Cambridge University Press, 2011), 58, 60; G. W. F. Hegel, *Lectures on the Philosophy of World History: Introduction* (Cambridge University Press, 1975), 154–5, 172–8 (author's italics).

[6] E. C. Semple, *Influences of Geographic Environment: On the Basis of Ratzel's System of Anthropo-Geography* (Henry Holt & Company, 1911), 633; F. Thomas, 'Some representative contributions of anthrogeography to modern political theory' in C. E. Merriam and H. E. Barnes (eds.), *A History of Political Theories: Recent Times* (Macmillan, 1935),

Whereas this classic Enlightenment strand of eco-determinist thought is oriented to explaining Europe's rise, and espouses an essentially progressivist understanding of history, a much more pessimistic or melancholic variant of eco-determinist thinking predominates today – one where the central concern is not with the nature and causes of social progress but with its limits, as well as with the possibility of societal collapse and the cyclical, even tragic, fate of civilisations and humanity. This 'counter-Enlightenment' variant of eco-determinism might initially seem the polar opposite of its Enlightenment forebear. And indeed, in their understandings of the arc of human history – of the past as well as the future – the two are very much at odds. Crucially though, the one is derivative of the other and shares many of the same colonial and racial premises. Above all, their conceptions of geography are essentially the same: narratives of ecological limits may dispense with the optimism of Hegel and followers, but the global colour line lives on.

Consider for example Malthus who, far from being the parochial commentator on pre-industrial English society that he is often taken to be, was inherently European and global in his work, with a worldview shaped by the French Revolution and new world exploration and settlement. Indeed, just as Malthus' 'law of population' was borrowed, however erroneously, from Benjamin Franklin, so his broader theory was structured around a 'universalising history of civilization' according to which different lands and peoples, and not just different classes, were differentially governed by the constraints of nature. Malthus viewed primitive and uncivilised societies as essentially subject to the natural 'positive checks' of war, malnutrition, disease and famine. Pastoralist societies were singled out, in particular, as red in tooth and claw, characterised by rapid population growth and a ceaseless search for new pastures which together propelled them into 'great tides' of emigration, 'continual battles' and wild oscillations in their numbers; 'war', claimed Malthus, is 'the predominant check to the population' of such 'savage nations'. Modern Europe, by contrast, was depicted as a site of slow but steady population growth, in which war had 'certainly abated', and where demography was shaped by 'preventive checks' – social and economic considerations leading people to delay marriage and have fewer children – as well as 'moral restraint'. Europeans, in Malthus' framework, were

---

457–507; E. Jones, *The European Miracle: Environments, Economies, and Geopolitics in the History of Europe and Asia* (Cambridge University Press, 1981); D. Landes, *The Wealth and Poverty of Nations: Why Some Are So Rich and Some So Poor* (Norton, 1998); J. Diamond, *Guns, Germs and Steel: A Short History of Everybody for the Last 13,000 Years* (Vintage, 1997), 401.

social, economic and moral beings, and capable of exercising foresight; others, by contrast, were essentially subject to nature.[7]

If we fast forward two centuries, we find the very same premises at work. In his prize-winning tome *Collapse*, for instance, Jared Diamond identifies human environmental impacts, climate change and overpopulation as responsible for the civilisational and societal 'collapses' that – from Easter Island's supposed collapse and descent into cannibalism, right through to the 1994 Rwandan genocide – he sees as recurring features of human history. The 'political trouble spots of the modern world' and its 'environmental trouble spots' are one, he says, before explaining this correspondence on the grounds that:

Today, just as in the past, countries that are environmentally stressed, overpopulated, or both become at risk of getting politically stressed, and of their governments collapsing. When people are desperate, undernourished, and without hope, they blame their governments, which they see as responsible for or unable to solve their problems. They try to emigrate at any cost. They fight each other over land. They kill each other. They start civil wars. They figure that they have nothing to lose, so they become terrorists, or they support or tolerate terrorism... Hence the best predictors of modern 'state failures' – i.e. revolutions, violent regime change, collapse of authority, and genocide – prove to be measures of environmental and population pressure.

All this leads Diamond to the strikingly Malthusian conclusions that our contemporary global civilisation is potentially on the brink of collapse and that its environmental problems '*will* get resolved, in one way or another'. Yet Diamond also insists, as indicated by his book's subtitle ('How Societies Choose to Fail or Survive') that it is the 'choices' made by societies which are crucial in determining whether they collapse: some societies, he claims, manage to exercise agency and solve their environmental problems through 'pleasant means of [their] own choice', while others, lacking such agency, remain beholden to the whip of ecological necessity and find their problems resolved 'by unpleasant and unchosen means, such as the ones that Malthus initially envisaged'. Unsurprisingly, it is modern Haiti and Rwanda, alongside various pre-modern civilisations, that are portrayed as governed by the latter. 'Rwandan and Haitian peasants... think only of food for the next day', says Diamond in his one explanatory comment on why they failed to avoid eco-collapse – while Europe and North America, above all, are assumed to be lands of choice.[8]

---

[7] Bashford and Chaplin, *New Worlds*, 11; Malthus, *Essay*, 1st ed., 26–34; Malthus, *Essay*, variorum ed., Vol. I, 304–5.
[8] Diamond, *Collapse*, 497, 516, 498 (authors italics), 313, 434.

Lest this be thought a feature of popular texts only, consider also Thomas Homer-Dixon's paradigm-defining work on environmental security. Homer-Dixon's framework centres on the ill-defined notion of 'environmental scarcity', as already discussed (Chapter 2), and on the claim that this scarcity 'causes violent conflict'. In addition, however, he claims that some societies can manage to adapt to scarcities providing they possess the necessary social and technical 'ingenuity'. Homer-Dixon's theorisation of this issue is, as Peluso and Watts comment, 'astonishingly banal'; he characterises 'ingenuity' merely as '*ideas* applied to solve practical social and technical problems', and as limited by four factors: market failure, constraints on science, shortages of capital and social friction. But we can put this shortcoming to one side. More to the point here is that this dual emphasis on ecological scarcity plus human ingenuity leads Homer-Dixon to endorse a standard colour-line reading of eco-collapse: in the near future 'we will see', he maintains, a 'bifurcation of the world into societies that can maintain an adequate supply of ingenuity in the face of rising scarcities and those that cannot'. Ingenuous Western societies will adapt; many African and Asian societies, by contrast, may enter 'a downward and self-reinforcing spiral of crisis and decay'.[9]

Nowhere is such thinking better encapsulated than in Robert Kaplan's well-known essays on geography and the environment. Many of Kaplan's claims are startlingly eco-determinist, as discussed in Chapter 2. But Kaplan also insists on the power of ideas, liberal ideas in particular. And the result is a by now familiar dualism – a hierarchical differentiation of both peoples and space. As Kaplan writes in 'The coming anarchy':

> We are entering a bifurcated world. Part of the globe is inhabited by Hegel's and Fukuyama's Last Man, healthy, well fed, and pampered by technology. The other, larger, part is inhabited by Hobbes's First Man, condemned to a life that is 'poor, nasty, brutish, and short'. Although both parts will be threatened by environmental stress, the Last Man will be able to master it; the First Man will not.

Some may object that this passage is metaphorical, and does not identify particular peoples or societies as inherently inferior to others. But the subtext is not difficult to discern. Kaplan, for instance, characterises 'the battle over land between Israelis and Palestinians' as 'a case of utter geographical determinism' and depicts Gaza in particular as home to

---

[9] Homer-Dixon, 'Environmental scarcities', 39; T. Homer-Dixon, 'The ingenuity gap: can poor countries adapt to resource scarcity?', *Population and Development Review*, 21:3 (1995), 587–612; Homer-Dixon, *Environment, Scarcity and Violence*, 109, 114–24 (author's italics), 44; Peluso and Watts, *Violent Environments*, 22.

'multitudes of angry lumpen faithful' where 'one can easily see the conflicts over scarce resources that Malthus predicted coming to pass'. 'Zionism', by contrast, is characterised as showing 'the power of ideas'. Israelis exemplify Kaplan's Last Man; Palestinians his First.[10]

That *all* such colour-line thinking is deeply problematic is hopefully too evident to warrant lengthy corroboration. The imagined division of the world into a zone of 'geography versus demography', and another of civilisation and progress, is colonial thinking par excellence, recalling Frantz Fanon's observation that '[t]he colonial world is a world cut in two' and 'inhabited by two different species'. Indeed, this imagined division is essentially racialised. Sometimes this is only implicit, as for instance in the widespread neglect or underestimation of the role of technical innovations, environmental adaptations, political economic dynamics and political contestation and debate within African and Asian societies. To give but one telling example, while Diamond and others claim that the 1994 Rwandan genocide was, in part, a result of extremely high rural population density and that this was a function of natural demographic pressures, in reality the latter was a result of politics – specifically a government-imposed anti-urban ideology that effectively trapped people on the land even amidst a spiralling agrarian crisis. Elsewhere, however, the racism is hard to miss. Consider, for instance, Huntington's view that because of the stagnating influence of the tropical climate, the black race has characteristics akin to 'those which unspecialized man first showed when he separated from the apes and came down from the trees. It is not to be expected that such people should ever rise very high in the scale of civilization'. Or recall Paul Ehrlich's famous portrayal of the emotional 'feel of overpopulation' as experienced during 'one stinking hot night in Delhi ... The streets seemed alive with people. People eating, people washing, people sleeping. People visiting, arguing, and screaming. People thrusting their hands through the taxi window, begging. People defecating and urinating ... People, people, people, people ... Would we ever get to our hotel?'.[11] Sometimes explicit and other times much less so, the racialised logic underpinning eco-determinist thought is clear.

---

[10] R. Kaplan, 'The coming anarchy: how scarcity, crime, overpopulation, tribalism, and disease are rapidly destroying the social fabric of our planet', *The Atlantic* (02/1994); Kaplan, 'The revenge of geography', 98.

[11] F. Fanon, *The Wretched of the Earth*, trans. C. Farrington (Grove Weidenfeld, 1991 [1961]), 38–40; Diamond, *Collapse*, 434; P. Verwimp, 'Development ideology, the peasantry and genocide: Rwanda represented in Habyarimana's speeches', *Journal of Genocide Research*, 2:3 (2000), 325–61; K. Boudreaux, 'Land conflict and genocide in Rwanda', *Electronic Journal of Sustainable Development*, 1:9 (2009), 61–71; E. Huntington, 'Environment and racial character', in M. Rutherford Thorpe (ed.), *Organic Adaptation*

This abhorrent politics aside, global colour-line thinking also presents an overly homogenised picture of peoples and their environments, obscuring their multiple internal particularities and axes of difference. It typically juxtaposes the most vulgar materialism and the most naïve idealism, reading history, as Kaplan does, as a battle between 'mountains' and 'ideas', and in the process marginalising discussion of the political economy of human–nature relations. Its progressivist variant assumes a 'diffusionist' model of world history, according to which 'the European miracle' is a product of internal – in this case internal environmental – attributes alone, and not at all a result of borrowings from non-European societies, still less of colonial settlement and exploitation. Its tragic variant, conversely, not only locates eco-collapse within non-European societies, but also depicts it as primarily the result of internally generated pressures within those societies and assumes that from there it may spill over, taking the West down with it. When Diamond, Kaplan and others talk of coming anarchy and collapse, then, they are not only making claims about historical directionality. In addition, in the suggestions that Easter Island is 'a metaphor ... for what may lie ahead of us in our own future', that contemporary West Africa provides a 'natural point of departure' for understanding twenty-first-century world politics, that '[t]o look to Sudan is to look to the future' and that US senators should 'come to Africa' to learn about what might await us (among many such claims), they are engaged in acts of imaginative geography, narrating African and Asian peoples as uncivilised Others whose numbers threaten to reverse and overwhelm Europe's gifts to the world.[12]

Lest this be imagined as a historical or theoretical issue only, the point here is this: that both modern eco-determinism, and contemporary water wars and climate chaos narratives specifically, are inextricably tied to these racial and colonial inheritances. Following in Hegel and Malthus' footsteps, most academic and policy discourse on water and climate security implicitly operates with a bifurcated model of the world, assuming that environmental conflicts could only conceivably occur in, and emanate from, the developing world; as Ragnhild Nordås and Nils Petter Gleditsch note, this focus is 'not a great point of controversy in the literature'. Except for the odd reference to California, 'water wars' are almost always imagined as taking place in the developing world, above all in that long-established zone of European fears and fantasies, the Middle East. 'Climate conflicts', likewise, are usually envisaged as outside the

to Environment (Yale University Press, 1924), 292; Ehrlich and Ehrlich, *Population Bomb*, 1.
[12] Kaplan, 'The revenge of geography', 101; Blaut, *Colonizer's Model*, 8–17; Diamond, *Collapse*, 119; Kaplan, 'The coming anarchy'; Welzer, *Climate Wars*, 13; T. Friedman, 'Come to Africa', *New York Times* (28/01/1996).

West, with a conspicuously high proportion of analyses and reports focusing on Hegel's primal continent, Africa (the only major exception to this rule is provided by a select few global quantitative studies). Moreover, as illustrated by the cases of Darfur, Syria and Lake Chad discussed in the previous chapter, such conflicts are routinely analysed without any regard to their political economy, as if decisions to migrate and fight were the product of natural or environmental forces alone. One also finds barely any mention within climate or water crisis discourse of Southern environmental movements, mobilisations, activism or agency; within such discourse Africans and Asians merely consume, while environmental consciousness raising is, by its absence, effectively imagined as the facility of 'Western' authors and experts alone. And perhaps most obviously of all, contemporary fears of climate migrants 'flooding' Europe and North America are shot through with racist premises, as a number of critical analyses have documented.[13]

Hence, when, for instance, a UK Secretary of State envisages 'a Hobbesian world, where life for far more people is "nasty, brutish, and short"' as a result of climate change, he is evidently not thinking of this as a universal condition, but instead imagining a global geopolitical order akin to those of Huntington and Diamond:

Picture a map of the world. Picture the areas we're most concerned about; where poverty, instability, and conflict meet. Parts of the Middle East and North Africa. Pockets of sub-Saharan Africa. Delicate borders on the Asian subcontinent. Now picture the areas where climate change will strike hardest. The overlap is uncanny – and unnerving.

As in Hobbes himself, 'the literal state of nature is reserved for non-whites'; for whites, by contrast, 'the state of nature is hypothetical', a racialised object lesson which functions to inculcate fear and to help legitimate assorted political projects.[14] Rooted in European colonialism and subsequent Euro-American dominance of the post-colonial world, such colour-line thinking is essentially a by-product of 'Western' self-understandings, preoccupations and desires – and tells us little about the actual likelihood of water or climate conflict.

One should not imagine, though, that such thinking is, or has historically been, inconsequential. Everywhere in the hands of European empires, Orientalist reasoning contributed to defining patterns of colonial

---

[13] Nordås and Gleditsch, 'Climate change and conflict', 635; Hartmann, 'Converging on disaster'; Adams et al., 'Sampling bias'; A. Baldwin, 'Racialisation and the figure of the climate-change migrant', *Environment and Planning A*, 45:6 (2013), 1474–90.
[14] C. Huhne, 'The geopolitics of climate change'; C. W. Mills, *The Racial Contract* (Cornell University Press, 1997), 64–6.

rule, to shaping post-colonial identities and to transforming local economies and environments – with legacies that live on to this day. Under British rule specifically, colonial governance was, as Mahmood Mamdani observes, '[m]ore than anything else ... about identity formation', revolving around the organisation and reorganisation of power and the imperative of maintaining political order through techniques not of 'divide and rule', but 're-identify and rule'.[15] Of course, this was never a matter of inventing new identities out of nothing on a tabula rasa of modernist fantasy: practices of colonial identity-making always took account of pre-existing local identities, and always reflected specific geopolitical and economic interests. But racial imaginaries were crucial nonetheless. Britain's starkly contrasting models of identify formation and rule in Sudan, Cyprus and Palestine provide clear cases in point.

A Bifurcated State on the Nile

The 1898 Battle of Omdurman is as appropriate a place as any to begin, heralding as it did the fall of the Mahdist state and onset of Anglo–Egyptian rule in Sudan. 'Nothing like the battle of Omdurman will ever be seen again', Winston Churchill said in his biography. 'It was the last link in the long chain of those spectacular conflicts whose vivid and majestic splendour has done so much to invest war with glamour'. With the use of gunboats, artillery batteries, Maxim guns and newly patented expanding dum-dum bullets – the use of which was prohibited between civilised states – an estimated 11,000 Mahdists were mown down in just one morning, against just 40 or so Anglo–Egyptian casualties. Around 16,000 wounded Sudanese soldiers were either executed or left to die on the battlefield. General Kitchener remarked that he thought the enemy had been given 'a good dusting', and went home to receive a hero's welcome. Churchill, for his part, lauded the 'fascinating thrills' of 'this kind of war' and, as the slaughter ended, asked his second sergeant 'if he had enjoyed himself'. 'Within the space of five hours', as he observed in more sober terms elsewhere, 'the strongest and best-armed savage army yet arrayed against a modern European Power had been destroyed and dispersed, with hardly any difficulty, comparatively small risk, and insignificant loss to the victors'.[16]

---

[15] Mamdani, *Saviors and Survivors*, 146; M. Mamdani, *Citizen and Subject: Contemporary Africa and the Legacy of Late Colonialism* (Princeton University Press, 1996), 23–4.

[16] W. Churchill, *A Roving Commission: My Early Life* (Charles Scribner's Sons, 1930), 171, 180, 193; G. W. Steevens, *With Kitchener to Khartum* (Dodd, Mead and Company, 1915), 284–7; Churchill, *River War, Vol. II*, 162, 164.

The system established by Britain in the wake of Omdurman was essentially the product of two contradictory sets of ideas. On the one hand, Britain approached Sudan, as it did its other colonies in Africa, informed by the clearly racialised and paternalistic belief that the subject races of the world were not yet ready for modernity. As adumbrated most famously in Frederick Lugard's *The Dual Mandate in British Tropical Africa*, the standard British elite view was familiarly Hegelian: that history had left Africa 'untouched'; that Africans were 'the child races of the world'; that the African mind was, compared to that of the European or Asiatic, 'far nearer to the animal world'; and that these racial characteristics were so great as to require methods of governance very different from those possible within modern Europeanised societies. Yet, on the other hand, Britain was committed – however inconsistently and shallowly – to the universalisation of liberal economic and political principles and thus to promoting private property, economic modernisation and individual and civil rights. In addition, the Mahdist state had forged strong trans-local political alliances and a national identity across northern Sudan, such that forestalling the resurgence of these were overriding British priorities. And Britain had specific and limited economic interests in Sudan – namely, establishing a presence which would thwart French designs on Egypt, and expanding local cotton production to serve its textile mills back home – that were essentially limited to the central Nile Valley.[17]

To square these contradictory ideological plus strategic commitments, Britain established in Sudan in the wake of Omdurman what Mamdani characterises as a form of late-colonial 'bifurcated state'. On the one hand, this comprised, in the Nile states around the urban centres of Khartoum and Omdurman, a modernising capitalist and to some extent liberalising region where individual rights, private land ownership, civil laws and an administrative state apparatus were all consolidated, and where the colonial authorities supported the nascent political and economic elite through education and training, integration into the colonial administration, lucrative state contracts and permits for agricultural development (alongside gradually suppressing the trade in slaves which had burgeoned under Turko–Egyptian rule). In Khartoum, Omdurman and Port Sudan, municipal councils were even established for advisory purposes in the early 1920s, with quotas for 'native' Sudanese. And yet,

---

[17] F. Lugard, *The Dual Mandate in British Tropical Africa* (William Blackwood and Sons, 1922), 3, 72, 69, 6, 197–229; G. N. Sanderson, *England, Europe and the Upper Nile 1882–1899* (Edinburgh University Press, 1965); T. Tvedt, 'Hydrology and empire: the Nile, water imperialism and the partition of Africa', *Journal of Imperial and Commonwealth History*, 39:2 (2011), 173–94.

on the other hand, across the vast bulk of Sudan's territory Britain pursued what Mamdani appropriately labels 'decentralised despotism'. This was not simply a system of indirect rule, with governance being left as far as possible to local actors. More precisely, it involved the creation of a highly decentralised patchwork of Native Administrations in which regional tribal leaders were vested with enormous powers over their tribes and associated lands and allowed to govern by their own mores rather than in accordance with 'Western' ways or those of the continent's more educated classes. And it also involved a process of 're-tribalisation' – the invention of tribal categories and traditions which, in reality, were much less clear-cut than Britain's colonial administrators and anthropologists supposed. While in the central Nile Valley Britain supported economic and political modernisation, across most of Sudan it did the exact reverse. And while indirect native rule was practised across Africa, in Sudan it was taken to extreme, with the Sudan Political Service 'rarely reaching one hundred and twenty-five officials on the ground to administer almost a million square miles'.[18]

Far from being external to these processes, representations of the Sudanese environment both informed, and were used to justify and legitimise, these racial ideas and the associated bifurcated system of rule. Northern Sudan was widely characterised and dismissed as little more than 'utter waste and desolation' where 'weary wind blew continuously... over the miles and miles of sand and rock, as if conscious of its own uselessness'. The south was also dismissed but in different terms – as a disease-infested, insect-ridden swamp inhabited, as one leading colonial administrator put it, by 'marsh-dwellers' whose 'mode of life... was as barbaric as the land in which they lived'. Indeed, Britain's Native Administration policy in Sudan was partly justified on these grounds: as Harold MacMichael, the administrator who more than any other was key to the production of colonial knowledge about the country, observed of southern Sudan, 'climate, geography, communications, language, culture, beliefs and background, being entirely different, what more natural than that policy should also differ?'. Overall, Sudan's environment and peoples alike were represented as essentially without history, and without a future. Only the River Nile and its agents of

---

[18] Mamdani, *Citizen and Subject*, 8, 16–18; Mamdani, *Saviors and Survivors*, 145–70; T. Niblock, *Class and Power in Sudan: The Dynamics of Sudanese Politics 1898–1985* (MacMillan Press, 1987); M. El Zain, 'Ruling elite, frontier-caste ideology and resource conflicts in the Sudan', *Journal of Peacebuilding and Development*, 3:1 (2006), 36–47; M. 'Abd Al-Rahim, *Imperialism and Nationalism in the Sudan* (Clarendon Press, 1969); A. H. M. Kirk-Greene, 'The Sudan political service: a profile in the sociology of imperialism', *International Journal of African Historical Studies*, 15:1 (1982), 21.

development were viewed as having the potential to bring history to Sudan; only through them, and only in specific riverine regions, could Sudan be 'redeemed from the great waste by the life-giving water'. As one British commentator put it, '[t]he whole importance of the country naturally depends on the river'.[19]

Politically, the consequence of all this was a system of colonial rule that was not just internally decentralised and differentiated, but deeply racialised and, across most of Sudan, tribalised too. Thus in Darfur, southern Sudan and other peripheral regions, tribal leaders were granted near total executive, administrative and judicial powers, including the authority to define 'customary laws', collect taxes and register and lease land on behalf of the colonial administrators. Tribal 'homelands' were officially mapped, delimited and institutionalised. And legal and political hierarchies were established among both races and tribes. For, not only did British elites distinguish between a more 'civilised' and 'Arab' north and a more 'backward' 'African' south, with the former being portrayed as better suited to 'Western' mores and forms of rule (Churchill once again: '[t]he Arab was an African reproduction of the Englishman; the Englishman a superior and civilised development of the Arab'). In addition, while some races and tribes were designated as 'native', others were categorised as 'settlers' – with 'native' tribes being allocated 'customary' land rights and 'non-native' ones being denied them. No matter that both 'race' and 'tribe' were imaginative colonial constructions; irrespective of this, 'race' and 'tribe' would henceforth become primary markers of identity and determinants of political rights and power across Sudan.[20]

Sudan's decentralised and bifurcated colonial system also, as a corollary, embedded a strikingly uneven pattern of socio-economic development and gross inter-regional – or what are often today called 'horizontal' – inequalities. Broadly speaking, Sudan's core riverine states – the three Ks of Kosti, Kassala and Khartoum – benefitted from modernising colonial investments and reforms, while its various peripheries were neglected if not actively marginalised and de-developed. The latter applied, for instance, to both southern Sudan and Darfur, albeit in different ways and for distinct reasons. Southern Sudan's development was restricted on essentially racial grounds, especially during the period of the 'Southern Policy' from 1930 to 1946,

---

[19] Churchill, *River War, Vol. II*, 35; H. C. Jackson, *Behind the Modern Sudan* (MacMillan and Co., 1955), 141–2; H. MacMichael, *The Sudan* (Ernest Benn, 1954), 116; J. K. Giffen, *The Egyptian Sudan* (Fleming H. Revell, 1905), 216–21; H. L. Tangye, *In the Torrid Sudan* (John Murray, 1910), 12.
[20] W. Churchill, *The River War: An Historical Account of the Reconquest of the Soudan*, Vol. I (Longmans, Green, and Co., 1899), 25; Mamdani, *Saviors and Survivors*.

when the region was declared to be so racially and socially distinct as to require closing off from the rest of the country. Darfur, by contrast, was marginalised for political reasons: having been key to the Mahdist project of unifying northern Sudan, it afterwards became no less central to the British objective of destroying all traces of Mahdism and thus became 'the heartland of indirect rule', supervised by just a few colonial officials and reduced to a labour reserve. The consequences were consistent and profound. While Khartoum and environs benefitted from relatively widespread government education provision, in many peripheral regions education was intentionally restricted with the consequence that in 1950–1, for example, there were just six government elementary schools in the whole of Darfur, three in Bahr el Ghazal and none in Upper Nile, in contrast to forty-nine in Khartoum province, seventy in Northern province and eighty-five in Blue Nile. Transport infrastructure, likewise, was concentrated overwhelmingly in the central Nile states, with railways extending from Khartoum north to Wadi Haifa and east to Port Sudan and Kassala, but south only as far as Sennar and west only as far as el Obeid – and not reaching Darfur or Bahr el Ghazal until after independence.[21]

These highly uneven political and economic geographies were paralleled by no less uneven environmental transformations – and the establishment of what we may perhaps think of as a 'bifurcated political ecology' across Sudan. As explored more fully in the next chapter, large-scale agricultural projects were concentrated around Khartoum and the Nile and 'useful' parts of Kordofan just to the west and Kassala just to the east, enabling the accumulation of wealth and power in these core riverine states. Cotton production flourished in these areas – most importantly through the government-funded Gezira scheme launched in the 1920s – as they were inserted, however violently, into the expanding world economy. Yet peripheral regions of Sudan presented a different story altogether. As illustration, in 1943–4 there were a total of 42 government irrigation pumps in Northern province and 27 in Khartoum province, against just 3 in Upper Nile, 1 in Equatoria and none at all in Darfur, while in the same year there were 355 water wheels in Khartoum against just 1 in each of Upper Nile and Darfur.[22]

---

[21] G. S. Symes, 'Appendix IV', in *Report on the Administration, Finances and Condition of the Sudan in 1937* (H. M. Stationary Office, 1938), 136–7; Mamdani, *Saviors and Survivors*, 161–4; R. Howe, 'Provinces', in *Report on the Administration of the Sudan in 1950/51* (McCorquodale and Co., 1955), 114–78; P. Woodward, *Sudan 1898–1989: The Unstable State* (Lynne Rienner, 1990), 117.

[22] W. N. Allen and R. J. Smith, 'Irrigation in the Sudan', in J. D. Tothill (ed.), *Agriculture in the Sudan* (Oxford University Press, 1948), 617, 629.

The pertinence of all this within the context of this chapter is that the political, economic and socio-ecological divides established in Sudan during the period of British rule – divides that were to a large degree constructed around a racialised imaginary – have reverberated through to today, both through their continuation and, as a counter-tendency, through periodic post-colonial projects of unification and centralisation. The post-independence military regimes of Ibrahim Abboud (1958–64), Jaafar al-Nimeiri (1969–85) and Omar al-Bashir (1989–2019) each pursued such projects, framing these as attempts to counter divisions created by colonial rule. Equally, during the 1980s the SPLM/A under John Garang advocated the creation of a 'New Sudan' that would transcend regional, ethnic and racial cleavages through a restructured central state and autonomous regional governments. In none of these cases, however, could such political ambitions overcome Sudan's multiple divides. The Khartoum-led projects of national integration were always pursued with the aim of assimilating southerners into northern culture under the banners of either 'Arabisation' or 'Islamisation', underpinned by assumptions about racial and cultural superiority (as, for instance, in the country's 1998 adoption of an Islamic constitution that identified Arabic as its official language and Islamic law as its source of legislation).[23] And Garang's 'New Sudan' came to nought. For, as detailed in Chapter 8, the North–South peace agreement of 2005, far from unifying the country, further consolidated both tribal and despotic rule in the south as well as the racialised state in Khartoum, leading ultimately to South Sudan's independence – and bringing to conclusion a pattern which had first been institutionalised almost a century before by Britain.

Sudan's conflict-ridden politics is often represented as structured around the divide between North and South, and between Arab and African identities, alongside a primary focus on the civil wars of 1955–72 and 1983–2005 and the country's subsequent partition.[24] Yet to depict North versus South as the primary, or primordial, axis of conflict in Sudan is, in truth, misleading. For, in contrast to the other cases examined in this chapter, both of which are divided along one main axis – Israelis versus Palestinians in the one case, Greek Cypriots versus

---

[23] Mamdani, *Saviors and Survivors*, 171–205; J. Garang, *The Genius of Dr. John Garang: Speeches on the War of Liberation*, ed. P. Wël (Paanda Publishers, 2013); D. H. Johnson, *The Root Causes of Sudan's Civil Wars: Old Wars and New Wars*, expanded 3rd ed. (James Currey, 2016), 62–5.

[24] A. M. Lesch, *The Sudan: Contested National Identities* (Indiana University Press, 1998); J. Copnall, *A Poisonous Thorn in Our Hearts: Sudan and South Sudan's Bitter and Incomplete Divorce* (Hurst, 2013).

Turkish Cypriots in the other – in Sudan there exist multiple ethnic or horizontal inequality conflicts, all rooted in the legacies of the colonial bifurcated state and subsequent responses to it. Among the cases considered here, only in Sudan did European colonialism bequeath a legacy of division between a core modernising region and multiple marginalised peripheries. And as we shall see in chapters 5, 6 and 7, one corollary of this has been extensive water-related dispossession, displacement, violence and insecurity across the country – from South Sudan to the central Nile states to Darfur. Thanks in large measure to its racialised colonial inheritance, Sudan is a multiply divided environment in a category all of its own.

Liberalism and its Contradictions in Cyprus

Britain's approach towards Cyprus diverged from that towards Sudan in two crucial respects. Ceded by the Ottomans in 1878, Britain had initially viewed Cyprus as of great geopolitical importance, with Benjamin Disraeli famously describing it as 'the key of Western Asia'. Cyprus, it was hoped, would become a *place d'armes* and a crucial refuelling station for trade with India. However, Famagusta harbour proved inadequate to these purposes and in any case Britain already possessed suitable ports in Malta and, after 1882, Alexandria. Unlike Sudan or indeed most British colonies, Cyprus thus became what Andrekos Varnava calls 'an inconsequential possession' – one that, until the Cold War at least, was neither of strategic nor great economic value.[25]

In addition, British elites imagined Cyprus very differently from Sudan – specifically as white and as European. In their imaginative geography, Cyprus lay in that uncertain zone between Orient and Occident, but nonetheless possessed an essentially European population unfortunately degraded by three centuries of Ottoman rule. Thus in Disraeli's 1847 work *Tancred* – an Orientalist fiction that envisaged and prefigured Cyprus' occupation forty years before it actually occurred – the island was imagined as a key to unifying Oriental spiritualism and the materialism of the West in a grand imperial synthesis (guided, among other things, by the twin maxims that '[a]ll is race; there is no other truth', and that '[t]he East is a career'). And, shortly after this fiction had been actualised, Gladstone answered his own question – 'What then is the case of Cyprus?' – as follows: 'It is the case of an island inhabited by people who have been civilised for centuries when we were barbarians, and nothing but

---

[25] B. Disraeli to Queen Victoria, 27/05/1878, in A. Varnava, *British Imperialism in Cyprus, 1878–1915: The Inconsequential Possession* (Manchester University Press, 2009), 85.

barbarians, and who have never lost the essentials of civilisation, except indeed, their form. They had been an oppressed people ... despotically governed by the Turk.'[26] British colonial policy in Cyprus followed inexorably from a combination of this thoroughgoing Orientalism and the island's limited strategic value. It was Orientalism which determined that it never became a settler colony, for despite being occasionally considered for European settlement, presumed racial hierarchies stood in the way: when Theodor Herzl proposed the establishment of a Jewish colony in Cyprus, the idea was rejected by Colonial Secretary Joseph Chamberlain on the grounds that he should instead try to find 'a spot in the English possessions where there were no white people as yet'. Hence Cyprus became a 'non-settler' rather than a settler colony. Decentralised despotism, as practised in Sudan and other darker corners of the Empire, was also discounted. Instead, as befitting a European territory, Britain pursued liberal constitutional rule in Cyprus. A partially elected legislative council was swiftly created. Church and state were formally separated. And detailed maps were drawn, censuses held and land ownership, taxation and justice systems transformed – all along lines considered appropriate for a liberal European society. As Rebecca Bryant puts it, the British project in Cyprus was above all ideological, articulated around a 'moral mission' to set Cyprus 'aright under British rule' and 'to prove the fertilizing, fecund influence of British reason'.[27]

The island's new administrators approached not just its people, but also its environment, through this lens. Forests, as we have seen, were a particular concern, but so too were goats and swamps: well-stocked forests, for their beauty, wood, assumed positive impacts on rainfall and their purported current state of disrepair; goats, for grazing the supposedly unregulated commons and degrading the forest; and swamps, for their claimed unproductivity and the undoubted problems with malaria. Samuel Brown, who oversaw the first British public works, put the problem thus: 'everywhere ruin, decay, desolation, the population decimated, and the fruitful land became a wilderness'. And the first High Commissioner, Garnet Wolseley, characterised Cyprus similarly:

The wells are few, and scarcely a river or what is marked on the map as a river, has more than a pool here and there of stagnant water on it. Where are the forests we

---

[26] B. Disraeli, *Tancred: Or, The New Crusade* (A. and W. Galignani, 1847), 96, 101; Varnava, *British Imperialism*, 45, 58–60; W. E. Gladstone, speech, *Times* (30/12/1879), in Varnava, *British Imperialism*, 152.

[27] *The Complete Diaries of Theodor Herzl*, Vol. IV, ed. R. Patai (Herzl Press, 1960), 1361; P. Panayiotopoulos, 'The emergent post-colonial state in Cyprus', *Journal of Commonwealth and Comparative Politics*, 37:1 (1999), 34; R. Bryant, *Imagining the Modern: The Cultures of Nationalism in Cyprus* (I.B. Tauris, 2004), 24.

thought Cyprus was covered with? This is in everyone's mouth, yet no one can give a very satisfactory answer. Like everything else that made this country a splendid one in ancient times, the forests have disappeared under the influence, the blighting influence, of the Turk.[28]

Whereas Sudan's various peoples and environments were viewed above all as without history, in Cyprus they were viewed very differently: as essentially degraded.

One crucial context to all this was Philhellenism and the ensuing emergence of Greek and other subsequent ethnic nationalisms within the Balkans. Within European and especially British Enlightenment thought, Greece was widely viewed as 'the missing link for the rejuvenation of European civilization'; to this extent, as Varnava, notes, the Enlightenment created the idea of modern Greece. The 1820s Greek uprising against Ottoman rule was thus welcomed by Britain's liberal elites, viewed as a revolt of 'Republican Hellenism' against 'Oriental Despotism' and part of the broader antagonism between West and East. That neither the uprising itself nor the resulting Kingdom of Greece conformed much to this image – the former being only in part motivated by nationalism and the latter being essentially a product of European realpolitik, as well as dynastic rather than liberal-constitutional in form – is beside the point. For, as far as Cyprus was concerned, the consequences were twofold. On the one hand, the newly established Greek state came to embrace ethno-nationalism, and thus started promoting Greek national identity via educational institutions and the orthodox clergy to minority Greek ('Rum') communities throughout the Ottoman empire, including within Cyprus. And, on the other hand, in the British imperial imagination there now existed a divide between 'Greeks' and 'the Turk', which found its mirror image in Cyprus. Never mind that the island's majority Orthodox and minority Muslim populations were in most respects mixed, with extensive intermarriage, mixed villages, a common language and customs and elites that were based on communities of faith (so-called *millets*) but nonetheless both well integrated into the Ottoman state. Irrespective, in typical colonial fashion Cyprus' population was reimagined, categorised and through this constructed as comprising neat, binary 'Greek Cypriot' and 'Turkish Cypriot' ethnic groups. 'The two races that inhabit the island are very distinct types', noted the young Kitchener, while commenting on a cartographic survey of the island which he had led. Through these various

---

[28] S. Brown, *Three Months in Cyprus during the Winter of 1878–1879* (Edward Stanford, 1879), 20; S. E. Harris, Colonial Forestry and Environmental History: British Policies in Cyprus, 1878–1960 (PhD thesis, University of Texas at Austin, 2007), 408; G. Wolseley, *Cyprus 1878: The Journal of Sir Garnet Wolseley* (Cyprus Popular Bank Cultural Centre, 1991), 22.

processes the seeds were planted of an ethno-national divide which had hitherto barely existed.[29]

That this divide became so fraught, leading eventually to the island's partition, was for a combination of socio-economic, political and geopolitical reasons. Britain for the most part invested little in Cyprus, reflecting its limited strategic value. Moreover, Cyprus' huge Tribute under the 1878 Convention with the Ottomans – which was formally a payment to Constantinople, but in practice a transfer from Cypriot government coffers to Ottoman bondholders in London – ensured that Cyprus under British rule was, as under the Ottomans, a net exporter of capital; as Winston Churchill observed, in a much less celebratory tone than on the Battle of Omdurman, this arrangement amounted to no less than 'the oppression of a small community by a great Power for the purpose of pecuniary profit'. The burden of paying the Tribute fell disproportionately upon the island's peasantry, with taxation reformed and individualised to facilitate this, and the main tax being on cereals. And the upshot of all this was that, by the 1920s, Cyprus had become home to a deepening agrarian crisis. Rural indebtedness became critical, especially during the interwar period: a 1930 government survey found that 82 per cent of peasant producers were indebted, making Cyprus' debt problem worse than that of the Punjab. This indebtedness resulted, in turn, in a significant consolidation of landholdings – with 40 per cent of cultivated land changing hands during the interwar period – as well as extensive rural to urban migration, the creation of an urban working class and transformation of the remaining rural poor from part-subsistence smallholders into landless wage labourers. Simultaneously, the establishment, demarcation and enclosure of State Forests led to persistent conflict between the upland peasantry and the British authorities – there were an average 7,000 forest offences per year during the 1930s and 1940s, more than in the entire United Provinces of India – which in turn led to widespread upland depopulation, including through the removal of villages. With parallel problems in urban areas, the result was that by the early 1930s there was widespread social unrest and anti-colonial protest.[30]

---

[29] R. Kasaba, 'The Enlightenment, Greek civilization and the Ottoman Empire: reflections on Thomas Hope's *Anastasius*', *Journal of Historical Sociology*, 16:1 (2003), 11; Varnava, *British Imperialism*, 56; J. Wallace, 'We are all Greeks? National identity and the Greek war of independence', *Byron Journal*, 23 (1995), 36–49; C. Hoffmann, The Eastern Question and the Fallacy of Modernity (DPhil thesis, University of Sussex, 2010), ch. 5; Y. Katsourides, 'The rise of Greek Cypriot nationalism to hegemony: agency, particularities, and popularization', in T. Kyritsi and N. Christofis (eds.), *Cypriot Nationalisms in Context* (Palgrave Macmillan, 2018), 25–46; H. Kitchener, 'Notes from Cyprus', *Blackwood's Edinburgh Magazine*, 126 (1879), 150–7.

[30] G. Hill, *A History of Cyprus: The Ottoman Province, The British Colony* (Cambridge University Press, 1952), 443–87; Panayiotopoulos, 'The emergent post-colonial state', 36; B. J. Surridge, *A Survey of Rural Life in Cyprus* (Government Printing Office, 1930),

Although it initially transcended the island's still nascent ethnic divide, this protest eventually became almost exclusively Greek Cypriot–led and organised along ethno-national lines – as a result, above all, of the influence of the Orthodox Church. Under the Ottomans the Church had held a powerful social position as leader of the Rum *millet*, the Christian Orthodox community, including through its extensive landownership and rights to collect taxes on behalf of the Sultan. Britain, and particularly its tax reforms, threatened these incumbent privileges such that despite the Orthodox clergy's elite position within Cypriot society it quickly took on leadership of the struggle against British rule and turned the national struggle into a cross-class one. This brought together old landed elites, the rising bourgeoisie, urban middle classes, trade unions and the Communist Party all under the banner not of independence but of 'Enosis', or unification, with Greece. The subsequent emergence of a 'Turkish Cypriot' political identity was partly a reaction to this and partly a corollary of the rise of post-Ottoman Turkish nationalism on the mainland. By the mid-1950s, Greek Cypriot paramilitaries had forced their compatriots to resign en masse from Cyprus' police forces, which thus became predominantly staffed by Turkish Cypriots, sharpening the inter-communal divide still further. And on top of all this were the impacts of dictatorship and Cold War militarisation in both Athens and Ankara, plus US Cold War machinations – the culmination of which were the Greek Junta's 1974 military coup in Cyprus, Turkey's invasion, and with it the island's partition.[31]

What, then, at the risk of repetition, is the case of Cyprus? It is the case of a small island in the eastern Mediterranean that is divided and partitioned along one main axis, making it less a multiply divided environment, as Sudan is, than a singularly divided one. It is a case where this single main axis of division was created and consolidated through a series of dividing processes: regional Cold War enmities, which underpinned

---

37, 43; Y. Katsourides, *The History of the Communist Party in Cyprus: Colonialism, Class and the Cypriot Left* (Bloomsbury Academic, 2014), 31; M. Given, 'Maps, fields, and boundary cairns: demarcation and resistance in colonial Cyprus.' *International Journal of Historical Archaeology*, 6:1 (2002), 1–22; Harris, *Colonial Forestry*, 248–67; Morgan, *Sweet and Bitter Island*, 105–10.

[31] U. Bozkurt and N. Trimikliniotis, 'Rethinking the postcolonial Cypriot statehood: the Cyprus problem, class struggles, and ethnic conflict', in N. Trimikliniotis and U. Bozkurt (eds.), *Beyond a Divided Cyprus: A State and Society in Transformation* (Palgrave Macmillan, 2012), 49; I. Marovich-Old, 'Nationalism as resistance to colonialism: a comparative look at Malta and Cyprus from 1919 to 1940', in N. Christofis and T. Kyritsi (eds.), *Cypriot Nationalisms in Context* (Palgrave Macmillan, 2018), 264; A. R. Novo, 'Friend or foe? The Cyprus Police Force and the EOKA insurgency', *Small Wars and Insurgencies*, 23:3 (2012), 414–31; C. Hitchens, *Hostage to History: Cyprus from the Ottomans to Kissinger* (Quartet, 1984).

the island's descent into crisis during the 1970s; before that, colonial exploitation plus resistance to it, which concretised the island's twin ethnic nationalisms; and, prior to all this, those racialised imperial and local imaginings which brought both modern Greece and Cyprus' two communities into being. And it is a case that, despite the numerical disparity between these two communities, is in many ways symmetrical in its divide. Hostage to a series of broader symmetries – from neither community being viewed by Britain as natural auxiliaries to colonial development as the Zionists were in Palestine, to those parallel processes of nation- and state-building and Cold War militarisation in the Greek and Turkish 'motherlands' just to the north – Cyprus' own 'internal' divide inevitably took on a certain symmetry too. The fact that Cyprus has been de facto partitioned for the best of part of fifty years reflects this. Cyprus, we may conclude, presents a case of a singularly and relatively symmetrically divided environment – presenting a form of division very different from that in Sudan, as well as that found just 200 miles to the south-east. And, as we shall see in Chapters 5 and 8, Cyprus' water politics reflect this too.

Palestine: A Settler-Colonial Encounter

Whereas in Sudan and Cyprus colonialism was organised around decentralised despotism and liberal constitutionalism respectively, in Palestine it was essentially settler-colonial in form, structured around an encounter between a European settler-colonial movement on the one hand, and an indigenous non-European population on the other. It is no doubt true that Zionist settlement was not pursued in the name of an imperial European state; true that Jews were modern Europe's greatest internal victims; and true also that the Zionist movement often understood or at least represented itself as an anti-colonial project. Yet Zionism was led by European immigrants, was founded on European ideologies, aimed to establish a European-style society, made wide use of industrialising Europe's superior economic, political and scientific resources, and was implemented behind and with the active support of the 'British imperial shield'.[32] These basic truths defined the structure and character of the

---

[32] F. Sayegh, 'Zionist colonialism in Palestine (1965)', *Settler Colonial Studies*, 2:1 (2012), 206–25; M. Rodinson, *Israel: A Settler-Colonial State* (Monad, 1973); Y. Shafir and Y. Peled, *Being Israeli: The Dynamics of Multiple Citizenship* (Cambridge University Press, 2002); J. Massad, 'The 'post-colonial' colony: time, space, and bodies in Palestine/ Israel', in *The Persistence of the Palestinian Question: Essays on Zionism and the Palestinians* (Routledge, 2006), 13–40; B. Wasserstein, *The British in Palestine: The Mandatory Government and the Arab–Jewish Conflict, 1917–1929* (Blackwell, 1991), 157.

early Zionist movement, and continue to shape the politics and political ecology of the Israeli–Palestinian conflict to this day.

There are at least three respects in which this is so. First, thanks to the settler-colonial origins of the Israeli–Palestinian conflict, Orientalist ideological formulations came to assume particular prominence within it. Theodor Herzl, the founding father of Zionism, hoped for a secular German-speaking Jewish state in Palestine that would function as 'a portion of the rampart of Europe against Asia, an outpost of civilisation as opposed to barbarism'. Mark Sykes, co-author of the 1916 Sykes–Picot Agreement which divided the Middle East into British and French spheres, held that '[i]t might be the destiny of the Jewish race ... to bring ... the vitality of Europe to Asia'. The Zionist movement was routinely imagined as Britain's local agent of colonial development. Conversely, in both British colonial and early Zionist thinking, Palestine and Arab society were depicted as 'stagnant' or 'abandoned', and in need of a *mission civilatrice*: '[w]e live in the twentieth century, they – in the fifteenth', said David Ben-Gurion, Israel's first prime minister. But Muslim Arab society was not Zionism's only Other. Middle Eastern and North African Jews were a further one, subject to both re-education and second-class status within an overwhelmingly Ashkenazi-dominated polity. Moreover, in both ideology and practice, Zionism sought to negate diaspora Jewish identity, by creating a hyper-masculine, pioneering 'new Jew' with typical European characteristics; Zionism was essentially conceived as a means for Jews – or more precisely, European Jews – to become more European, not less. 'We do not want Israelis to become Arabs', said Ben-Gurion, 'We are duty bound to fight against the spirit of the Levant, which corrupts individuals and societies, and preserve the authentic Jewish values as they crystallized in the [evidently European] Diaspora.'[33]

Second, the settler-colonial origins of the Israeli–Palestinian conflict were constitutive of the very distinction between Palestinian and Jewish-Israeli nationalisms and of Palestine's subsequent territorial partition. Most settler-colonial encounters and accompanying super-charged racial imaginaries have historically not led to separate nationalisms or territorial

---

[33] T. Herzl, *The Jewish State: An Attempt at a Modern Solution of the Jewish Question*, trans. S. D'Avigdor (Henry Pordes, 1993 [1896]), 30; S. Leslie, *Mark Sykes: His Life and Letters* (Cassell and Company, 1923), 271–2; J. Norris, *Land of Progress: Palestine in the Age of Colonial Development, 1905–1948* (Oxford University Press, 2013), ch. 2; A. Shlaim, *The Iron Wall: Israel and the Arab World* (Penguin, 2000), 96; S. Swirski, *Israel: The Oriental Majority* (Zed, 1989); M. Nordau 'Jewry of muscle' (1903), in P. Mendes-Flohr and J. Reinharz (eds.), *The Jew in the Modern World: A Documentary History* (Oxford University Press, 1980), 434–5; S. Smooha, *Israel: Pluralism and Conflict* (University of California Press, 1978), 88.

division – but this one did, and for quite particular reasons. Turn of the century Palestine was not on the whole an attractive destination for European settlers, since it was already heavily settled with a relatively poor Arab population and thus large reserve pool of cheap labour. Early Jewish investors, like Baron de Rothschild, preferred to employ Arab labourers on typical colonial plantations. Unwilling to compete directly with their Arab counterparts, Zionist workers thus faced intractable problems of getting work and started campaigning for the exclusion of Arabs from the Jewish labour market. And this is precisely what occurred. To circumvent the exigencies of the market, the Jewish community in Palestine, the Yishuv, developed an exclusionary political economic model which, unlike the inclusionary settler-colonialism practised, for instance, in South Africa, involved 'colonizing the land while excluding the people'. A 'split labour market' was created, with a discrete Jewish economic sector largely, though not completely, separated and protected from the parallel Arab sector. Land was nationalised and managed through cooperative settlements, kibbutzim and moshavim. And powerful central institutions – most importantly the quasi-governmental Jewish Agency and the Histadrut labour federation – were established to maintain and oversee these 'greenhouse conditions'. These exclusionary arrangements led in turn to the consolidation of distinct Palestinian and Jewish-Israeli nationalisms, and to Palestine's subsequent political and geographical division. Moreover, they also had formative institutional and societal impacts, on the one hand giving shape 'precisely to those aspects of their society which Israelis pride themselves on being most typically Israeli' – most notably its interventionist state institutions and powerful labour aristocracy – while on the other setting Palestinian society on a path towards exclusion and fragmentation, with a weak labour movement and limited institutionalisation.[34] As the above suggests, conflict in the divided environment of Palestine has not so much been a secondary effect of the presence of rival nations, or even nationalisms, but has instead been constitutive of these nations' very existence, and an ever-present influence on their 'internal' institutions, identities and societies.

The settler-colonial origins of the Zionist movement are also, third, what lie behind the Israeli–Palestinian conflict's striking asymmetry. Late Ottoman Palestine, while far from static, was nonetheless a predominantly

---

[34] Shafir and Peled, *Being Israeli*; M. Younis, *Liberation and Democratization: The South African and Palestinian National Movements* (University of Minnesota Press, 2000), 75; E. Bonacich, 'A theory of ethnic antagonism: the split labour market', *American Sociological Review*, 37:5 (1972), 547–9; G. Shafir, *Land, Labour and the Origins of the Israeli-Palestinian Conflict, 1882–1914* (Cambridge University Press, 1989), 19, xii.

rural and feudal society, led by a small class of largely absentee notables and with a peasantry that was increasingly indebted thanks to Ottoman tax and land reforms. The Zionist movement, by contrast, had its intellectual and institutional genesis in urban, upper middle-class Europe, and proved readily able to marshal resources for its settlement project – most importantly, British government support and Jewish capital for land purchases. Palestine's remarkable demographic and economic transformation between 1900 and 1950 was a direct product of this disparity. In 1893, Jews accounted for around 4 per cent of its estimated 537,000 population, and a negligible proportion of landholdings; by 1948, by contrast, they accounted for a third of the population of Palestine and their share of cultivable land had risen to 22 per cent. The corollary of this immigration and settlement was Palestinian dispossession, first through the eviction of tenant farmers and later – and more crucially – through the war of 1948–9 when 726,000 Palestinians were displaced and refused the right to return.[35] Unlike in Cyprus, where a racialised but non-settler colonial imaginary paved the way for a relatively symmetrical ethno-national conflict, as discussed above, in Palestine settler-colonialism dictated the opposite. And, as we shall see in subsequent chapters, this asymmetry has been cast in stone ever since.

Crucially, the Zionist movement's 'fight against the Levant' was directed not only against its 'spirit' but also against its supposedly degraded physical environment. Britain and the Zionist movement alike imagined Palestine as 'a ruined and empty country, without a tree or a shadow'. The 'desert' was a particular obsession: 'the desert is a reproach to mankind. It is criminal waste' wrote Ben-Gurion. Indeed, in the colonial imaginary the poverty of Palestine's arid lands was conflated with that of its Arab inhabitants: '[n]owhere has the interaction between the deterioration of a land and the degradation of its people been so clear as in Palestine', claimed Walter Lowdermilk (to whom we return in Chapter 5); 'the Moslem Arab invasion was an invasion of the Sown by the Desert', observed prominent British journalist Philip Graves. The pioneering transformation of Palestine's physical environment – 'making the desert bloom' – thus became central to the Zionist project. 'Desert' here was not just a climatic category referring to the Negev Desert and lower Jordan Valley, but also often a metaphor for the entire 'wasteland' of Palestine. Thus swamps were drained, uneven and unproductive lands

---

[35] J. McCarthy, *The Population of Palestine: Population History and Statistics of the Late Ottoman Period and the Mandate* (Columbia, 1990), 10, 23; UN Special Committee on Palestine, *Report to the General Assembly, Vol. 1* (1947); K. Stein, *The Land Question in Palestine, 1917–1939* (University of North Carolina Press, 1984), 3–4; UNCCP, *Final Report of the United Nations Economic Survey Mission for the Middle East* (1949), 22.

were levelled, cultivation was intensified and mass afforestation schemes were pursued – the latter eventually seeing the Jewish National Fund (JNF) plant 220 million trees, even if these were often European firs unsuited to arid and semi-arid climates. At the same time, these environmental transformations were imagined as key to national self-actualisation. In Europe, Jews had been landless city-dwellers, alienated, so Zionist thought had it, not only from the promised land, but from land and labour in general. Working and transforming the land was thus imagined as central to the redemption of the Jewish people: 'is it at all possible not to dream while you are ploughing the land of Israel and see around you Jews ploughing', wrote Ben-Gurion in his diary, 'Is it not a dream?'[36]

As with the Zionist project more broadly, this eco-modernising mission is usually characterised as resulting from the unique ideological commitments of Labour Zionism. But this is misleading. In the first place these commitments, as illustrated above, were themselves rooted in and dependent upon racialised ideas about East and West, self and other, past and future: as a set of ideas they were far from straightforwardly progressive. Moreover, underpinning them was a set of deeply practical, material considerations. Not only was agricultural expansion a key Zionist economic priority. In addition, the ideological commitment to redemption through agriculture meshed with Zionism's colonising and territorial objectives, since using and improving the land were viewed as instrumental to claiming, controlling and establishing ownership over it.[37] Zionist agricultural development was – and indeed remains – a means of taking possession of the land. And as an inevitable corollary, given that the land was not empty, it was also a means of facilitating Palestinian dispossession.

It is within these contexts that water became so central to early Zionist politics and practice. Irrigation water supplies were deemed a key input to agricultural intensification, economic growth and the colonisation of land and labour; good municipal supplies were crucial to meeting

---

[36] Jewish National Fund, 'Forests and the environment – past and future' (1990), 1; A. De-Shalit, 'From the political to the objective: the dialectics of Zionism and the environment', *Environmental Politics*, 4:1 (1995), 85, 74; Ben Gurion, *Recollections* (Macdonald, 1970), 146; W. C. Lowdermilk, *Palestine: Land of Promise* (Harper and Brothers, 1944), 53; P. Graves, *Palestine, The Land of Three Faiths* (Jonathan Cape, 1923), 26; B. Wasserstein, *Israel and Palestine: Why They Fight and Can They Stop?* 3rd ed. (Profile, 2008), 76; Y. Galai, 'Narratives of redemption: the international meaning of afforestation in the Israeli Negev', *International Political Sociology*, 11:3 (2017), 273–91.

[37] M. Lowi, *Water and Power: The Politics of a Scarce Resource in the Jordan River Basin*, 2nd ed. (Cambridge University Press, 1995), 51–2; A. Rouyer, *Turning Water into Politics: The Water Issue in the Palestinian–Israeli Conflict* (Macmillan, 2000), 8–9; Selby, *Water, Power and Politics*, 68–9.

expectations of modern 'European' urban life; and swamps became important sites for land reclamation. Water, in turn, became not only a key focus of Zionist environmental transformations and infrastructural development, as detailed in the next chapter, but also a factor in the Zionist movement's ultimate choice of settlement destination and in the demands it made of European states. With Cyprus having been rejected as a settlement destination, El-Arish in the Sinai was offered to and accepted by Herzl instead, but was never colonised for want of adequate water supplies: Palestine, we may conclude, was eventually favoured partly because it was not a desert. Water, moreover, was a key element in Zionist lobbying over the delineation of Palestine's boundaries. The Zionist Organisation's submission to the 1919 Paris Peace Conference presented the inclusion of major regional water resources, especially the sources of the Jordan River, not just as of 'vital importance' to the future economic life of Palestine but as more important than any other factor. Water subsequently also became central within the Anglo–French negotiations over Palestine's borders – mainly, it seems, because of Zionist priorities. By contrast, water resources barely featured either in Arab demands at Paris or in Britain's approach to the borders question (and, contrary to Zionist demands, the Litani and Yarmouk Rivers and two of the three headwater springs of the Jordan all came to lie outside of Mandate Palestine).[38] Unlike in Cyprus, where water has rarely been politically that important, in Israel–Palestine Zionism's settler-colonial ambitions made it hugely so.

The significance of all this within the context of this chapter, and the book, is that the racialised constructions and associated settler-colonial project advanced in early twentieth-century Palestine established a series of patterns that continue to this day. Differentiation of Jewish-Israeli and Palestinian identities, and the resulting construction of the Israeli–Palestinian conflict as comprising two distinct nations fighting over a singular piece of land (in which respect the Israeli–Palestinian conflict stands out from most other settler-colonial ones). Striking asymmetries in power and resources between Zionism and its indigenous Other. Asymmetries also in water use, development and supply, as explored in Chapters 5 and 6. The abiding importance of agriculture

---

[38] W. Laqueur, *The History of Zionism* (Trinity Press, 1972), 120–2; J. C. Hurewitz, *The Middle East and North Africa in World Politics: A Documentary Record. Vol. 2 British-French Supremacy, 1914–1945*, 2nd ed. (Yale University Press, 1979), 140; A. T. Wolf, *Hydropolitics along the Jordan River: Scarce Water and Its Impact on the Arab–Israeli Conflict* (UN University Press, 1995), 23, 18; A. Garfinkle, *War, Water and Negotiation in the Middle East: The Case of the Palestine–Syria Border, 1916–1923* (Moshe Dayan Center for Middle Eastern and African Studies, Tel Aviv University, 1994), 39.

and hence water within Zionist strategy (as noted in an official Knesset enquiry on water as recently as 2002; for Israel 'agriculture has a Zionist-strategic-political value, which goes beyond its economic contribution'). And, not least, the enduring prevalence of deeply Orientalist, indeed racist, narratives within mainstream Zionist thought. Still now, an ostensibly left-of-centre and peace-seeking Israeli Prime Minister, Ehud Barak, can characterise the Palestinians as 'the products of a culture in which to tell a lie ... creates no dissonance. They don't suffer from the problem of telling lies that exists in Judeo-Christian culture. Truth is seen as an irrelevant category' – without, it may be added, any note of complaint from the leading Israeli 'new historian' interviewing him. The Israeli Ministry of Defence official who negotiated the Oslo II water agreement with the Palestine Liberation Organization (PLO) (discussed in Chapter 8) can depict the Palestinians as 'liars' who 'want us to bring them water and to live at our expense', who do 'nothing' because 'it's easier to cry' and 'to be miserable', and who in any case don't have the same 'cultural' need for water as Israelis do. More broadly, right across Israeli public and policy discourse, water regularly functions as a site for the performance of Jewish-Israel's 'developed' and 'Western' self-identity, and for Orientalist constructions of cultural difference. As one Israeli expert has put it: 'Israelis have the same tastes as Europeans and Americans; they like green lawns and flowers around their homes. In Arab towns and villages one sees few green lawns .... Like Westerners Israelis take frequent showers in the summer months; this is not an Arab tradition.'[39] While many Palestinians no doubt have preconceptions and prejudices of their own, it is only on the Zionist side that narratives of national identity and otherness are so systematically Orientalist, and so neatly aligned with both a civilising-cum-redemptive moral vision and national and international apparatuses of military, political and economic power. Such is the nature of settler-colonialism – including, as we shall see, in relation to water.

Identity and Alterity in an Age of Climate Change

The chapter has sought to explore representations and constructions of peoples, their environments and human–nature relations, within both modern 'Western' political thought and some of the colonial projects it

---

[39] *Report of the Parliamentary Committee of Inquiry on the Israeli Water Sector* (2002), 12; B. Morris, 'Camp David and after: an interview with Ehud Barak', *New York Review of Books* (13/06/2002); Selby interview with Noah Kinnarty (28/07/1998); G. Fishelson in Rouyer, *Turning Water*, 30.

spawned and legitimised. Its core premises have been fourfold: that such representations and constructions are through and through imaginative and hence in large degree fictional; that they are always relational, defined through opposition; that they are also always hierarchical both in their European provenance and in content, always implicitly or explicitly ranking some peoples as by some measure superior to others; and that from Immanuel Kant to Robert Kaplan, and Winston Churchill to Ehud Barak, all such representations and constructions may therefore be considered racialised, indeed racist.[40] Building upon recent work on environmental Orientalism, in particular – and departing from the narrowly humanist emphases of Said and others – the chapter has sought to demonstrate that ideas about environments, landscapes, resources and climates, and peoples' relations with them, have always lain near the heart of such racist colonial imaginaries. And it has sought to argue, above all, that the legacies of these colonial imaginaries are still very much with us today, within both thought and practice. Thus contemporary eco-determinisms, including water wars and climate collapse narratives, remain structured – just like the European Enlightenment and counter-Enlightenment traditions to which they are indebted – around an assumed global environmental colour line in which some peoples are viewed as essentially subject to natural or environmental forces, while others, especially those of a white European lineage, are held to have transcended nature to inhabit the realm of ideas. Many contemporary pseudo-scientific concepts, such as 'desertification', are racialised colonial-era myths that, like Malthusianism more broadly, refuse to die despite all the evidence to the contrary.[41] And at the level of practice, the politics and political ecologies of contemporary Sudan, Cyprus and Israel–Palestine are each essentially structured, in their own unique ways, around racialised conceptions of identity and alterity – and underpinning these, conceptions of time and space, humanity and nature – inherited from the high era of European colonialism. To this extent, the imaginaries examined in this chapter are foundational both to contemporary climate and water security discourse and to the material patterns of water-related (in)security examined in subsequent chapters.

The latter can be summarised as follows. In Sudan as elsewhere in Africa, Britain supposed that the population mostly required despotic and traditional rule, and this combined with British interests in the Nile and in the dismemberment of the vestiges of Mahdism led to the creation of a classically bifurcated state – paving the way for multiple core–periphery conflicts,

[40] I. X. Kendi, *How to Be an Antiracist* (Bodley Head, 2019), 9.
[41] D. S. G. Thomas and N. J. Middleton, *Desertification: Exploding the Myth* (Wiley, 1994); J. Swift, 'Desertification: narratives, winners and losers', in Leech and Mearns, *Lie of the Land*, 73–90.

including water-related conflicts and insecurities, which have plagued Sudan (and what is now South Sudan) ever since. In Cyprus, by contrast, a supposedly white European but degraded population was deemed to require liberal constitutional governance, which alongside the parallel rise of Greek and Turkish ethnic nationalisms and their extension into Cyprus, paved the way – once again in combination with key material factors, above all colonial exploitation and the Cold War – for the creation of a relatively symmetrical 'bi-communal' conflict and relatively symmetrical partition (with the island's water politics reflecting this). And by contrast again, in Palestine a European settler-colonial project – which, like its forebears, overflowed with racist ideas – paved the way for a highly asymmetrical conflict characterised by huge Israeli–Palestinian inequalities, mass Palestinian displacement and dispossession and attendant inequalities in water security. Of course, racial and colonial legacies are not always as neatly determinant of present-day patterns of conflict and (in)security as the above might be taken to imply (to name but one, Syria's current spectrum of divisions are barely indebted to colonialism at all). Yet it is striking nonetheless that the contrasting patterns of conflict and water (in)security across these three cases can each be traced to distinct racialisations. In three different contexts – one imagined as African, one as European and one as in the Levant – the same colonial master pursued three very different models of rule, bequeathing distinctly racialised legacies which in key respects endure to this day.

To be clear, our argument is not, as in blunt versions of idealist and constructivist thought, that free-floating ideas about peoples and their environments created the national and ethnic divisions that endure to this day. For, not only were these ideas always enabled by and dependent on various material hierarchies and interests, from the general fact of European dominance to the particular strategies pursued and structures established in each colonial context. In addition, all of the identity constructions discussed above were infused with, and by-products of, those understandings of progress – of some peoples as being more 'civilised' or 'advanced' or 'developed' than others – that rose to prominence as corollaries of the European Enlightenment, European imperialism and capitalist modernity. Our thesis, in short, is not just that contemporary patterns of conflict and water (in)security are historical legacies of specific constructions of identity, but that these racialised constructions were in turn corollaries of capitalism, colonialism and the conceptions of development that went with them (as Cedric Robinson emphasised, capitalism tends 'not to homogenise', as per Marx, 'but to differentiate – to exaggerate regional, subcultural, and dialectical differences into "racial" ones'). Moreover – and with apologies for not discussing this issue in detail – it bears emphasising that gender figures crucially here too, modern gender divisions, representations, hierarchies and relations having been foundational both to the emergence of capitalism and colonial rule, and for this reason affecting patterns of environmental access, control, responsibility and vulnerability to this day (as well as making water an important site for the reproduction of gendered relations and injustices).[42] The political ecology of environmental (in)security is rooted in the contradictions of capitalist development, all the way down.

What, though, might all this tell us about the conflict and security implications of climate change? We suggest two main things, paralleling the chapter's part-critical and part-reconstructive purposes. First, the analysis above hopefully provides even further reason, lest the previous chapters were not evidence enough, for moving beyond eco-determinist and mainstream climate security reasoning – since not only is this reasoning theoretically flawed, empirically baseless and questionably congruent with a range of right-wing and authoritarian state agendas, as argued in Chapters 2 and 3; it is also, we now find, essentially racist in structure too. The problem with this, to be clear, is not just that this racism is morally objectionable, but also that it is substantively misleading, ignoring or at best understating the complexity, heterogeneity, historicity, agency and universally hybrid character of peoples and their environments, and thus offering a poor guide to understanding and analysis. Put more simply, the imagined geographies underpinning mainstream climate security thinking – of climate conflict as essentially restricted to the global South, in particular Africa; of Southern peoples as lacking agency vis-à-vis their environments; and of the global North as being left to pick up the pieces in the form of 'climate migration' and 'security' interventions – are fundamentally flawed. Just as famine is not, despite regular portrayals to the contrary, a predominantly African phenomenon, so climate change–induced conflict is unlikely to become so either.[43] Whatever the future holds, we can be pretty sure that it will not conform to such racialised stereotypes.

And yet, second, the enduring power of racial ideas – the potency both of explicitly racist views and of how they have become embedded and naturalised within national identities and racial inequalities – is such that race itself may well become one of the central axes of future climate-related conflict. Indeed, a case could be made that this future is already

---

[42] C. Robinson, *Black Marxism: The Making of the Black Radical Tradition* (Zed, 1983), 26; S. Federici, *Caliban and the Witch: Women, the Body and Primitive Accumulation* (Autonomedia, 2004); A. McClintock, *Imperial Leather: Race, Gender and Sexuality in the Colonial Contest* (Routledge, 1995); M. Lugones, 'Heterosexualism and the colonial/modern gender system', *Hypatia*, 22:1 (2007), 186–219; F. Sultana, 'Fluid lives: subjectivities, gender and water in rural Bangladesh', *Gender, Place and Culture*, 16:4 (2009), 427–44.
[43] de Waal, 'The end of famine?', 188–9.

with us. Global greenhouse gas emissions have historically been deeply racialised, with the United States and EU states alone contributing almost 50 per cent of cumulative global emissions, and emissions from Asia, in particular, having historically been suppressed. Conversely, climate change's most severe consequences are likely to fall disproportionately elsewhere, above all within Asia. Contrasting North–South interpretations of responsibility for cutting emissions to a large degree reflect these historical and projected inequalities.[44] Equally, in many countries, attitudes towards climate change are deeply divided along racial lines (as well as by class and gender). In the United States, climate change denial has become, in the words of sociologists Aaron McCright and Riley Dunlap, 'almost an essential component of conservative white male identity' – a defensive reaction against long-entrenched race, gender and class privileges, and arguably also a response to declining US hegemony. Worldwide, right-wing populists overwhelmingly combine anti-immigrant policies and rhetoric with fierce climate change denial. On the far right, eco-fascist white supremacists are advancing simultaneously racist and patriarchal narratives of overpopulation, population control and (white) masculinist 'protection' of nature, and calling for the killing of Latinos and Muslims as a route to saving the planet.[45] Meanwhile, countless Western liberals and left-green activists and commentators envisage Africa returning to a state of nature, and northern latitudes being forced to absorb billions of climate refugees – with these liberals and eco-socialists for the most part remaining completely innocent and unaware of how these racialised narratives mirror colonial and right-wing premises, and may end up feeding extreme right-wing agendas. This is

---

[44] B. Müller et al., 'Differentiating (historic) responsibilities for climate change', *Climate Policy*, 9:6 (2009), 593–611; A. Ghosh, *The Great Derangement: Climate Change and the Unthinkable* (University of Chicago Press, 2016), 96–111; Mann and Wainwright, *Climate Leviathan*, 43; J. T. Roberts and B. C. Parks, *A Climate of Injustice: Global Inequality, North–South Politics, and Climate Policy* (MIT Press, 2007); D. Ciplet et al., *Power in a Warming World: The New Global Politics of Climate Change and the Remaking of Environmental Inequality* (MIT Press, 2015).

[45] A. McCright and R. Dunlap, 'Cool dudes: the denial of climate change among conservative white males in the United States', *Global Environmental Change*, 21 (2011), 1168; J. Selby, 'The Trump presidency, climate change, and the prospect of a disorderly energy transition', *Review of International Studies*, 45:3 (2019), 471–90; M. Lockwood, 'Right-wing populism and the climate change agenda: exploring the linkages', *Environmental Politics*, 27:4 (2018), 712–32; P. Beinart, 'White nationalists discover the environment', *The Atlantic* (05/08/2019); B. Hartmann, 'The ecofascists', *Columbia Journalism Review* (Spring 2020); J. Dyett and C. Thomas, 'Overpopulation discourse: patriarchy, racism, and the specter of ecofascism', *Perspectives on Global Development and Technology*, 18: 1–2 (2019), 205–24; M. Darwish, 'Nature, masculinities, care, and the far-right', in P. M. Pulé and M. Hultman (eds.), *Men, Masculinities, and Earth: Contending with the (m)Anthropocene* (Palgrave Macmillan, 2021), 183–206.

not the place for systematic discussion of these issues (indeed, for the most part they lie outside the scope of our analysis: we return to them explicitly only in the book's Conclusions). But the implication is hopefully clear. If the above provides any guide, then the future landscape of climate conflicts may revolve less around resource scarcities than around racist conceptions of selfhood and otherness and the hierarchies and representations that sustain them. If the above is right, our windows on the future may be neither Sierra Leone nor Darfur – but Christchurch, New Zealand and El Paso, Texas.

5   Hydraulics

From Greenock to Tennessee

In 1824, the Scottish engineer Robert Thom proposed a huge new water infrastructure scheme for the town of Greenock on the Clyde, with the promise that this new scheme would assure the town's water supply and revolutionise its industrial fortunes. The Thom scheme – comprising a large reservoir for multi-year storage, a six-and-a-half-mile aqueduct, self-acting sluices, a second regulating reservoir and two artificial mini-rivers with artificial waterfalls – came into operation in 1827, the largest such water conveyance works in Britain. The water flowing through this system, it was hoped, would be sufficient to turn Greenock into a second Glasgow, powered in this case by water rather than coal and steam. Thirty-three cotton mill sites were initially prepared, with the expectation that more would follow. With Greenock the inspiration, similar projects were proposed across northern Britain – for Manchester's River Irwell, for the River Tame just to the south, and even for Glasgow – all dedicated to using hydraulic engineering and waterpower as motors of industrial development.

Yet these plans all came to nothing. The majority of Greenock's mill sites were never occupied. Those which were did not produce cotton. And the Irwell, Tame and Glasgow water diversion schemes were never implemented. This was not, however, as Andreas Malm argues in *Fossil Capital*, because of any natural disadvantages of water power: water was cheaper than coal; it flowed abundantly down the windward slopes of northern Britain; and Thom's new techniques provided constant, assured supplies, even during periods of drought. The problem was rather one of collective action: that within the context of early industrial capitalism, where inter-capitalist competition was fierce and working-class demands were rising steadily, mill owners simply could not agree on how the costs and benefits of large-scale water infrastructures should be distributed. As Malm puts it, such schemes posed 'problems of coordination and resource distribution of which an extra steam engine, a large boiler, one more ton of coal per day were blissfully oblivious'.[1] The dream of using water to power Britain's industrial revolution ultimately came to nothing and Thom himself, unlike his compatriot James Watt, became little more than a footnote to history.

Fast forward a century, and cross the Atlantic, and one encounters an even more ambitious hydraulic scheme – though with a very different outcome. Born through an act of Congress in 1933, the Tennessee Valley Authority (TVA) scheme involved, and still involves today, a system of dams and locks to ease river navigation, additional dams for electricity generation, an electricity transmission network right across the Valley, assorted flood control measures and associated land use, conservation, education and agricultural and industrial development programmes. Centred on the state of Tennessee but also serving parts of six other southern states, the TVA was, and remains, the largest regional planning authority in the United States. Moreover, for four decades through to the 1970s the TVA came to serve as a worldwide exemplar of US-style integrated regional development. During this period the TVA was what James Scott called 'the granddaddy of all regional development projects.' As one US outlet put it, 'TVA' had become 'magic letters the world over'.[2]

Why the Greenock and Tennessee projects had such contrasting fates – the one still-born and forgotten, the other swiftly propelled to global fame – is not difficult to see. The TVA was a product, and indeed has often been presented as emblematic, of Franklin Roosevelt's New Deal, the enormous programme of federal public works projects and regulatory reforms launched by Roosevelt in 1933 to support recovery from the Great Depression. More broadly, the establishment of the TVA reflected and embodied the seismic early twentieth-century transformation in the powers of the capitalist state, away from the limited form of classical liberal imagination to one with hugely enhanced infrastructural, epistemological and administrative functions and capabilities. Like many other major US water development schemes, the TVA also owed much to military conflict and power: its material beginnings lay in a hydropower dam and nitrate plant for First World War explosives production; it expanded massively during the late 1930s and 1940s to feed critical war industries, most notably the Manhattan Project at Oak Ridge Tennessee;

---

[1] Malm, *Fossil Capital*, 110.
[2] J. Scott, *Seeing Like a State: How Certain Schemes to Improve the Human Condition Have Failed* (Yale University Press, 1998), 6; R. G. Lynch, 'TVA's world offspring', *Milwaukee Journal* (22/03/1959), in D. Ekbladh, '"Mr TVA": grass-roots development, David Lilienthal, and the rise and fall of the Tennessee Valley Authority as a symbol for US overseas development, 1933-1973', *Diplomatic History*, 26:3 (2002), 335.

and it acquired its global fame as an arm of US Cold War overseas development and counter-insurgency strategy. The contrast between Greenock and Tennessee can be overplayed, no doubt: the TVA model, while globally influential, was not replicated anywhere else in the United States; and large-scale regional water schemes had long been developed by European states in their colonies, most notably by Britain in Egypt and India. But the disjuncture between the two hydraulic schemes is instructive nonetheless: whereas in the mid-nineteenth century the political ecology of water in the 'Lockean heartland' of capitalist development was shaped principally around competition between individual capitalists, a century later institutions of state acting in the name of a population and over vast areas of claimed national territory had become key.[3]

This chapter takes this immense shift as its starting point to explore how, over the last hundred or so years, modern state capitalist projects of hydraulic development have transformed forms and patterns of water use and supply, in turn variously uprooting, constituting and helping to reconfigure nature, populations, economies and states. 'Water security' has always, of course, been an infrastructural or at least technological accomplishment: whether through dew ponds or pots and pans, artesian wells or *qanats*, sewage channels or land drainage schemes, clothes or dwellings, technologies in the broadest sense have always been foundational to human well-being and development vis-à-vis water. Yet over the last century this has become the case in unprecedented ways. Hydraulic infrastructures have exploded in scale: whereas in 1900 there were no dams higher than 15 metres, by 2000 there were an estimated 50,000 – collectively amounting, in Christopher Sneddon's apt phrase, to a 'concrete revolution' comparable in impact to the better-known Green Revolution. Hydraulic developments have vastly increased in ambition too, intersecting with and often being foundational to a host of parallel techno-social projects, from agricultural intensification and rural electrification to public health, land reform, regional integration and more. Hydraulics have arguably been instrumental both to what Henri Lefebvre theorised as the signature of capitalist modernity, 'abstract space', and to modern conceptions of abstract time, in which water and other commodities are constantly available at the flick of a switch or the turn of a tap, hostage to the vagaries of neither the seasons nor the

---

[3] Ekbladh, 'Mr TVA'; P. K. Conkin, 'Intellectual and political roots', in E. C. Hargrove and P. K. Conkin (eds.), *TVA: Fifty Years of Grass-Roots Bureaucracy* (University of Illinois Press. 1983), 3–34; K. M. O'Neill, 'Why the TVA remains unique: interest groups and the defeat of new deal river planning', *Rural Sociology*, 67:2 (2002), 163–82; K. van der Pijl, *Transnational Classes and International Relations* (Routledge, 1998).

weather.[4] Modern hydraulic development, moreover, has become a constant revolution, an important ally in capitalism's general drive towards the ever-increasing mobilisation, consumption and exploitation of natural resources.

There already exists a large literature on these themes. Thus geographers and historians of water politics, in particular, have mapped out how natural 'waterscapes' have been transformed through hydraulic development; have theorised the rise of a new 'hydro-social cycle' traversing the nature–society divide; have interrogated the politics of 'hydraulic bureaucracies' and their 'hydraulic mission' – the missionary 'conviction that every drop of water flowing to the ocean is a waste and that the state should develop hydraulic infrastructure to capture as much water as possible for human uses'; have analysed also the links between water development and state formation, building on Karl Wittfogel's classic, if Orientalist, work on the subject; have investigated the roles of diverse scientific knowledges in these processes; and much else besides.[5] In this chapter we build upon this extensive body of work to explore the political ecology of modern hydraulic development, attending both to its various contexts and motivations and to some of its major consequences.

More specifically, we seek to make three main contributions. We explore, first, the ways in which modern projects of hydraulic development and the new waterscapes and patterns of water use and supply resulting from them, reflect and help consolidate specific state-building and national development agendas – our central argument being that modern hydraulics is always politics by other means, configured around particular state interests, ideologies and capabilities. We demonstrate, second, and contrary to liberal premises, that because of their essentially techno-political character, hydraulic projects repeatedly involve conflict and everywhere create new forms of insecurity for some, alongside 'water security' for others. And we argue, third, that the renewal of hydraulic missions in response to real or

---

[4] C. Sneddon, *Concrete Revolution: Large Dams, Cold War Geopolitics, and the US Bureau of Reclamation* (University of Chicago Press, 2015); H. Lefebvre, *The Production of Space*, trans. D. Nicholson-Smith (Blackwell, 1991 [1974]); E. P. Thompson, 'Time, work-discipline, and industrial capitalism', *Past and Present*, 38 (1967), 56–97.

[5] T. Karpouzoglou and S. Vij, 'Waterscape: a perspective for understanding the contested geography of water', *WIREs Water*, 4:3 (2017), 1210; Linton and Budds, 'The hydrosocial cycle'; F. Molle et al., 'Hydraulic bureaucracies and the hydraulic mission: flows of water, flows of power', *Water Alternatives*, 2:3 (2009), 328–49; P. Wester, 'Capturing the waters: the hydraulic mission in the Lerma-Chapala Basin, Mexico (1876–1976)', *Water History*, 1:1 (2009), 10; K. Wittfogel, *Oriental Despotism: A Comparative Study of Total Power* (Yale University Press, 1957); E. Swyngedouw, *Liquid Power: Contested Hydro-Modernities in Twentieth-Century Spain* (MIT Press, 2015); D. Gilmartin, 'Scientific empire and imperial science: colonialism and irrigation technology in the Indus basin', *Journal of Asian Studies*, 53:4 (1994), 1127–49.

imagined fears of climate change will, whatever the merits of such investments, inevitably also presage a resurgence of hydraulic development–related conflicts and insecurities. We develop these arguments in relation to our four once-unitary cases, Cyprus, Israel–Palestine, Syria and Sudan. It should be noted that across the chapter our focus is on core processes of hydraulic development and state-building; consideration of territorial frontiers and peripheries is held over to Chapter 6.

Engineering Ethno-Nationalism in Israel and Cyprus

'When nations wandered in historic times, they let chance carry them, draw them, fling them hither and thither, and like swarms of locusts they settled down indifferently anywhere', wrote Herzl in *Der Judenstaat*. 'But this modern Jewish migration must proceed in accordance with scientific principles.... Everything must be systematically settled beforehand.... By these means a country can be occupied and a State founded in a manner as yet unknown to history.'[6] From the very beginning, before Palestine was even chosen as the location for the new Jewish homeland, Zionism imagined itself as a political project devoted to the colonisation of abstract space.

As indicated in the previous chapter, agriculture and in turn water became central to this space- (and time-) dominating political project. Agriculture became a political imperative, valued more in strategic-territorial terms than as an input to economic development – as Israeli state institutions and policymakers regularly observe. 'The rural and agricultural sector in Israel discharges a national and social responsibility in dispersing population [and] populating frontier regions', the Ministry of Agriculture and Rural Affairs reported in 1997. 'Agriculture has a Zionist-strategic-political value, which goes beyond its economic contribution', a Knesset enquiry on water issues noted in 2002. Or as a leading Israeli water expert once put it to one of the authors, when explaining the philosophical basis for this:

The whole philosophy of the Zionist movement was that you maintain control of the land, over your country, by working there and being there. There's no doubt that if they move out of the border with Lebanon, somebody else will be there, and that somebody is Arabs, not Jews, and the government of Israel doesn't want Arabs to be there on the border, because then the border will move further and further south. The same is true in the Negev.[7]

---

[6] Herzl, *Jewish State*, 67.
[7] Ministry of Agriculture and Rural Development, 'Agriculture economic report for 1996' (1997) (in Hebrew), in A. Kartin, 'Factors inhibiting structural changes in Israel's water

In turn, the scientific management and re-engineering of Palestine's water resources became a key aspect of Zionist state-building (in keeping with Herzl's vision of a Jewish 'Altneuland' whose 'real founders' would be hydraulic engineers). Management of the water sector was centralised, well-funded and kept in line with national policy and planning priorities. The para-statal water company Mekorot was established as early as 1937, quickly becoming a Yishuv-wide provider of water supplies and, to this day, Israel's only national water utility. This was followed, post-independence, by the passing of a unified Water Law which defined all water resources, groundwater and rainwater included, as public property, 'subject to control by the state'. The water sector benefitted from high levels of state investment amounting, from 1950 to 1970, to 3–5 per cent of gross capital formation. A national Water Commission was created under the authority of the Ministry of Agriculture. And there was a direct line between water sector management and the apex of political power: it is no coincidence, for instance, that Israel's third Prime Minister, Levi Eshkol, was previously Mekorot's first director general (as well as the first director general of Israel's Ministry of Defense, head of the Jewish Agency's Settlement Department and Minister of Agriculture).[8]

Before 1948, the main focus of Zionist concern was not any overall scarcity of water but rather ensuring that it was put to productive use and, where necessary, removed (as Samer Alatout points out, a discourse of 'water abundance' rather than 'absolute scarcity' dominated during this period). Thus, on the one hand various hydro-electric schemes were proposed to power the Yishuv's developing urban centres while, on the other and more significantly, there was extensive land reclamation. Much of northern and coastal Palestine then consisted of marshland, to the extent that Herzl could record, at the beginning of his one journey through Palestine, that '[o]nly large-scale drainage operations and the elimination of swamps . . . could make the country habitable'. In turn, during the 1920s and 1930s alone around 150,000 acres of Palestine's wetlands were reclaimed in support of the interlinked goals of Jewish settlement and agricultural modernisation. Revealingly, British concessions for swamp

---

policy', *Political Geography*, 19:1 (2000), 108; *Report of the Parliamentary Committee of Inquiry on the Israeli Water Sector*, 12; Selby interview with Arlosoroff.

[8] T. Herzl, *Old-New Land*, trans. L Levensohn (Markus Wiener, 1997 [1902]), 238; I. Galnoor, 'Water planning: who gets the last drop?', in R. Bilski et al. (eds.), *Can Planning Replace Politics? The Israeli Experience* (Martinus Nijhoff, 1980), 137–215; S. Arlosoroff, 'The Israeli water law concept: summary', in E. Feitelson and M. Haddad (eds.), *Joint Management of Shared Aquifers: The Fourth Workshop* (Palestine Consultancy Group and Harry S. Truman Institute for the Advancement of Peace, 1998), 111–15; E. Feitelson, 'Implications of shifts in the Israeli water discourse for Israeli–Palestinian water negotiations', *Political Geography*, 21:3 (2002), 300–3.

drainage and land reclamation all went to Zionist entrepreneurs: not a single such concession was granted to members of Palestine's majority Arab population.[9]

Post-independence, the focus shifted to ameliorating water supply scarcities through large-scale water conveyance infrastructures. Prior to this, the rising (Jewish and Arab) demand for water had all been met locally, mainly from wells sunk into Palestine's shallow Coastal Aquifer. In the face of serious local over-pumping and some salinisation, however, Zionist planners started considering out-of-basin water transfer from the upper Jordan River. The template for this was provided by the American soil conservationist Walter Lowdermilk's 1944 study *Palestine: Land of Promise*, which portrayed Palestine as a land of abundant supplies but poor use and distribution, and in turn advocated the diversion of the upper Jordan River for irrigation, the simultaneous construction of a Mediterranean–Jordan Valley seawater canal (both to generate hydroelectric power and to compensate for the diversion of the upper Jordan River), and the reclamation and repopulation of the Negev through the capture and transfer of floodwaters – all of which, Lowdermilk argued, could enable the absorption of an additional four million European Jews. Inspired by this and related proposals, during the 1950s and 1960s Israel established a highly integrated national water supply network which, broadly speaking, performed (and still performs) two functions. At the most general level, this hydraulic network and the system of water subsidies and common water-pricing supporting it, made water available and affordable for Jewish agriculture and settlement right across the territory of the new state, supporting the vision of state-building through permanent productive presence on the land outlined above. And, more specifically, this system enabled the transfer of water from the upper Jordan basin in the north-east of the country, as well as from springs around the edge of the West Bank, to the more heavily populated coastal plain and then southwards towards the Negev. At the centre of this was the country's National Water Carrier, a combination of huge underground pipelines, open canals, tunnels, reservoirs and pumping stations which from 1964 onwards transferred around 400 mcm/y of water from the Lake Kinneret, 220 metres below sea level, up and over into the coastal plain.[10]

---

[9] Alatout, '"States" of scarcity'; R. Shamir, *Current Flow: The Electrification of Palestine* (Stanford University Press, 2013); T. Herzl, *The Complete Diaries of Theodor Herzl*, Vol. 2, ed. R. Patai, trans. H. Zohn (Herzl Press, 1960), 740; J. Broich, 'British water policy in Mandate Palestine: environmental orientalism and social transformation', *Environment and History*, 19:3 (2013), 257, 271.

[10] Wasserstein, *Israel and Palestine*, 79; Lowdermilk, *Palestine*, 121–42, 154–61; D. Sitton, 'Development of water resources', Israel Ministry of Foreign Affairs (20/10/2002).

With the establishment of this politically configured hydraulic system, the natural hydro-geography of Israel–Palestine was turned upside down. 'Pristine nature' did not rule before this, of course: nineteenth and early twentieth-century European travellers to Palestine often expressed surprise at the extent of Arab irrigation. But the transformations effected after 1948 were of a different order altogether. Israel became the Jordan River's major consumer, despite being a downstream riparian and marginal contributor to river flows. The northern Negev Desert became the main terminus of the upper Jordan River, replacing the Dead Sea. The former was turned into a site of intense and expanding irrigated agriculture, as noted in Chapter 2. By contrast, the lower Jordan River was transformed into a glorified sewage outlet, its average flow cut by a whopping 98 per cent. And the Dead Sea inexorably declined in turn, its waters retreating by around a metre per year (while Syria and Jordan have also been factors in this through their abstractions from the Jordan River's major tributary, the Yarmouk, Israel has been responsible for the bulk of abstractions).[11] Notwithstanding the many proposals for unified river basin development modelled on the success of the TVA, in practice the Jordan River was re-engineered around a distinctively national political project: that of integrating and unifying the abstract space of the Israeli state – while simultaneously dispossessing, excluding and otherwise discriminating against Palestine's indigenous population, as we shall see, especially in Chapter 6.

In recent years this system has been augmented in two crucial respects. First, Israel has invested heavily in wastewater treatment and reuse, such that the wastewater of metropolitan Tel Aviv, for example, is injected into a nearby aquifer basin before being collected through recovery wells and conveyed to the Negev for reuse in agriculture, through a designated effluent line. In addition, since 2004 Israel has engaged in a major desalination drive, constructing five major plants with a combined capacity of around 600 mcm/y. The combined effect of these developments, including the reuse of water originally produced through desalination, has been to increase Israel's overall water budget by over 1,000 mcm/y, more than 40 per cent. This in turn has enabled a further intensification of agriculture in the northern Negev, plus initiatives like the new lake in Beer Sheva, described in Chapter 2. Moreover, these developments have further transformed the geography of water supply. No longer is Israel's national supply network organised from north-east to south, with Lake Kinneret its major storage facility; instead, the Mediterranean is now

---

[11] Broich, 'British water policy', 264; C. Messerschmid and J. Selby, 'Misrepresenting the Jordan River Basin', *Water Alternatives*, 8:2 (2015), 258–79.

Israel's single most important water resource. Reflecting this shift, Mekorot is currently constructing a 'reverse carrier' to transfer desalinated Mediterranean seawater to Lake Kinneret as and when needed. Israel's natural waterscape has, in short, been multiply upended. A highly unified hydro-political territory has been created, criss-crossed and sustained by multiple supply networks. And simultaneously, this space has been rendered largely invulnerable to seasonal and weather variations. Owing in particular to desalination, Israel's water policymakers can now claim that the country is 'mostly immune to adverse weather conditions', to the extent that 'Israel can now withstand even a long drought'.[12]

Cyprus' modern hydraulic history presents, in many respects, a quite similar story. As in Israel, land reclamation and groundwater development were the main focuses under British rule – with the former beginning under Joseph Chamberlain and continuing thereafter, and the latter being so extensive that by 1960 there were over 10,000 boreholes across the island. However, Britain invested little in Cyprus' hydraulic – or other – development overall, viewing it, as already discussed, as a largely inconsequential possession. A government Water Supply and Irrigation Department was established only in 1938. No dams were built until the 1940s (aside from some initial failures around the turn of the century). And while groundwater development partially compensated for this lack of surface water storage – providing ready access to water all year round and enabling some limited perennial irrigation – by the early 1950s groundwater levels were in significant decline, with some seawater intrusion.[13]

Following independence, however, this colonial pattern of limited hydraulic investment and groundwater-led development fundamentally changed, as one corollary of the new Cypriot state's post-colonial development agenda. This agenda had in essence three aims: righting the chronic poverty which still stalked the island after nearly a century of British exploitation; doing so through state-led planning and investment (as in our other cases); and supporting Greek Cypriot development

---

[12] A. Kanarek and M. Michail, 'Groundwater recharge with municipal effluent: Dan Region Reclamation Project, Israel', *Water, Science and Technology*, 34:11 (1996), 227–33; E. Feitelson and G. Rosenthal, 'Desalination, space and power: the ramifications of Israel's changing water geography', *Geoforum*, 43:2 (2012), 272–84; Z. Rinat, 'National water project faces changes: inside the efforts to refill Israel's dwindling Lake Kinneret', *Ha'aretz* (24/02/2020); S. M. Siegel, *Let There Be Water: Israel's Solution for a Water-Starved World* (Thomas Dunne, 2017), 123.

[13] Varnava, *British Imperialism*, 142–5; D. J. Burdon, 'Groundwater in the island of Cyprus', *Association of International Hydrological Science General Assembly*, 2: 37 (1954), 322–3; D. Christodoulou, *The Evolution of the Rural Land Use Pattern in Cyprus* (Geographic Publications, 1959), 114–15; D. J. Burdon, *The Underground Water Resources of Cyprus* (Government Printer, 1953).

specifically within a state which, though constitutionally bicommunal, was in practice dominated by those cross-class Greek Cypriot forces that had waged the struggle against British rule. Agriculture – which was simultaneously the largest economic sector and Greek Cypriot–dominated, with Greek Cypriots making up a disproportionate share of the island's rural population – became, in turn, the backbone of the country's post-independence development strategy. Agriculture became the major recipient of state investment, accounting as late as 1982–6 for 26 per cent of the Republic of Cyprus' (RoC) total development budget. The Ministry of Agriculture and associated institutions became Greek Cypriot–dominated. Agricultural investment and extension services were overwhelmingly targeted at majority Greek Cypriot areas, while rural poverty in Turkish Cypriot majority areas was ignored. And water development became the major component of this, accounting during 1982–6, for example, for 60 per cent of agricultural investment.[14] As within Israel, ethno-nationalism translated into strategies of land control which, in turn, translated into a privileging of agricultural interests and a project of politically configured large-scale hydraulic development.

Although this dynamic took shape well before 1974, Turkey's invasion and occupation of northern Cyprus deepened and extended it in several key respects. Within that larger part of the island which remained under RoC control both the ambition and pace of hydraulic developments increased, mostly thanks to a series of emergency plans and increased state interventionism, but also partly with the aim, in the Paphos region in particular, of accelerating development plans for the now abandoned Turkish Cypriot land.[15] Somewhat later, Turkey's State Hydraulic Works (DSI) and the authorities in northern Cyprus commenced their own hydraulic development programme. Mirroring the broader symmetries of the conflict, Cyprus became home to two rival, unilateral projects of hydro-territorialisation.

In RoC territory this translated most notably into a large dam-building programme devoted to capturing the abundant winter rains flowing from the Troodos, in line with the post-independence motto 'not a drop of

[14] R. Bryant and M. Hatay, *Sovereignty Suspended: Political Life in a So-Called State* (University of Pennsylvania Press, 2020), 10; A. Melamid, 'The geographical distribution of communities in Cyprus', *Geographical Review*, 46:3 (1956), 364–5; P. Hocknell, *Boundaries of Cooperation: Cyprus, De Facto Partition and the Delimitation of Transboundary Resource Management* (Kluwer Law International, 2001), 179; P. I. Panayiotopoulos, 'Cyprus: the developmental state in crisis', *Capital and Class*, 19: 3 (1995), 29.

[15] P. Hocknell, 'Contested "development": a retrospective of the UN development programme in Cyprus', in O. Richmond and J. Ker-Lindsay (eds.), *The Work of the UN in Cyprus: Promoting Peace and Development* (Palgrave Macmillan, 2001), 163; RoC Planning Bureau, *The Second Emergency Economic Action Plan 1977–1978* (1977), 48.

water to the sea'. More than a hundred dams were constructed around the Troodos and its foothills, turning Cyprus into an 'island of dams' with one of the highest per capita water storage capacity rates in the world. Water transfer tunnels and pipelines were constructed to connect these dams together, and especially to feed the new Southern Water Conveyor constructed to transfer water along Cyprus' southern coast and, through that, to support irrigation projects, urban expansion and growing tourist demand in the island's drier south-east. Although the latter project was conceived before 1974, it now became instrumental in consolidating the country's water supplies without reliance on aquifers under Turkish control. Moreover, since the 1990s a series of desalination plants with a combined capacity of around 115 mcm/y have been constructed along the southern coast, mainly to meet growing domestic and, especially, tourist demand. As a result, total freshwater storage capacity within the RoC has increased from a mere 6 mcm in 1961 to more than 300 mcm today, with total water consumption increasing likewise such that political leaders can claim that 'water shortages ... will not be experienced again by Cypriot society'.[16] Moreover, the country now has a hydraulic system that functions as an infrastructural corollary of the Greek Cypriot state and ethno-nationalist project.

While in these and many other respects there are important parallels with Israel, at a slightly more fine-grained level it is the differences which stand out. The RoC's hydraulic system is nowhere near as integrated as Israel's, just as its hydro-geography has not been so comprehensively de-naturalised either – this reflecting both differences in physical geography and, more than this, the contrast between a settler-colonial society with a high-modernist scientific-plus-pioneering ethos and centralised state institutions, and what was at least initially a 'peasant state' dedicated to servicing farmers on their own inherited land. Furthermore, whereas in Israel irrigation water use has continued to expand even as its agricultural sector has decreased in economic importance – reflecting the political-strategic imperatives discussed above – in Cyprus, farmland and orchards have been paved over and turned into golf courses, and agricultural water use has significantly declined (between 1995 and 2009, for example, Cyprus' irrigated area and irrigation water use both declined by around

[16] A. Sofroniou and S. Bishop, 'Water scarcity in Cyprus: a review and call for integrated policy', *Water*, 6:10 (2014), 2901; P. Pyla and P. Phokaides, 'An island of dams: ethnic conflict and the contradictions of statehood in Cyprus', in F. Menga and E. Swyngedouw (eds.), *Water, Technology and the Nation-State* (Routledge, 2018), 115–30; WDD, 'Desalination plants', www.moa.gov.cy/moa/wdd/wdd.nsf/page23_en/page23_en?open document; WDD, 'Water balance', www.moa.gov.cy/moa/wdd/wdd.nsf/page10_en/page10_en?opendocument; WDD, *Dams of Cyprus* (2009); D. Eliades, 'Water shortages a thing of the past', *Cyprus Mail* (15/09/2010).

25 per cent). A state that was initially dominated by traditional landed interests is no more.[17]

On the other side of the Cyprus divide a separate process of state-led hydraulic development and territorialisation has unfolded, once again reflecting and structured by a broader political project. Following Turkey's 1974 invasion, northern Cyprus was in many respects turned into an extra-legal if nominally independent appendage of the Turkish state. An estimated 25,000 Anatolian peasants were settled and granted citizenship there, accompanied by a significantly larger number of temporary workers, army personnel and university students who were not granted citizenship. School curricula were transformed along Kemalist lines to highlight connections with the 'motherland'. And northern Cyprus became highly dependent on Turkey for trade, aid and subsidies (for example, over 60 per cent of the government's budget was typically covered by Ankara). This extreme dependency was further extended post-2005, in the wake of the failure of the Annan peace process and the rise of Turkey's Justice and Development Party. An electricity connection was laid to the mainland, large Seljuk mosques were built, Islamic youth groups were funded and more – all of this oriented to advancing control of northern Cyprus in accordance with an Islamist-nationalist agenda.[18]

The hydraulic corollary of this political project – indeed, its most striking emblem, labelled 'the project of the century' by Turkish President Recep Tayyip Erdogan – is a new floating transmarine water supply line linking northern Cyprus and Turkey. Inaugurated in 2015, this 107 km-long, 60-inch diameter pipeline, suspended 250 metres below the surface of the Mediterranean, conveys water from the Anamur River and Alaköprü Dam in Turkey to behind a newly heightened dam in Kyrenia (Girne), in northern Cyprus, from where its waters are distributed throughout the territory via new and upgraded supply networks (with additional infrastructure works still to come). With a theoretical capacity of 75 mcm/y – as compared to the territory's prior

---

[17] Panayiotopoulos, 'Cyprus', 26; C. Zoumides et al., 'Quantifying the poorly known role of groundwater in agriculture: the case of Cyprus', *Water Resources Management*, 27:7 (2013), 2501–14.

[18] M. Hatay, *Beyond Numbers: An Inquiry into the Political Integration of the Turkish Settlers in Northern Cyprus* (PRIO Cyprus Centre, 2005), viii; Y. Papadakis, 'The good, the bad and the ugly: Greek Cypriot and Turkish Cypriot schoolbooks on the "history of Cyprus"', in V. Lytra (ed.), *When Greeks and Turks Meet: Interdisciplinary Perspectives on the Relationship since 1923* (Routledge, 2014), 125–40; N. Kliot and Y. Mansfield, 'The political landscape of partition: the case of Cyprus', *Political Geography*, 16:6 (1997), 515; A. Dayioğlu et al., 'Turkish nationalism and the Cyprus question: change, continuity and implications for engagement with Northern Cyprus', *Ethnopolitics* 20 (2019), 450–466.

annual water consumption of 125 mcm/y – the pipeline has already transformed municipal water availability, if not yet agriculture. If used instead of groundwater, this water could significantly alleviate pressure on local aquifers (for, as in the south, groundwater within northern Cyprus is heavily depleted). More than this, though, the project was used by Turkey to implement a broader disciplining of northern Cyprus' water sector. Under terms agreed in 2016 and due to hold until 2050, water from the pipeline is to be distributed by a single operator, with all existing water infrastructure in northern Cyprus, together with all surface and groundwater rights, being transferred to it. In addition, a new water Turkish–northern Cypriot regulatory committee has been established, bypassing the existing water office of the local Ministry of Agriculture and Natural Resources and granting its Turkish representatives veto powers over the management of northern Cyprus' water sector. Whatever the benefits in terms of water supply, the pipeline and associated regulatory changes deepen northern Cyprus' incorporation into and dependency on Turkey and the island's hydro-political partition. Indeed, Turkish leaders have been explicit about this: 'here we declare to the world that Turkey and Cyprus have been intertwined together, and cannot be separated anymore', Prime Minister Ahmet Davutoğlu said at the project's inauguration.[19] As in both Israel and the RoC, what we witness here is the attempted hydraulic re-engineering of the space of the state.

River Development and Beyond in Syria and Sudan

Inherited and artificially created out of the ashes of the Ottoman Empire by France, Syria, like Cyprus, was not particularly valued by its new European rulers, being neither a source of valuable resources nor strategically significant – and having been taken above all for reasons of prestige (those Mandate territories that France did value were carved out to form the state of Lebanon). Syria had witnessed gradual and in some respects dramatic economic changes during the final decades of Ottoman rule, especially

---

[19] Presidency of the Republic of Turkey, 'May not only the lands of Cyprus but also the hope for peace benefit from this water coming from Anatolia!' (17/10/2015); G. Elkiran, 'Water resources management and trend of water use in North Cyprus', *Desalination and Water Treatment*, 177 (2020), 272–3; *The Intergovernmental Agreement between the Republic of Turkey and the Turkish Republic of Northern Cyprus Regarding the Delivery and Management of Water* (2016) (in Turkish); R. Bryant and M. Mason, *Water Technology and Sustainability in North Cyprus: Climate Change and the Turkey–North Cyprus Water Pipeline* (PRIO Cyprus Centre, 2017); C. Hoffmann, 'From small streams to pipe dreams: the hydro-engineering of the Cyprus conflict', *Mediterranean Politics*, 23: 2 (2018), 277–9; A. Conker and H. Hussein, 'Hydraulic mission at home, hydraulic mission abroad? Examining Turkey's regional "pax aquarum" and its limits', *Sustainability*, 11:1 (2019), 13.

thanks to the construction of railways and consequent expansion in trade. But, after 1920, investment and economic development stalled, despite French rhetoric to the contrary. Instead, Syria's political economy became dominated and held back by a small class of landowning notables, initially acting in collaboration with the colonial state, who siphoned off the bulk of the profits from local production and made scant investment in public works. Only from the late 1940s did Syria's agricultural sector start to witness a rapid expansion (but not intensification), such that in 1958, when Syria entered into a brief and unhappy union with Egypt, the country still had no large dams, irrigation on only around a tenth of its cultivated land, and very limited agricultural exports.[20]

However, the 1963 military coup which brought the Ba'ath Party to power changed all this. Led by radical young officers hailing mostly from rural areas, this coup – amounting to what Ray Hinnebusch labelled a 'green uprising' – resulted in Syria becoming committed to an essentially agrarian socialist political and economic programme. Henceforth and over the next three decades, the Syrian state pursued a high-modernist vision of rapid catch-up economic development through agrarian modernisation. State farms were established across the country, sitting alongside private land ownership. A state Higher Agricultural Council oversaw a compulsory cropping regime. Rural areas and the country's middle peasantry became central to the regime's political support base, especially through the establishment of a powerful Peasants Union. Regime rhetoric reflected this: 'I am first and last ... a peasant and the son of a peasant', proclaimed President Hafez al-Assad. Self-sufficiency in food staples became a key policy objective (partly for ideological reasons, but partly also owing to concerns about the strategic vulnerability of food imports within the context of the Cold War, Arab–Israeli conflict and, later, Iraqi sanctions regime).[21] Agriculture thus came to assume a prominence within Syria's political

---

[20] M. E. Yapp, *The Near East since the First World War* (Longmann, 1991), 86; R. Owen, *The Middle East in the World Economy 1800–1914*, rev. ed. (I.B. Tauris, 1993), 244–64; E. Williams, 'Contesting the colonial narrative's claim to progress: a nationalist's proposal for agrarian reform', *Review of Middle East Studies*, 44:2 (2010), 187–95; J. Barnes, 'Managing the waters of Ba'ath country: the politics of water scarcity in Syria', *Geopolitics*, 14:3 (2009), 524; S. M. Kaylani, The Role of Agriculture in the Economic Development of Syria, 1948–1962 (MA thesis, American University of Beirut, 1964), 53, 71.

[21] R. Hinnebusch, *Peasant and Bureaucracy in Ba'thist Syria: The Political Economy of Rural Development* (Westview Press, 1989), 19; Barnes, 'Managing the waters', 522–4; H. Batatu, *Syria's Peasantry, the Descendants of Its Lesser Rural Notables, and Their Politics* (Princeton University Press, 1999), 193; R. Springborg, 'Baathism in practice: agriculture, politics, and political culture in Syria and Iraq', *Middle Eastern Studies*, 17:2 (1981), 192; E. Woertz, *Oil for Food: The Global Food Crisis and the Middle East* (Oxford University Press, 2013), 103.

economy that was exceptional by regional standards. Whereas in Israel, and to a lesser degree Cyprus, agriculture was privileged for ethnopolitical reasons, despite it being of declining importance in strictly economic terms, in Syria the prioritisation of agriculture was at once politically and economically motivated.

State investment in agricultural and irrigation development reflected this. Agriculture received 20–25 per cent of all public investment. The prices of government-designated 'strategic crops' were held at well above international market value. Agricultural inputs, most importantly fuel and fertilisers, were generously subsidised: still in 2004, fuel subsidies amounted to 11 per cent of GDP. State farms had their losses wiped clean each year. Irrigation, specifically, received 60–70 per cent of the total annual agricultural budget. Low interest loans were provided for private well drilling. And major hydraulic projects were allocated extraordinary levels of state spending, with the Euphrates development project, most notably, accounting for 25 per cent of total government expenditure over a period of twenty years.[22]

As a result, having had no large dams prior to 1960, by the turn of the twenty-first century Syria had more than seventy, covering all of the country's major watercourses: the Euphrates and its tributaries; the Orontes, which flows from Lebanon through Syria and into Turkey; the Yarmouk River, the main tributary of the Jordan; and the major internal rivers, most notably the Barada which feeds Damascus. Foremost among these is the Tabqa Dam on the Euphrates, constructed with Soviet support during the early 1970s and to this day providing, through the 610 square km Lake Assad lying behind it, the clear majority of the country's surface water storage capacity. Linked to this and other reservoirs are extensive irrigation and water conveyance networks, including, from Lake Assad, a network of canals for irrigation on both sides of the Euphrates Valley, plus a large canal feeding the Balikh Valley to the east (which alone carries 570 mcm annually, more than Israel's National Water Carrier) and the main water supply source for the city of Aleppo. By the 1990s, all of the country's major surface water resources had been dammed, developed and transformed. In addition, from the late 1980s onwards Syria saw a startling expansion in groundwater development and use: whereas in 1988 the country had an estimated

---

[22] Barnes, 'Managing the waters', 524; Z. D. Cuyler, 'Fuel subsidy policy and popular mobilization in Syria', *MERIP* (16/03/2015); World Bank, *Syrian Arab Republic Irrigation Sector Report* (2001); C. Firilllo and J. Vercueil (eds.), *Syrian Agriculture at the Crossroads* (FAO, 2003); Gül et al., 'Economic analysis', 290; M. Ababsa, 'Agrarian counter-reform in Syria (2000–2010)' in R. Hinnebusch et al., *Agriculture and Reform in Syria* (St Andrews Centre for Syrian Studies, 2011), 88.

50,000 irrigation wells, within a decade this number had close to tripled.[23]

The knock-on social, economic and environmental consequences were profound. Surface and groundwater hydraulic development fed rapid increases in Syria's irrigated and cultivated areas, with hitherto open pastureland becoming farmed and farmland increasingly reaching into the marginal steppe and desert lands, the *badiya*. In turn, during the three decades from the late 1960s there was a steady expansion in Syrian agricultural production, and especially in the production of strategic crops, most notably cotton and wheat. Much of north-east Syria, in particular, became a land of wheat monoculture, such that by the mid-1990s it had become the country's breadbasket, and Syria a net wheat exporter. Moreover, thanks to the extensive subsidies, a high proportion of Syria's population was effectively maintained – even trapped – on the land, such that Syria came to have one of the lowest levels of urbanisation in the Middle East. Yet for all this, agriculture did not become a motor of economic development or diversification and did not prompt much technological or value upgrading. There was little investment in drip irrigation, for example, and little move towards increasing high-value agricultural production. In net terms, agriculture became more of a drain on state finances than a source and generator of capital.[24]

Moreover, as explored more fully in the next chapter, this agriculture-centred economic regime was mired in contradictions and soon destined to collapse. State investment in new hydraulic infrastructure declined from the late 1980s: the heyday of Syrian dam building was over. Groundwater abstraction rose rapidly in turn and aquifers were soon being gravely over-exploited, with groundwater levels plummeting across the country – the World Bank describing this as Syria's 'single most important water management challenge'. Major springs started to disappear and social conflicts between small and large landowners, and between rural and urban areas, became widespread. Adding to this mix, the liberalisation of Syria's economy following the death of Hafez al-Assad in 2000 removed many of the props, most importantly crop price controls and input subsidies, which till then had underpinned Syrian agriculture. As a result, from the early 2000s agricultural production started going into decline and Syria's rural population began moving en masse to the country's burgeoning cities. Despite possessing relatively

---

[23] FAO, Geo-Referenced Database on Dams; UN-ESCWA, *Inventory of Shared Water Resources in Western Asia* (2013), 88; Gül et al., 'Economic analysis', 290.

[24] Selby, 'Climate change and the Syrian civil war, part II'; World Bank, 'Urban population (% of total)', World Development Indicators DataBank; Hinnebusch, 'The Ba'ath's agrarian revolution (1963–2000)', in Hinnebusch, *Agriculture and Reform*, 14.

plentiful water supplies – in per capita terms Syria has much higher natural water availability than either Israel or Cyprus, as demonstrated in Chapter 2 – within the space of little more than thirty years a programme of rapid catch-up agrarian and hydraulic development had run into the ground.[25]

Sudan presents us with a different pattern again. Unlike Syria, or for that matter Israel and Cyprus, Sudan has a long history of large-scale hydraulic development, tied to the Nile and external interests in it, in particular. Thus, under the Turkiyya (the period of Turkish–Egyptian rule, from 1820 to 1881) water wheels and irrigation canals proliferated, especially along the Blue Nile, to support increased sugar cane and, later, cotton production, while under the Mahdiyya these infrastructures were extended as part of the Mahdi's centralising and expansionist state-building project. It was under Anglo-Egyptian rule, however, that the basic contours of the political ecology of modern Sudan were established. British colonial interests in Sudan were overwhelmingly economic, dominated by the facts that Egyptian cotton production, so crucial to the imperial economy, was dependent upon and limited by summer Nile flows and that upstream water development would be required if these flows were to be increased; Britain sought to secure, in the words of Lord Cromer, Consul-General of Egypt, 'the effective control of the waters of the Nile from the Equatorial Lakes to the sea'. Indeed, unlike each of the other cases examined in this book, a case can be made that the colonisation of Sudan was 'essentially hydrological' in motivation.[26]

British investments, made especially during the 1920s and 1930s, reflected this. The Jebel Aulia Dam on the White Nile, the largest dam in the world at the time, was constructed to regulate Nile flows and ensure water for cultivation in Egypt. The Gezira scheme – a 100,000-hectare irrigated area with 4,300 km of canals and ditches at the confluence of the Blue and White Niles, which was financed, developed and run by the London-listed Sudan Plantations Syndicate, and which was at the time the largest irrigation scheme in the world – simultaneously transformed Sudan into a major cotton exporter. The 3 km-wide Sennar Dam was constructed just upstream on the Blue Nile, to provide the Gezira with constant water supplies. And various swamp clearance and canalisation

---

[25] FAO, Geo-Referenced Database on Dams; World Bank, *Syrian Arab Republic Irrigation*, 18; F. De Châtel, 'Waterless Wadi Barada', *Middle East Report*, 271 (2014); Selby, 'Climate change and the Syrian civil war, part II'.

[26] P. M. Holt, 'Modernization and reaction in the nineteenth century Sudan', in W. Polk and R. Chambers (eds.), *Beginnings of Modernization in the Middle East: The Nineteenth Century* (University of Chicago Press, 1968), 410–12; E. Baring, Earl of Cromer, *Modern Egypt*, Vol. 2 (Macmillan, 1908), 110; Tvedt, 'Hydrology and empire', 174.

works were also undertaken. Together, these hydraulic investments not only integrated the Nile valley, improving transport and communication between Khartoum and the south. More importantly, by regulating Nile flows and the spring flood in particular, they enabled a wholesale switch from inundation to perennial irrigation and a concomitant intensification of agricultural production. Sudan henceforth became a cash crop and, especially, cotton export economy, with agricultural exports that accounted for the overwhelming majority of foreign earnings and government revenues. At the same time, Sudan became heavily indebted – a pattern that began with its Gezira loans and cotton production subsidies (a 1931 government budget note described the Gezira scheme as an economic 'liability') and continues to this day. Moreover, Sudan also became much less agriculturally self-sufficient than previously: where the Gezira region, for example, had formerly been 'the subsistence granary of the northern Sudan', under Britain the region was transformed to service the mills of Lancashire, leaving Sudan's population acutely dependent on the vagaries of international markets.[27]

Economically motivated as they were, these hydraulic developments were geographically highly concentrated, being almost entirely located in the environs of Khartoum and Omdurman on the Nile (as briefly discussed in the previous chapter). This is not to suggest, of course, that they were politically inconsequential or of little bearing for Sudanese politics or society. To the contrary, the Gezira and associated works became the backbone of Sudan's agricultural export-oriented cash-crop economy and, in turn, its class structures and Nile Valley–focused form of state. Sudan became dominated by two main classes: its powerful riverine merchant class, which had emerged on the back of the Turkiyya-era slave trade and now became the key intermediary in the country's import-export trade, and an expanding class of large landowners, who had been granted property and cultivation licences by the colonial authorities. Representatives of these classes accounted for nearly all of the positions in the Advisory Council and subsequent Legislative Assembly established under British rule and dominated Sudan's first post-independence government.[28]

---

[27] R. O. Collins, *The Waters of the Nile: Hydropolitics and the Jonglei Canal, 1900–1988* (Oxford University Press, 1990), 78–88; T. Barnett, *The Gezira Scheme: An Illusion of Development* (Frank Cass, 1977); R. H. B. Condie, 'Cotton exports and economic development in the Sudan', *Sudan Notes and Records*, 37 (1956), 70–8; C. Warren, *The Agricultural Economy of the Sudan* (USDA, 1962), 1; S. Mollan, 'Business, state and economy: cotton and the Anglo-Egyptian Sudan, 1919–1939, *African Economic History*, 36 (2008), 114; A. Gaitskell, *Gezira: A Story of Development in the Sudan* (Faber and Faber, 1959), 26.

[28] Niblock, *Class and Power*, 51–81, 94–6; A. S. Sidahmed and A. Sidahmed, *Sudan* (RoutledgeCurzon, 2005), 20, 123–6.

Britain's control, during the first half of the twentieth century, of nearly the whole of the Nile River basin (Ethiopia being the important exception) enabled it to pursue integrated river basin development – the one instance of this among our assorted cases. Thus, Britain's hydraulic and irrigation projects in Sudan were accompanied by others in Uganda and Egypt, as well as by plans for drainage of the Sudd and for further dams upstream, all oriented to optimising the overall storage capacity and productive potential of Nile flows. Following the 1952 Free Officers' coup in Egypt, and Sudanese independence in 1956, however, this integrated development model was dropped, with Egypt in particular favouring a much more state-based and internalist model of hydraulic development. The 1959 Nile Waters Agreement between Egypt and Sudan was both a consequence and an enabler of this – being inspired principally by Egypt's interest in constructing the Aswan High Dam (Sudan's consent for this was necessary simply because Lake Nasser, formed by the Dam, stretches into Sudanese territory) and simultaneously paving the way, through the allocation of a healthy Sudanese quota for Nile withdrawals, for Sudan's own independent water development.[29]

Thus followed, after independence, a series of major water and agricultural development initiatives within Sudan. During the 1960s these focused on infrastructure, most notably the Roseires Dam on the Blue Nile (which, in turn, allowed a twofold expansion of the Gezira scheme) and the Khashm el-Girba Dam on the Atbara River further east, constructed with World Bank and West German funding. By contrast, the 'Breadbasket Strategy' of the 1970s, as promoted by President al-Nimeiri, sought to utilise Gulf investment, international aid and extensive state subsidies to pursue a rapid expansion of mechanised rain-fed agricultural production, and through that to transform Sudan into the breadbasket of Africa and the Middle East. The immediate consequence of this strategy was that, during the late 1970s alone, Sudan's cultivated area expanded by almost a quarter. For all this, however, the country's agricultural exports declined in both volume and value – partly because of the declining value of primary commodities on international markets; partly owing to the 1970s oil shocks (Sudan's oil bill increased by 325 per cent between 1976–7 and 1980–1 alone, forcing it to reduce the oil subsidies which had underpinned agricultural expansion); partly because government investment had been diverted away from established irrigation schemes; partly because of Sudan's growing mountain of foreign debt;

---

[29] H. E. Hurst et al., *The Nile Basin, Vol. VII: The Future Conservation of the Nile* (Government Printing Office, 1946); UAR and Sudan, *Agreement*.

and, to cap it all, because of the extreme drought of the early 1980s.[30] We return to the Breadbasket Strategy in the next chapter, but suffice here to say that it was a monumental failure.

A second major phase of hydraulic development began in the early 2000s, as the Islamist-military regime of Omar al-Bashir sought to reinvent itself, both internationally and domestically, following the 9/11 attacks on New York and Washington and its years as an international pariah. Modernisation through dam building and agricultural expansion became a key plank of the regime's policy agenda. A new Dams Implementation Unit (DIU) was established, reporting directly to the president and his High Political Committee. Dams became a key focus of investment: the DIU became the largest recipient of oil revenues in the country after the army and intelligence service, the 9 km-long Merowe Dam in northern Sudan receiving 40 per cent of all official national investment during 2005–8. Foreign capital also poured in, with China, Saudi Arabia, Kuwait, Qatar and the United Arab Emirates all providing funding for the Merowe Dam and for the heightening of the Roseires Dam above the Gezira. A complex of large dams was also constructed on the Upper Atbara, with a series of further dams planned elsewhere. In addition, China along with Egypt, Jordan, Saudi Arabia and the Gulf states invested heavily in land leases and associated irrigation schemes – among them, a fifty-year lease to Saudi Arabia of over 800,000 hectares in Red Sea state; a ninety-nine-year lease to Saudi Arabia of over 400,000 hectares in the Upper Atbara and Setit Dam area; a lease to Kuwait of 400,000 hectares for agricultural and livestock production; and a lease to China of land in the El-Rahad agricultural project between Gedaref and Gezira states. Linked to these and other investments, an Agricultural Revival Programme was launched which, among other things, resulted in a huge surge in cotton production. And simultaneously, electricity production and supply also rose rapidly. Between 2000 and 2016, hydro-electricity production in Sudan increased sixfold, so that it came to account for more than half of the country's total electricity production. And the proportion of Sudanese households with electricity rose from 23 per cent in 2000 to 45 per cent in 2014 (though coverage remains highly uneven, ranging from 80 per cent or more in Khartoum and neighbouring states, to less than

---

[30] Collins, *Waters of the Nile*, 247–75; R. P. C. Brown, *Public Debt and Private Wealth: Debt, Capital Flight and the IMF in Sudan* (MacMillan, 1992), 111–23; I. Elnur, *Contested Sudan: The Political Economy of War and Reconstruction* (Routledge, 2009), 51–8; P. M. Holt and M. W. Daly, *A History of the Sudan: From the Coming of Islam to the Present Day*, 6th ed. (Pearson, 2011), 139–40; B. C. D'Silva, *Sudan: Policy Reforms and Prospects for Agricultural Recovery after the Drought* (USDA, 1985), 26.

20 per cent in Darfur and Kordofan).[31] Unlike the case of Syria, in Sudan hydraulic development is still very much in its expansionary phase. Yet these expansionary hydraulic developments have nonetheless still had their contradictions, as will be explored below.

What should we draw from this brief survey of the main contours of hydraulic development in Israel, Cyprus, Syria and Sudan? For now, we suggest one main thing: that however much our modern apparatuses of dams, pipelines, irrigation channels and so on are, at the most general level, consequences of capitalist imperatives – particularly the requirement that abstract space and time be organised to facilitate ever-rising production and consumption – they are also always shaped by and subordinate to specifically political factors, above all those associated with the state. Across each of the four cases above, the modern hydro-political infrastructures that have been created reflect and embody specific political or political–economic projects, from the Zionist project of colonising land by farming it, through to the Ba'athist vision of agrarian socialist modernisation. In each case, moreover, these new systems are specifically post-colonial, reflecting the interests and agendas of post-colonial states (sometimes departing from and sometimes continuing agendas inherited from their erstwhile colonial masters). And in each case, the new apparatuses are, in one way or another, organised around logics of territorial space. Within both research and practice, the river basin is typically represented as the central geographical reference point for integrated water management and development (sometimes being viewed as a 'natural' unit, and elsewhere as a socially constructed one).[32] Yet if the above is any guide, this is not so – since in none of our cases has the

---

[31] H. Verhoeven, *Water, Civilisation and Power in Sudan: The Political Economy of Military-Islamist State Building* (Cambridge University Press, 2015), 130, 143–4; H. Verhoeven, 'African dam building as extraversion: the case of Sudan's dam programme, Nubian resistance, and the Saudi-Iranian proxy war in Yemen', *African Affairs*, 115:460 (2016), 562–73; K. A. El Amin, 'The state, land and conflicts in the Sudan', *International Journal of Peace and Conflict Studies*, 3:1 (2016), 7–18; 'Bill allows Saudi Arabia to cultivate Sudan lands', *Radio Dabanga* (19/07/2016); 'Sudanese farmers call for transparency on China land lease', *Radio Dabanga* (31/01/2018); C. Henderson, 'Gulf capital and Egypt's corporate food system: a region in the third food regime', *Review of African Political Economy*, 46:162 (2019), 599–614; Republic of Sudan, *Executive Programme for Agricultural Revival* (2008); USDA Foreign Agricultural Service, 'Production, supply and distribution dataset', https://apps.fas.usda.gov/psdonline/app/index.html#/app/adv Query; US Energy Information Administration, 'International electricity data', www.eia.gov/international/data/world; Federal Ministry of Health et al., *Multiple Indicator Cluster Survey, 2000 Sudan Final Report* (2000), 39; CBS and UNICEF Sudan, *Sudan Multiple Indicator Cluster Survey 2014: Final Report* (2015), 16.

[32] C. J. Barrow, 'River basin development planning and management: a critical review', *World Development*, 26:1 (1998), 171–86; F. Molle, 'River basin planning and management: the social life of a concept', *Geoforum*, 40:3 (2009), 484–94.

unitary river basin been the primary scalar object or target of modern hydraulic development. Just as nature does not preordain contemporary political boundaries, so neither does it determine the space, scale or form of modern hydraulics.

The Five Contradictions of Hydraulic Development

Now, one corollary of the centrality of politics within modern hydraulic development – of the fact that hydraulic development is always 'techno-political' in form – is that, whatever its benefits, it is always also fraught, contested and generative of conflicts and insecurities. We may summarise these alliteratively as the five contradictions of hydraulic development: dislocation, destruction, discrimination, danger and dependency.

To start with, hydraulic development always involves spatial dislocation. It involves the dislocation, most obviously, of water, as surface flows are channelled, diverted and drained, as valleys are flooded and as groundwaters are mined. It always involves a form of appropriation, of resource capture, as water is relocated from some spaces and some uses to others. It also, as a result, regularly involves the forced displacement of livelihoods and communities, especially of those racialised as irrelevant, or worse as obstacles, to logics of state-building and development. Even when there is no direct or immediate displacement, hydraulic development can easily be read as a threat of displacement to come. And violent conflicts, or threats thereof, often follow.

One can find illustrations of this at multiple scales across our cases. In Mandate Palestine, Arab communities were routinely uprooted during the course of Zionist land reclamation schemes. In 1921, for example, the Palestine Jewish Colonization Association was granted a concession for reclamation of a 4,000-hectare expanse of the Kabbara region, the largest swamp on Palestine's coastal plain. The British authorities recognised that the scheme would lead to the displacement of the region's native Arab population, but considered this acceptable given that the project would improve the land and be in what they considered the general long-term interest. The upshot was that at least 840 villagers were displaced and removed to a nearby hilltop, losing their wetland livelihoods in the process (but not before being hired to undertake the drainage work themselves). The resulting Arab settlement of Jisr al-Zarqa remains to this day one of Israel's poorest and most neglected communities.[33]

---

[33] Broich, 'British water policy', 272–4; M. Rapoport, 'A classic Zionist story', *Ha'aretz* (10/06/2010).

On a much larger scale, in the late 1970s Sudan's Nimeiri regime put into action long dormant plans for drainage of the vast Sudd swamps in southern Sudan through excavation of a 360 km-long canal, the Jonglei Canal. The objective of the project was resource capture sold as modernisation: the canalisation of the White Nile during its course through the Sudd in the hope of reducing evapo-transpiration 'losses' and increasing Nile flow by up to 7 bcm/y, with this volume to be shared between Egypt and Sudan including in support of the latter's Breadbasket Strategy. The project immediately disrupted cattle migration routes and livelihoods, causing local resentment and protests. Moreover, its longer-term consequences would have been still more profound: 1.5 million hectares of land were set to be reclaimed from the Sudd and turned over to cash crop production, while an estimated 450,000 people, mostly Dinka, Shilluk and Nuer, were to be either displaced or see their livelihoods destroyed. Not all southerners, it should be said, were wholly opposed to the canal: the SPLM/A leader-to-be John Garang, for instance, initially praised it in his doctoral thesis for its potential to open 'up hitherto dormant opportunities for socio-economic development' and through that promote national and regional integration. However, the project became widely mistrusted, viewed as a means of turning the Sudd over to mechanised farming controlled by northern elites and facilitating northern troop movements – as well as, in John Garang's telling formulation, 'a hole that sucks water all the way to the north'. The result was that canal installations and the "Bucketwheel" digger being used to excavate them were among the SPLM/A's earliest targets after the outbreak of the second civil war in 1983, with the project coming to a halt shortly afterwards.[34]

Further north on the Nile, as well as in Syria, mega-dam construction has repeatedly led to mass displacement. Construction of the Roseires Dam in the 1960s led to the displacement of around 80,000 people, while another 110,000 were displaced during its recent heightening. The filling of Lake Nasser behind Egypt's Aswan Dam involved the displacement of

---

[34] P. P. Howell, 'The impact of the Jonglei Canal in the Sudan', *Geographical Journal*, 149:3 (1983), 291; Collins, *Waters of the Nile*, 311–17, 387–401; A. M. Ahmad, 'Post-Jonglei planning in southern Sudan: combining environment with development', *Environment and Urbanization*, 20:2 (2008), 578; G. T. Lako, 'The impact of the Jonglei scheme on the economy of the Dinka', *African Affairs*, 84:334 (1985), 19; M. Suliman, 'Civil war in the Sudan: from ethnic to ecological conflict', *Ecologist*, 23:3 (1993), 107; J. Garang de Mabior, Identifying, Selecting, and Implementing Rural Development Strategies for Socio-Economic Development in the Jonglei Projects Area, Southern Region, Sudan (PhD thesis, Iowa State University, 1981), 219–20; M. Suliman, 'Civil war in Sudan: the impact of ecological degradation', *Contributions in Black Studies*, 15 (1997), 99–121; J. Garang, *The Genius of Dr. John Garang: Speeches on the Comprehensive Peace Agreement*, ed. P. Wël (Paanda Publishers, 2013), 77.

around 70,000 Nubian Sudanese (plus similar numbers within Egypt). The Merowe reservoir submerged 900 villages and displaced 60,000. Construction of the Upper Atbara and Setit Dam Complex led to the displacement of around 30,000 families. The filling of Lake Assad in Syria displaced 60,000 people. And planned dams are likely to have similar consequences: the proposed Kajbar Dam on the upper Nile, for instance, is predicted to submerge up to nearly 100 villages and displace more than 10,000 people.[35]

Brute displacement numbers aside, mega-dams also typically have a range of further consequences. Their victims are, to a disproportionate degree, poor and marginalised populations and 'tribes' considered obstacles to History. In Sudan, for instance, dam building has particularly affected Nubian communities, to the extent that it has often been viewed as a form of demographic engineering and Arabisation. Consultation, resettlement conditions and compensation have frequently been dire, and local political mobilisation and protests, together with state repression, have often followed. In 1960, for example, protests against the decision to resettle Nubian communities displaced by the filling of Lake Nasser in a government agricultural development project 800 miles away resulted in the Wadi Halfa area being placed under martial law. Within the context of widespread disaffection with the country's military government, demonstrations quickly spread to Khartoum and other parts of the country, where Nubian protestors were joined by students, intellectuals and senior civil servants who together formed a coalition calling for an end to military rule; these events have often been credited as key precursors to the 1964 revolution that toppled the Abboud regime. Moreover, Nubian protests and demands for compensation have continued through to the 2000s, with thousands of families yet to be compensated for the loss of their homes in the 1960s.[36]

---

[35] M. H. S. Ebrahim, 'Irrigation projects in Sudan: the promise and the reality', *Journal of African Studies*, 10:1 (1983), 9; M. T. Alrajoula et al., 'Hydrological, socio-economic and reservoir alterations of Er Roseires Dam in Sudan', *Science of the Total Environment*, 566 (2016), 945; H. Dafalla, *The Nubian Exodus* (Hurst, 1975), xvii; G. M. Sørbø, *Tenants and Nomads in Eastern Sudan: A Study of Economic Adaptations in the New Halfa Scheme* (Nordiska Afrikainstitutet, 1985), 12, 59; A. Dirar et al., *Displacement and Resistance Induced by the Merowe Dam: The Influence of International Norms and Justice* (University of East Anglia, 2015), 1; 'People displaced by Sudanese dam project complain of poverty, unemployment', *Radio Dabanga* (01/02/2017); G. Meyer, 'Rural development and migration in northern Syria', in M. Salem-Murdock and M. M. Horowitz (eds.), *Anthropology and Development in North Africa and the Middle East* (Westview, 1990), 249; 'Nubians demonstrate against northern Sudan dams', *Radio Dabanga* (15/02/2016).

[36] M. J. Hashim, 'The dams of Northern Sudan and the policy of demographic engineering', *International Journal of African Renaissance Studies*, 5:1 (2010), 148–60; Y. Oron et al., 'The Republic of Sudan', *Middle East Record*, 1 (1960), 416–17;

Similarly, an array of indigenous rights, women's rights and dam-specific groups and committees – the Leadership Office of the Hamadab Affected People, the Women's Committee of the Manasir Affected People, the Amri Dam Affected Peoples Committee, the Resist Kajbar Dam Committee, Rescue Nubia and more – have mobilised to contest the regime's post-2000 wave of dam building, with protests that have spread from the dam sites themselves to Khartoum, Washington, London and the UN in New York. The Sudanese state, in turn, has repeatedly responded with mass arrests, arbitrary detentions and violence. During one incident in 2006, for example, three people were killed and more than fifty injured when a community meeting was attacked by militia associated with the Merowe Dam Authority. During another incident in 2007, four were killed and at least twenty injured during protests against the construction of the Kajbar Dam. As the Committee of Anti Dal-Kajbar Dam observed of this dam-building, '[n]o consultation has been made with the affected people; no heed is given to their just demands; and the reaction of the government of Khartoum has been ruthless to any manifestation of public protest and rejection'. The pattern here will be sadly all too recognisable to anyone familiar with the 2000 report of the World Commission on Dams or any of the numerous anti-dam movements worldwide. Indeed, it is conservatively estimated that, globally, 30–60 million people have been displaced by mega-dam construction over the past century, the majority of them forcibly removed.[37]

Dam building and diversions have also often generated significant international tensions. Rival Israeli and Arab League schemes for the diversion of the upper Jordan River sparked three years of intermittent military hostilities

---

Y. F. Hasan, 'The Sudanese revolution of October 1964', *Journal of Modern African Studies*, 5:4 (1967), 491–509; M. Janmyr, 'Nubians in contemporary Egypt: mobilizing return to ancestral lands', *Middle East Critique*, 25:2 (2016), 127–46; N. Magdi, '11.7K Nubians entitled to post-Aswan Dam construction compensations', *Egypt Today* (01/07/2019).

[37] T. Schmidinger, 'Spatial control, "modernization" and assimilation: large dams in Nubia and the Arabization of northern Sudan', in A. Fischer-Tahir and S. Wagenhofer (eds.), *Disciplinary Spaces: Spatial Control, Forced Assimilation and Narratives of Progress since the 19th Century* (Transcript, 2017), 165–86; M. Zeitoun et al., 'A "justice" reading of the trans-national struggle of the people displaced by the Merowe Dam', *Local Environment*, 24:2 (2019), 129–45; N. Hildyard, 'Neutral? Against what? Bystanders and human rights abuses: the case of Merowe Dam', *Sudan Studies*, 37 (2008), 6; 'Dozens detained, wounded at Sudanese dam protest', *Radio Dabanga* (18/02/2016); 'Nubians demonstrate against northern Sudan dams'; Committee of Anti Dal-Kajbar Dams, 'The Sudanese government plan of demographic engineering of Nubia and the Chinese and Egyptian connection to it: a letter of protest and resistance' (12/01/2011), 1–2, 17; World Commission on Dams, *Dams and Development: A New Framework for Decision-Making* (Earthscan, 2000); P. McCully, *Silenced Rivers: The Ecology and Politics of Large Dams*, 2nd ed. (Zed, 2001), 7–8.

during the mid-1960s, even though as discussed in Chapter 2, this 'Jordan waters crisis' did not spark war. Dam building on the Euphrates has repeatedly contributed to, or at least legitimised, political–military tensions: Saddam Hussein sent troops to Iraq's border with Syria during the filling of Lake Assad in 1975; Turkey stopped the Euphrates' flow entirely during January–February 1990 to enable filling of the Atatürk Dam reservoir, this in turn leading to water and electricity shortages in Syria; and riparian protests and recriminations accompanied both this and the construction of Turkey's Birecik Dam in 1996. It was a stated need to generate revenues for construction of the Aswan High Dam that led Nasser to nationalise the Suez Canal Company in mid-1956, precipitating the Israeli–British–French invasion later that year. And the rhetoric which has emanated from Egypt since Ethiopia started building the GERD is well-known. More broadly, quantitative research has consistently found hydraulic development projects to increase inter-state political tension and conflict.[38] Within mainstream environmental and water security discourse, such tensions and conflicts are typically pinned on natural scarcities, as discussed in Chapter 2. In truth, however, they typically result from three things: competition between rival state-led agrarian development ambitions; projects of resource capture in support of these ambitions; and, most directly, the construction of large-scale hydraulic infrastructures. Like the instances of local displacement and resistance discussed above, such tensions are consequences not of scarcity but hydraulic state-building and development.

Recent hydro-political developments in northern Cyprus demonstrate this more clearly still. As noted earlier, under Turkey's terms for 'the project of the century' all existing water infrastructure as well as surface and groundwater rights in northern Cyprus were to be transferred to a single Turkish distributor. These terms were widely opposed within northern Cyprus, most notably by municipalities – who instead proposed a publicly owned distribution system, water charges being their largest source of income – and not least by the largest party in northern Cyprus' then coalition government, the Republican Turkish Party. The terms were also, it might be added, contrary to Turkey's own water laws, which prohibit monopoly distribution of water supplies. The dispute over the agreement, and by extension over relations with Ankara, resulted

---

[38] Lowi, *Water and Power*, 115–44, 58–9; A. Çarkoğlu and M. Eder, 'Domestic concerns and the water conflict over the Euphrates–Tigris river basin', *Middle Eastern Studies*, 37:1 (2001), 41–71; D. B. Kunz, *The Economic Diplomacy of the Suez Crisis* (University of North Carolina Press, 1991); Yoffe et al., 'Conflict and cooperation'; B. K. Sovacool and G. Walter, 'Internationalizing the political economy of hydroelectricity: security, development and sustainability in hydropower states', *Review of International Political Economy*, 26:1 (2019), 46–79.

in the collapse of the then government in 2016, and the formation of a new right-wing coalition which aligned the northern Cypriot leadership more closely with Turkey.[39] As this case illustrates, water transfer projects can generate conflict even when promising unprecedented abundance.

Besides causing multiple forms of spatial dislocation – the relocation, capture and transfer of water resources, the displacement of communities and livelihoods, and the struggles and conflicts which inevitably ensue – hydraulic development also, second, typically involves straightforward destruction, hydraulic projects being, just like capitalism in general, productive and destructive in equal measure. Foremost here is the widespread destruction of natural waterscapes and processes, and with them habitats and biodiversity: the drying up of natural watercourses and the disappearance of springs; the startling loss of wetland ecosystems, as marshes are drained and rivers straightened; the impacts of dams on fish migrations and on natural processes of siltation; and much else besides. Paralleling this, moreover, hydraulic projects have often led to or involved the destruction of long-set communities and patterns of livelihood. Sudan's Gezira scheme, for instance, was established via a 1921 Land Ordinance which effectively destroyed existing communal property relations, turning communal land into state property and in the process transforming over a 100,000 independent farmers and pastoralists into tenant farmers, sharecroppers and wage labourers; protests and rent strikes inevitably ensued. Equally, recent dam building in Sudan has been critiqued not just for causing displacement, but as cultural destruction. 'The Nubians like many other indigenous people in Sudan are facing a war of cultural cleansing and systematic marginalization, forcible relocations and deprivation by the Islamist government which is hostile to the Nubian culture and heritage', says the Rescue Nubia and Resist Kajbar Dam Committee. Or, in the words of Suad Ibrahim Ahmed, the late journalist and leader of the Nubian Alliance:

Allowing our land to be taken means the extinction of our distinct language and culture forever.... We told the government that we want development. We say 'yes to development, but no to the dam'. Any development that displaces us, destroys our language and our culture is rejected, and we prefer to live without development if that is the case.[40]

---

[39] Bryant and Mason, *Water Technology*, 18; A. Sözen and S. Sonan, 'Yet another case of electoral and government epidemic? The Turkish Cypriot legislative election of January 2018', *South European Society and Politics*, 24:1 (2019), 139–40; M. E. Bariş and A. A. Karadag, 'Water resources management issues in Turkey and recommendations', *Journal of Applied Sciences*, 7:24 (2007), 3903.

[40] V. Bernal, 'Colonial moral economy and the discipline of development: the Gezira scheme and "modern" Sudan', *Cultural Anthropology*, 12:4 (1997), 447–79; Mollan, 'Business, state and economy'; Rescue Nubia and Resist Kajbar Dam Committee,

Implicit here, third, is that hydraulic development is also always discriminatory, involving a valuing of certain lives and livelihoods above others, and enabling – or at least concretising – forms of differentiation that would not have been possible without it. Thus, mega-dams have disproportionately displaced and destroyed marginal and 'tribal' communities, as discussed. The irrigation schemes and agricultural development projects linked to them are typically designed with elite and corporate agro-industrial interests in mind, not those of small farmers or sharecroppers. Hydraulics involves a privileging of abstract knowledges over those tacit and practical ones through which waters were traditionally managed and understood. Hydraulic development is also an overwhelmingly gendered pursuit, empowering male-dominated organisations, disciplines and networks over and above female access or traditional female-led water management practices – as represented by the hyper-masculinity of the mega-dam and the two figures who typically accompany it, the male engineer and the usually male head of state. And last, hydraulic infrastructures can, like all infrastructural systems, be designed to disproportionately include and benefit some while disproportionately excluding and marginalising others. Between 1962 and 1969, Israel's predominantly Arab district of Nazareth had an average total water consumption of just 4 mcm/y, whereas the neighbouring Jewish district of Afula, with a smaller population, was consuming 285 mcm/y; and still today, Israel's 'non-Jewish' localities, with around 20 per cent of the country's population, use just 1.7 per cent of its agricultural water – in large part because of an ethnically inflected national water supply network and the legal, administrative and subsidy regimes surrounding it (and we will encounter further evidence to this effect in the next chapter).[41]

Two final contradictions can be summarised more briefly. So, fourth, the creation of modern hydraulic systems has generated wholly new sets of dangers, techno-scientifically produced hazards and unintended side effects which, as Ulrich Beck famously theorised them, are the signature tunes of the modern 'risk society'. Thus from the Nile valley to Cyprus and Palestine, colonial era hydraulic developments led to the proliferation of new diseases and health conditions, from the spread of more virile strains of malaria feeding off newly created bodies of water, to endemic

'Urgent press release: Kajbar massacre', *SudaneseOnline* (14/06/2007); N. Bol, 'Yes to development, no to a new dam, Nubians say', *IPS News* (08/01/1998).

[41] Mitchell, *Rule of Experts*, ch. 1; L. M. Harris, 'Water rich, resource poor: intersections of gender, poverty, and vulnerability in newly irrigated areas of southeastern Turkey', *World Development*, 36:12 (2008), 2643–62; Israeli Water Commission, *Water in Israel, Consumption and Production 1962–1968* (1970), 419 (in Hebrew); Israel CBS, 'Consumption of water in agriculture, by type of locality', *Statistical Abstract of Israel 2019* (2019).

parasite infections among agricultural labourers on perennial irrigation schemes. The widespread use of desalinated water within Israel has been linked to increased risks of heart disease and, when used in agriculture, crop defects. Integrated water supply systems and the reuse and recycling of river water have become facilitators of the circulation of modern agricultural, industrial and consumer waste, from pesticides, nitrates and hormones to heavy metals and micro-plastics. And, especially if not properly built or maintained, mega-dams have the potential to visit catastrophic destruction on lands and populations downstream. As these few examples illustrate, and as Mitchell demonstrates so powerfully in *Rule of Experts*, techno-political development is not an upward, linear process, but constantly generative of new problems, dangers and risks.[42]

Last, within modern economies the end-users of hydraulic systems come to expect the continuous, regulated delivery of water supplies, which in turn generates new forms and degrees of dependency – on water supply infrastructures, on the energy required to power them and on the technical expertise needed to manage them – as well as new vulnerabilities arising from risks of infrastructural disruption. We provide extensive evidence of such disruption in later chapters, especially that associated with war (Chapter 7). Suffice to say here that such disruption can arise for any number of reasons: out of intentional or accidental destruction, from environmental shocks, from sudden reductions in energy availability or access, or for technical reasons. In spring 2020, to give just one example, the Turkey–northern Cyprus water conveyance line snapped, leading to deepening water shortages in northern Cyprus until the supply was restored ten months later.[43] Modern hydraulic infrastructures are simultaneously vectors of resilience, dependency and insecurity.

Resurgent Hydraulic Politics

This chapter has sought to examine, however schematically, some of the wide-ranging consequences of modern hydraulic development: for water

---

[42] U. Beck, *Risk Society: Towards a New Modernity*, trans. M. Ritter (Sage, 1992 [1986]); J. L. Derr, *The Lived Nile: Environment, Disease, and Material Colonial Economy in Egypt* (Stanford University Press, 2019); Christodoulou, *Evolution of the Rural Land Use*, 114–15; S. Sufian, *Healing the Land and the Nation: Malaria and the Zionist Project in Palestine, 1920–1947* (University of Chicago Press, 2007); M. Shlezinger et al., 'Association between exposure to desalinated sea water and ischemic heart disease, diabetes mellitus and colorectal cancer; a population based study in Israel', *Environmental Research*, 166 (2018), 620–7; U. Yermiyahu et al., 'Rethinking desalinated water quality and agriculture', *Science*, 318 (2007), 920–1; O. Smith, 'Big dams, big damage: the growing risk of failure', *New Security Beat* (21/08/2018); Mitchell, *Rule of Experts*, ch. 1.
[43] P. Michael, 'Water supply from Turkey finally restarts in the north', *Cyprus Mail* (05/10/2020).

availability and use, for time and space, for nature, populations, economies and states, and in revolutionising patterns of water (in)security. Its central arguments, thus far, have been twofold. It has sought to show, first, that while modern mega-dams, tubewells, conveyance lines, irrigation systems and so on have been universally transformative, this has been in sharply different ways in different geographical contexts, shaped, among other things, by the particular interests and agendas of particular states. Such political factors are, we suggest, key to understanding contemporary patterns of hydraulic development and the radical reconfigurations of ecology and society that have followed. And, second, the chapter has sought to show that for all the positives associated with hydraulic development – provision of treated water to billions, support for increased food production and more – it nonetheless has a series of dark sides. These five contradictions – dislocation, destruction, discrimination, danger, dependency – are, we suggest, commonplace, continuing and unavoidable corollaries of modern hydraulics. Hydraulic state-building and development are, in this sense, productive of both water security and profound water-related insecurities and conflicts.

Lest this appears too obvious, it is worth emphasising that contemporary liberal discourse, including on environmental security, typically portrays the development–state–security relationship quite differently. Within mainstream development policy, for example, economic growth, strong institutions and security are widely assumed to be mutually reinforcing, establishing a virtuous circle of progress away from poverty and conflict – with the exact reverse being assumed to apply to the links between low development, state fragility and insecurity. In environmental and climate security discourse, likewise, 'constrained economic productivity', 'disrupted institutions', 'state weakness' and the like are widely identified as the key intervening variables between resource scarcity and conflict, with even the most sensitive of studies suggesting that it is 'when states contract' that violence over the environment becomes 'more likely'.[44] Yet this is profoundly misleading. Far from being antitheses of hydraulic development and state-building, conflict and insecurity are among their dark underbellies – nasty secrets that liberal, as well as nationalist, merchants of progress would rather ignore.

Indeed, the ideology of development is so all-pervasive that its dark sides are routinely obscured and ignored, including in relation to the

---

[44] R. MacGinty and A. Williams, *Conflict and Development*, 2nd ed. (Routledge, 2016); J. Selby and M. Tadros, 'Introduction: eight myths of conflict and development in the Middle East', *IDS Bulletin*, 47:3 (2016), 1–18; Homer-Dixon, *Environment, Scarcity and Violence*, 81–103; J. Barnett and W. N. Adger, 'Climate change, human security and violent conflict', *Political Geography*, 26:6 (2007), 647.

environment. Thus, as mentioned above, international water-related tensions are regularly pinned on 'scarcity' rather than on their much more direct causes, namely, dam building, water diversions and projects of agrarian development. Drops in river flow and groundwater levels are often attributed to climate change, even when, as in the cases of Syria, Lake Chad and the Jordan River, these changes are clearly above all a function of state-led agricultural expansion and intensification. Meanwhile, talk of 'climate refugees' and 'environmental migration' proliferates without any equivalent discussion of 'development refugees' – despite there being copious evidence of development directly causing mass displacement, among other things, via dam building.[45]

The implications of all this for an era of deepening global climate change are not difficult to envisage. The need for a rapid transition away from fossil fuels will inevitably generate pressures for increased hydro-electric power production and dam construction. Concerns about changing precipitation patterns and increased climatic variability, especially floods and droughts, will tend in the same direction, generating calls for all manner of adaptation measures – more dams, more irrigated agriculture, and improved and more extensive water supply infrastructures. In addition, food security concerns are likely to pave the way for increased land appropriation, as national governments, sovereign wealth funds and assorted corporate interests seek to increase agricultural production and secure supply chains, as well as profits. In these and various other respects, global and national climate mitigation and adaptation agendas will in all likelihood drive – or at the very least generate tendencies towards – a new era of large-scale hydraulic development. Waterscapes will be further reimagined and reconfigured to meet the real or imagined challenges of climate change. Governments, regimes and major elite and corporate interests will, as during the twentieth century, be at the forefront of this work. And dispossession, displacement, conflict and insecurity will inevitably ensue – as they have done historically.

Indeed, this future has already part-arrived. 'Green grabbing' – the appropriation of environmental resources by state and corporate interests under the banner of saving the planet – is already a well-established threat to livelihoods in the global South. Climate adaptation projects have repeatedly been implicated in livelihood insecurity and conflict. Since the turn of the millennium there has been an upsurge in international purchases of water-rich agricultural lands – 'land grabbing' and 'water

---

[45] M. L. Wine et al., 'Agriculture, diversions, and drought shrinking Galilee Sea', *Science of the Total Environment*, 651 (2019), 70–83; Wine, 'Climatization of environmental degradation'; T. Scudder, 'Development-induced relocation and refugee studies: 37 years of change and continuity among Zambia's Gwembe Tonga', *Journal of Refugee Studies*, 6:2 (1993), 123–52; Selby and Daoust, *Rapid Evidence Assessment*.

grabbing' – partly motivated by concerns about the impacts of climate change on food security. There has also been a renaissance in dam building – what Sneddon characterises as 'a new era of damning the planet's rivers' – following a period during the 1990s and early 2000s when far greater attention was paid to its social costs. Among our cases, Israel and Cyprus' desalination programmes, Israel's water transfer to Lake Kinneret, the northern Cyprus pipeline, the new and forthcoming dams in Sudan, the upstream dam-building programmes in Ethiopia and Uganda and agricultural investments in Sudan and South Sudan have all been partly inspired or legitimised by concerns about global warming and have all, in one way or another, caused, threatened or contributed to conflict. In central Africa, meanwhile, plans are afoot for a 2,000 km-long water transfer to 'refill' Lake Chad from the Oubangui River in the Congo basin, at a cost of up to $50 billion – despite the fact that the lake has not recently been shrinking. And to give one final example, in *Windfall: The Booming Business of Global Warming*, journalist McKenzie Funk reports on New York venture capitalist Phil Heilberg's attempts to lease one million acres of land from the government of South Sudan, for climate security–inspired food production. The land is 'mostly empty', Heilberg says; '[h]e hadn't surveyed it all himself, but seemed to believe this', Funk comments. However, the plot in question ultimately turned out to have a population of around 100,000. And while thankfully this particular purchase never went ahead, the broader point still stands. As the twenty-first century unfolds, hydraulic developments inspired by climate change–related interests and concerns are likely to have a profound bearing on patterns of displacement, destruction, discrimination, conflict and insecurity.[46]

---

[46] Fairhead et al., 'Green grabbing'; A. Dunlap and J. Fairhead, 'The militarisation and marketisation of nature: an alternative lens to "climate conflict"', *Geopolitics*, 19: 4 (2014), 937–61; S. H. Eriksen et al., 'Reframing adaptation: the political nature of climate change adaptation', *Global Environmental Change*, 35 (2015), 523–33; S. H. Eriksen et al., 'Adaptation interventions and their effect on vulnerability in developing countries: help, hindrance or irrelevance?', *World Development*, 141 (2021), 105383; P. Woodhouse, 'Foreign agricultural land acquisition and the visibility of water resource impacts in sub-Saharan Africa', *Water Alternatives*, 5:2 (2012), 208–22; J. C. Franco and S. M. Borras Jr., 'Grey areas in green grabbing: subtle and indirect interconnections between climate change politics and land grabs and their implications for research', *Land Use Policy*, 84 (2019), 192–9; Sneddon, *Concrete Revolution*, 133; A. B. Adeniran and K. A. Daniell, 'Transaqua: power, political change and the transnational politics of a water megaproject', *International Journal of Water Resources Development*, 37:2 (2021), 234–55; Daoust and Selby, 'Understanding the politics'; Funk, *Windfall*, 139–59.

6   Frontiers

Water and the Capitalist Resource Frontier

If the intensive re-engineering of nature and society is one of the defining features of capitalist social relations, then equally so, and no less importantly, is the extensive 'opening up', colonisation and exploitation of new frontiers. How else could it be otherwise? Whether led by states, corporations or bands of pioneering men and women, the incorporation and 'development' of hitherto untapped or under-utilised regions and their resources is one of the surest routes to the accumulation of capital and the extension of social power. Capitalism's central ideological tenets – its faiths in linear development via individual action, freedom and opportunity – likewise find no surer confirmation than in the progressive transformation of supposedly unused wilderness into productive land and property. Little wonder that Frederick Jackson Turner, high priest of the American frontier, could characterise it both as 'the meeting point between savagery and civilization' and as the source of America's most abiding, and singularly capitalist, qualities: dynamism, materialism, individualism and democracy. As Jason Moore has put it, 'capitalism' is in essence 'a frontier process' in which 'endless accumulation and endless geographical appropriation are joined at the hip'.[1]

While this seems clear in general terms, one thing that is much less certain is the place of water within these processes. Indeed, water barely figures within frontiers research. Turner's focus was the wildwood, reflecting the environmental challenges posed by the settlement of his native Wisconsin (American democracy 'came out of the American forest', he famously declared); as Donald Worster observes, the Turner thesis contains 'no water', and 'no aridity', as well as 'no technological dominance' and 'very little in it of the West as it is geographically defined

---

[1] F. J. Turner, *The Frontier in American History* (Holt, Reinhart and Winston, 1920), 3, 37; J. W. Moore, *Capitalism in the Web of Life: Ecology and the Accumulation of Capital* (Verso, 2015), 107.

today'. Moreover, this neglect of what we might call 'water frontiers' has continued to this day. The recent literature on new 'resource frontiers' contains next to nothing on the exploitation or appropriation of water specifically. There have been countless studies of the opening up of agricultural land through deforestation, but very little on the frontier dynamics within desert or dryland margins. And one searches in vain through the pages of recent hydro-politics research for analyses of frontier water conflicts. Explicit adoption of a frontier perspective on water conflict and insecurities is almost entirely absent.[2]

This chapter seeks to make good on these existing research gaps while simultaneously extending our analysis, begun in Chapter 5, of the political ecology of water development and state-building and their impacts on water-related conflicts and insecurities. The two chapters have quite different substantive foci: where the previous chapter concentrated on major water engineering projects and geographically core regions and populations, our attention here shifts to geographical peripheries and the profound socio-political conflicts that are typically found within them, and attends more to questions of land control than to large-scale hydraulics. Yet irrespective of these differences (which are in any case not absolute, but matters of degree) the two chapters' central arguments are essentially the same: that projects of state-building and development have repeatedly involved, or resulted in, large-scale environmental and social destruction, dispossession, violence and insecurity; that the water arena provides ample evidence to this effect; that water-related conflicts and insecurities are caused less by scarcity – among other things, because they often coincide with relative abundance – than by political and political-economic interests, structures and agendas; and notwithstanding all this, that patterns of water-related conflict and insecurity can take very different forms in different contexts, determined above all by core state-building and development agendas. If this is so of core regions, then it applies doubly, or more, at the frontier.

As we use it here, the term 'frontier' refers to spaces that are simultaneously geographically peripheral to existing centres of political and economic power and sites of outward expansion and incorporation by those existing centres – typically involving some combination of colonisation, commodification, extraction and appropriation. Understood thus,

---

[2] Turner, *Frontier*, 293; Worster, *Rivers of Empire*, 11; N. L. Peluso and P. Vandergeest, 'Political ecologies of war and forests: counterinsurgency and the making of national natures', *Annals of the Association of American Geographers*, 101:3 (2011), 587–608; M. B. Rasmussen and C. Lund, 'Reconfiguring frontier spaces: the territorialisation of resource control', *World Development*, 101 (2018), 388–99; Selby, 'Climate change and the Syrian civil war, part II'.

'frontier' is essentially a socio-political rather than environmental category; the location of frontiers changes across time ('the frontier is the outer edge of the wave', said Turner; it is 'a mobile front in continuous formation', say Mezzadra and Neilson) and is determined by political structures and developmental trajectories, not by the differential endowments of nature.[3] For obvious reasons, however, there usually is an environmental dimension to frontiers, and to this extent they are perhaps better understood as socio-ecological spaces. In character, they are deeply contradictory. They are commonly imagined by their colonisers as open, empty, under-used and uncivilised. They typically combine, or are thought to combine, abundances of free land and other resources, together with scarcities of capital and labour. They are spaces of freedom, potential and opportunity, if also hardship, and the regular focus of myths of nation-building and development. But they are also sites of lawlessness, violence and expropriation. They are typically characterised by attenuated state authority and legitimacy, if also highly coercive practices of control. Their local and indigenous populations, who are usually viewed as obstacles to or at best objects of frontier development rather than as its subjects – especially if they happen to possess their own collective national or ethnic identities – are typically neglected, marginalised, excluded and sometimes expelled. Frontiers, as a result, are lands of both movement and confinement, of simultaneous in-migration, out-migration and displacement, and, at the extreme, include enclaves for those peoples expelled in the name of progress. In turn, frontiers are always spaces of resistance, resilience and local adaptation. And for all these reasons they are also always sites of rapid, thinly regulated and short-termist economic development, with all of its attendant social and ecological consequences.

In what follows we illustrate and explore these patterns and tendencies with reference to Sudan, Palestine, Syria and Lake Chad, showing in each case how 'water frontiers' – by which we simply mean frontier regions where the development, appropriation or control of water resources is an important objective, without meaning to imply that the dynamics found within such regions can be reduced to water-related interests alone – are sites of extreme appropriation, inequality, degradation, conflict and insecurity, both in general and in relation to water specifically. After considering these four cases we then turn the tables somewhat, to explore frontier political agency – the diverse means through which frontier actors endure and resist internal colonial state power vis-à-vis the environment. And we close, as in previous chapters, by turning to climate change,

---

[3] Turner, *Frontier*, 3; S. Mezzadra and B. Neilson, *Border as Method, or, the Multiplication of Labor* (Duke University Press, 2013), 15.

noting that frontiers are widely misunderstood within climate crisis discourse and reflecting on the how they are actually likely to fare as the planet warms.

Abundance and Violence on Sudan's Multiple Frontiers

We have already seen in the previous chapter that Sudan has a particularly acute history of water-related dispossession and conflict that spans the colonial, post-colonial and contemporary eras. Swamp drainage, river canalisation, dam building, irrigation schemes, and more: all have had their victims, especially in the country's riverine core where the goals of hydraulic state-building and agricultural modernisation have been most consistently pursued. Yet for all this, the violence which has always stalked Sudan's core states pales into insignificance relative to that which has raged across the country's frontier regions since the 1980s, since when large-scale violence in Sudan has had a predominantly frontier character. And water, crucially, has been an important and recurring element in this.

For context, Sudan's first civil war of 1955–72 was confined to the south of the country and was essentially political in its causes, revolving around the under-representation of southern leaders and their demands for secession. By contrast, the second civil war of 1983–2005 and the 'network of internal wars' that Sudan has experienced since then have brought large-scale armed conflict to 'semi-peripheral' regions of the 'Muslim North' – to South Kordofan, Blue Nile, Kassala, Red Sea and, as already discussed in Chapter 3, Darfur – that is, to areas lying outside both the core riverine states around Khartoum and the equatorial zone to the south. The fundamental reason for this shift is that, since the 1980s, resource appropriation has become a primary motivation for and cause of conflict in Sudan, it being in the country's semi-peripheral regions that what David Harvey termed 'accumulation by dispossession' has been most marked. Within these regions, it is resource abundance – of land, oil, minerals, livestock and, yes, water – which has been the major correlate of conflict and source of vulnerability. Moreover, this new conflict geography has emerged in large part because of a specific post-1970 national development regime allied to Sudan's deepening integration into, and dependence on, the global economy. Right across the country's semi-periphery, new dynamics of capitalist development have underpinned the emergence of new conflict-ridden internal frontiers.[4] Here we consider two outstanding examples of this: South Kordofan and, once again, Darfur.

---

[4] D. H. Johnson, *The Root Causes of Sudan's Civil Wars* (James Currey, 2003), 63–87, 127, 130; Selby and Hoffmann, 'Beyond scarcity', 366–8; D. Harvey, 'The "new" imperialism:

The central Sudanese province of South Kordofan lies in the country's semi-arid Sahel zone, between the desert lands to the north and what is now South Sudan, just to the south. Centred on the Nuba Mountains which reach up to 1,300 metres above sea level, the province has few surface water resources beyond intermittent streams flowing east towards the Nile; yet it nonetheless receives relatively good summer rains – its lowland capital, Kadugli, has annual average precipitation of 700 mm – and its clay plains are suitable for both grazing and cultivation. Until the 1970s, the region had an essentially subsistence agricultural and livestock economy and a communal and customary land regime. The region was not without history, of course. Its Nuba peoples, in particular, had repeatedly been objects of coercive outside interventions, from slave-raiding under the Turkiyya and Mahdiyya periods through to forced resettlement by the British. However, the socio-economic changes within the region had been limited – and intentionally so. Under the British, the Nuba Mountains had been a 'closed district', with access, trade and external contact restricted in a bid to preserve tribal identity, and following independence this pattern was initially maintained.[5] Only from the 1970s was it upended: henceforth, a region that had long been an ethnic and ecological intermediate zone became in addition an open, dynamic, capitalist frontier.

The initial catalysts to this change were a series of World Bank–funded Mechanised Farming Projects that sought, as the Bank put it, to 'open up' supposedly uncultivated and 'almost uninhabited' land for sorghum and sesame production. The projects involved the creation of standard 630-hectare farm plots, land clearance, provision of credit for tractors and tractor-drawn equipment, new roads and the establishment of a government-owned Mechanised Farming Corporation (MFC) that would henceforth own and lease the new land and oversee the whole process. Mechanisation, here, was thought to offer the key to land use change, enabling the systematic development of clay soils that were too heavy for cultivation with simple hand tools, especially when wet. In turn, both the Bank and the Sudanese government advocated the development of rain-fed mechanised farming right across Sudan's transitional zone, with this ambition becoming central to the Nimeiri administration's Breadbasket Strategy. Mechanised rain-fed cultivation for export became a national priority. Gulf capital flowed in, in support. And Sudan's land

---

accumulation by dispossession', *Socialist Register*, 40 (2009), 63–87; A. J. Ayers, 'Sudan's uncivil war: the global-historical constitution of political violence', *Review of African Political Economy*, 37:124 (2010), 153–71.

[5] H. A. Kadouf, 'Marginalization and resistance: the plight of the Nuba People', *New Political Science*, 23:1 (2001), 45–63; G. K. Komey, *Land, Governance, Conflict and the Nuba of Sudan* (James Currey, 2010), 36–43.

regime was simultaneously transformed, the government's 1970 Unregistered Land Act abolishing customary land rights and converting nearly all unregistered – that is, communal and tribal – land into government property.[6] Here, external liberalisation – *infitah* – went hand in glove with the internal 'opening up' of new frontier land for development.

Between the late 1970s and mid-1980s, as a result, the area under mechanised cultivation in Sudan's central clay plains more than doubled, to over 3.7 million hectares; in the Nuba Mountains alone, nearly 1,400 mechanised farming schemes were established during this period. There was some diversification of Sudan's agricultural exports with sorghum, in particular, becoming the country's second largest export crop after cotton. Yet overall, the consequences of this expansion drive were dismal, as discussed in the previous chapter. And the problems were particularly acute in producer regions. In South Kordofan – one of the main targets of the MFC programme, possessing, by 1986, a quarter of the country's total mechanised agricultural land – mechanisation brought a potent combination of elite enrichment, local dispossession, ecological degradation and tribal conflict. Most of the newly opened land was leased or sold by the MFC to government officials, former military officers and wealthy riverine merchants (often at bargain prices and with the additional benefit of low-interest government loans, subsidised farm equipment, fuel quotas and trade licences); for example, in the largest scheme in the Nuba Mountains, in the Habila area, only 12 of 143 farms were allocated to local co-operatives, with the rest going to absentee merchants and government officials. Furthermore, since this land was far from unused or uninhabited, these development schemes necessarily involved appropriation, livelihood destruction and displacement, especially of Nuba communities. Farmers were driven from their lands without compensation, and transhumance routes were blocked. Large areas of land were seized outside of designated sites. People were frequently fined or imprisoned for 'trespassing' on land held by mechanised schemes. And grain harvests were burned in protest. The combination of widespread bush clearance, crop monoculture and continuous cropping also led to soil exhaustion and erosion, and declining crop yields. Thus followed typical frontier cycles of land development and abandonment, and repeated waves of

---

[6] World Bank, *Mechanised Farming Project: Sudan, Appraisal Report* (1968), i, 5; World Bank, *Development Credit Agreement (Second Mechanized Farming Project) between the Democratic Republic of the Sudan and International Development Association* (1972), 16; World Bank, *Project Performance Audit Report: Sudan Second Mechanized Farming Project* (1982), 1; G. Kibreab, 'Property rights, development policy and depletion of resources: the case of the central rainlands of Sudan, 1940s–1980s', *Environment and History*, 7:1 (2001), 57–108.

dispossession and displacement. As the Nuba Mountains Farmers Union observed in 1974, 'there is glaring inequality and injustice in all schemes ... the discrimination is basically against the local people of the Nuba Mountains, and favors certain people from other areas'.[7]

Sudan's second civil war broke out in 1983 and by 1985 the Nuba Mountains had become one of its key battlegrounds, with many Nuba joining the SPLM/A. This conflict geography was no accident: the Nuba had long been subject to racial oppression, the Sudanese state under al-Nimeiri had become committed to Islamisation and this combined with large-scale land appropriation and displacement made many deeply hostile to Khartoum. Moreover, land appropriation was not just one factor among others: in the assessment of Sudanese academic Mohamed Suliman, the encroachment of mechanised agriculture was '[t]he single most important issue behind the outbreak of the conflict in the Nuba Mountains ... This devastated the economic and social life of the Nuba and ultimately destroyed friendly relations with the Baggara.'[8]

For their part, between 1985 and 1993 Sudanese government forces and local Baggara militias conducted a brutal counter-insurgency war in the region. At least 100,000 Nuba, and possibly more than 200,000, were killed or disappeared, while a swathe of 'peace camps' were simultaneously established in fulfilment of the government's project of Islamising the region; by 1992, an estimated 167,000 Nuba had been relocated to them. Far from being simple educational facilities, these camps, as described by the NGO African Rights, were

> concentration camps in the true sense of the word .... Inmates are kept there against their will, they are forced to work for low wages or no wages, men are forced to become members of the PDF [Popular Defence Forces], women are raped, and children have their identities changed. It is all part of a programme for dismembering Nuba society .... every woman who has been in a peace camp has either been raped or threatened with rape.

---

[7] Suliman, 'Civil war', 105; A. M. Yahya and B. A. Mohammed, *The Future of Mechanized Schemes and Agricultural Investment in the South Kordofan State/Nuba Mountains* (Chr. Michelsen Institute, 2016), 10; African Rights, *Facing Genocide: The Nuba of Sudan* (1995), 40–1; G. K. Komey, 'Land factor in wars and conflicts in Africa: the case of the Nuba struggle in Sudan', in T. Falola and R. C. Njoku (eds.), *War and Peace in Africa* (Carolina Academic Press, 2010), 364–5; Komey, *Land, Governance, Conflict*, 43–9; Johnson, *Root Causes*, 132; H. B. Ibrahim, Agricultural Development Policy, Ethnicity and Socio-Political Change in the Nuba Mountains, Sudan (PhD thesis, University of Connecticut, 1988), 114–15.

[8] M. Suliman, 'The Nuba Mountains of Sudan: resource access, violent conflict', in D. Buckles (ed.), *Cultivating Peace: Conflict and Collaboration in Natural Resource Management* (International Development Research Centre and World Bank, 1999), 212.

Resettlement to these camps also enabled further land appropriation while providing a captive labour pool for farms, with some camps being directly attached and adjacent to farming schemes. In South Kordofan, during the three decades from 1970, agricultural development both paved the way for and was further facilitated by state-led appropriation and genocidal dispossession.[9]

Like South Kordofan, the western Sudanese region of Darfur has been a site of acute state-orchestrated and inter-ethnic violence since the 1980s. Indeed, we have already explored one reading of this violence, showing in Chapter 3 that the region's 2003–5 war should not be understood through a 'climate conflict' lens. But how, then, should it be understood? If climate conflict narratives of the war are flawed, does this mean that it was 'purely political' in its causes? Or in what sense, if at all, was this war a product of capitalist development 'at the frontier'?

The key to answering these questions lies in recognising that, within Darfur itself, the war centred above all on issues to do with land – and in understanding the economic and political reasons for this. That land issues were central to Darfur's slide into war has been extensively documented in both academic studies and humanitarian and conflict reporting on the subject. In the years prior to the war, there was a continuous increase in the number and intensity of local disputes over land, many of them violent. The capture and possession of land – and the dispossession and displacement of others from land they considered theirs – became one of the war's recurring motifs. The political discourse of local parties reflected this, whether in the form of justifications of violence ('we have no land!') or resistance against it ('they came to steal our land!'). Land issues have been central to subsequent attempts to craft a durable political settlement within Darfur, as well as an obstacle to its realisation. And land continues to be a major source of discord, as evidenced by the refusal of most displaced Darfuris to accept resettlement beyond internally displaced person (IDP) camps.[10] Moreover, and crucially, the fundamental reason why land conflicts have erupted with such

---

[9] M. Burr, Working Document II: Quantifying Genocide in Southern Sudan and the Nuba Mountains 1983–1998 (US Committee for Refugees, 1998), 33; African Rights, *Facing Genocide*, 3, 120–8, 267–75; Komey, *Land, Governance, Conflict*, 77–99; L. Wise, 'The genocide–ecocide nexus in Sudan: violent "development" and the racial-spatial dynamics of (neo)colonial-capitalist extraction', *Journal of Genocide Research*, 23:2 (2021), 201.

[10] M. A. Abdul-Jalil, 'Nomad-sedentary relations and the question of land rights in Darfur: from complementarity to conflict', in R. Rottenburg (ed.), *Nomadic-Sedentary Relations and Failing State Institutions in Darfur and Kordofan* (Martin-Luther-Universität Halle-Wittenberg, 2008), 14–15; J. Flint, *The Other War: Inter-Arab Conflict in Darfur* (Small Arms Survey, 2010), 12–16; J. Tubiana, 'Le Darfour'; Abdul-Jalil and Unruh, 'Land rights'; Darfur Peace Agreement (2006), 22, 29, 31; 'Displaced "reject Central Darfur model villages": sheikh', *Radio Dabanga* (04/07/2013).

force in Darfur is not drought but because, together with South Kordofan, the region has become a developmental frontier par excellence.

In one sense, Darfur has always been a frontier region, defined above all by its central Jebel Marra mountain range which rises to over 3,000 metres above sea level, the highest point in Sudan. At its core this highland area has a temperate climate and heavy seasonal rains, which in turn feed permanent and seasonal streams as well as groundwater resources that radiate outwards across the region. In these respects Darfur is unique within Sudan, centring on a relatively water-rich and fertile finger of land that juts out into the northern Sahel and the Sahara, far beyond other areas of the country. Indeed, across the whole of the eastern Sahel – from Lake Chad 1,000 km to the west, to the River Nile 1,000 km to the east – Darfur is the only region where verdant green land and savannah woodland can be found so far north (as can be seen from even a quick scan of Google Earth). Darfur is not only a watershed region, dividing the Nile basin from the endorheic Lake Chad basin in the centre of the continent; metaphorically speaking, it is also very much a regional oasis in the desert.

Under British rule and the early post-independence years these environmental advantages were barely exploited, except locally – owing, as in the Nuba Mountains, to Native Administration policies and restrictions on development. While Darfur was integrated into the Sudanese and colonial economies, this was mainly as a labour reserve for the Gezira scheme and its cotton production and exports. But in the second half of the twentieth century this pattern of limited economic incorporation changed, with two main infrastructural developments setting the stage. In 1959, Sudan's southern rail line was extended to Nyala, creating the conditions for the emergence of a regional export crop economy. And this was followed, from the 1960s, by the introduction and uptake of diesel-powered tubewells, which for the first time enabled systematic exploitation of Darfur's shallow but plentiful groundwater resources. The latter development, in particular, transformed the geography, intensity and overall political economy of agricultural production within Darfur. Newfound access to groundwater meant that, where cultivation had previously been largely confined to valley terraces, it now also became possible in the light *goz* soils of the open savannah – using a simple borehole, pump fuel and oil drums for overland distribution. Year-round access to constant underground stocks of water also made multi-season cropping possible, in turn revolutionising production processes.

Indicative of this, a 1988 survey found that low water availability was viewed by farmers as only the fifth biggest obstacle to irrigation expansion, with fuel and cash, the latter to buy fuel and hire labour, being much bigger constraints (and this was despite the fact that the survey was conducted following a year of poor rains).[11] By the 1980s, if not earlier, Darfur's agricultural sector had become distinctly modern – dynamic, export-oriented and highly capital- and fuel-dependent – and nothing like the traditional system of Orientalist imagination.

In turn, these technological and economic changes upended established land use and associated demographic patterns, paving the way for the land conflicts and violence discussed above. Much of this has already been detailed in Chapter 3. Open rangelands, which had hitherto been used mainly for communal grazing, became prime sites for cash crop development and were progressively demarcated, enclosed, irrigated, cultivated and settled. The cash crop sector became the main source of economic opportunity and dynamism within Darfur's rural economy. Unlike in South Kordofan where, as we have seen, agricultural development was a state-led project, in Darfur's transition small-scale, private-farmer horticulture predominated. Pastoralist communities moved decisively into farming as a result, often migrating and settling new lands in the process. This, in turn, led to a rapid expansion in the population of South Darfur, where the opportunities for irrigated horticulture were greatest.[12] There was also eastward cross-border movement into Darfur, with the region as a whole witnessing significant in-migration. Combining relative resource abundance, economic opportunity, processes of enclosure and commodification, extensive in-migration and settlement, as well as decided limits to the rule of law, post-1960s Darfur bore many of the classic hallmarks of a capitalist frontier.

Although these transformative processes revolved mainly around the cash crop economy, in the decade prior to the 2003–5 war there had also been a significant expansion in Darfur's livestock sector. Historically, livestock had not figured heavily within Sudan's development thinking. But following the economic crisis of the 1980s and the failure of the Breadbasket Strategy, this sector became viewed as crucial to the goal of

---

[11] Mamdani, *Saviours and Survivors*, 163–75; A. Abdelkarim, *Primitive Capital Accumulation in the Sudan* (Frank Cass, 1992), 40–1; Morton, *Agricultural Development*, 34–5; Jebel Marra Rural Development Project, 'Irrigation Survey 1988 (Khartoum)', in Morton, *Agricultural Development*, 81.

[12] Morton, *Agricultural Development*, 35; M. Adams, 'The Baggara problem: attempts at modern change in southern Darfur and southern Kordofan (Sudan)', *Development and Change*, 13:2 (1982), 270.

increasing export earnings. Reflecting this, Sudan's Comprehensive National Strategy of 1992 aimed to triple the size of the country's livestock population and increase exports twentyfold within the space of just ten years, via a systematic modernisation of feeding and ranching processes. To this end, the sector was comprehensively liberalised, starting with the replacement of the parastatal Livestock and Meat Marketing Corporation with commercial livestock banks. The consequences were dramatic. Livestock's contribution to Sudan's agricultural exports rose from just 6–7 per cent in the 1960s and early 1970s, to an average 27 per cent by the turn of the millennium. Sudan became the leading livestock exporter in the region, with most of its animals, above all sheep, being exported live to Saudi Arabia. Livestock became Sudan's third most important export commodity, after oil and gold; and the livestock sector became the biggest in Sudan's domestic economy, bigger even than oil. Moreover, Darfur was at the heart of this expansion, accounting for as much of a third of the country's total livestock production. Large-scale ranches were established, owned by wealthy urban elites and government officials (in South Darfur, for example, several private investors owned ranches of over 20,000 hectares).[13] As in farming so also in livestock: in both sectors the commercialisation, globalisation and liberalisation of Sudan's economy resulted, on the ground, in the progressive privatisation and appropriation of land.

But how did these developments feed into the 2003–5 war? Most climate conflict accounts of these processes simply assume, where they acknowledge them at all, that the expansion of cultivation and especially livestock numbers in Darfur led to increased competition over scarce and increasingly degraded land and water resources, which in turn resulted in violence between 'farmers' and 'herders'. Yet such tacit Malthusianism has little to commend it – partly for the theoretical reasons set out in Chapter 2; partly because of the evidence of greening and in-migration discussed in Chapter 3; and, perhaps most fundamentally, because narratives of Darfur as a scarce and degraded environment are as old as British colonial rule in the region, yet fly in the face of the huge recent expansions in its cash crop and livestock sectors. Instead, the intensifying conflict over land in the run-up to the 2003–5 war seems to have had three major structural causes. First, the developmental processes outlined above upended established divisions and hierarchies – between tribes,

---

[13] H. M. Nur, 'Ambitious plans and unresponsive sectors: new horizons for pastoral development in Sudan', *Nomadic Peoples*, 5:1 (2001), 134–54; M. Buchanan-Smith, et al., *On the Hoof: Livestock Trade in Darfur* (Tufts University and UNEP, 2012), 10, 19, 52; R. Behnke, *The Economics of Pastoral Livestock Production in Sudan* (Feinstein International Center, Tufts University, 2012), 2–3; Young et al., *Darfur*, 67–9.

between farmers and pastoralists, and between northern and southern Darfur – creating incentives for pastoralist groups, in particular, to migrate, seize land and through that assert or reassert their place within Darfur's social order (hence the reason why some of those who migrated, like the Zaghawa, were relatively wealthy, whereas others, like the Northern Rizeigat, had historically been *dar*-less and economically marginalised). Second, the above political and economic changes coincided with, and were in part facilitated by, no less acute transformations in local governance and authority structures following the dissolution of Sudan's Native Administration system in 1971; it was this combination of profound economic and political change and frontier lawlessness that enabled contestation over land to descend into mass violence. And third, there undoubtedly were a growing number of conflicts over access to *wadis*, water-rich land and transhumance routes in the years prior to the 2003–5 war. Crucially, however, these conflicts were not the result of a generalised condition of scarcity, but rather of frontier-style grabbing of land and water resources.[14]

In the late twentieth century, then, both Darfur and South Kordofan were water frontiers in the fullest of senses. In both cases, abundant, low-value land suddenly became a valuable resource and an object of competition and conflict. In both cases, the primary latent property of this land, which underpinned it becoming so valuable, was the relative richness of its water resources. In both cases, this land and its waters suddenly became valuable not out of scarcity but for reasons of political economy, especially technological diffusion and export-oriented national economic policies. In both cases, these profound political–economic changes sparked the widespread appropriation and enclosure of land and water; equally widespread dispossession and displacement; and extreme violence, both between local actors but also with and by the Sudanese state. It would be a mistake, no doubt, to describe the conflicts in South Kordofan and Darfur as 'water conflicts', not least because of their essentially political and political–economic causes. But as elsewhere on Sudan's semi-periphery, water – or more precisely water-rich land – was the primary target and objective of local development. And it was these regions' emergence as water resource frontiers which set the stage for the extreme violence that followed.

[14] Morton, *Agricultural Development*, 9, 30–2, 90, 96–7; de Waal, 'Is climate change the culprit?'; Mamdani, *Saviors and Survivors*, 235–40; Abdul-Jalil and Unruh, 'Land rights', 162–5, 169–71; UNEP, *Sudan*, 81–2.

Water Apartheid in the West Bank and Gaza

The West Bank and Gaza Strip present us with what is, in many respects, a polar opposite case. For, whereas on Sudan's semi-periphery it was the introduction of new technologies and economic liberalisation that paved the way for land appropriation and war, in the West Bank and Gaza what occurred was the reverse, with war opening up a new era of, and space for, state-led frontier colonisation. The war in question was the Six Day War of June 1967, by the end of which Israel had become the occupying military power across the West Bank and Gaza (plus in the Golan Heights and Sinai peninsula, which are not considered here). East Jerusalem was immediately annexed, while large swathes of land in both the West Bank and Gaza were redefined as state property and earmarked for Jewish settlement, with settlements being established first in the Jordan Valley and around Jerusalem, and then, after 1981, on a much larger scale across the Occupied Palestinian Territories. Simultaneously, Israel established a dual legal and institutional system, under which Palestinians were governed by a complex matrix of Military Orders, military courts and state violence, and were denied both citizenship and access to public funds, while settlers continued to exercise those political, social and economic rights available to all other Jewish Israelis. Of course, the Israeli–Palestinian conflict had always had a frontier character, set in motion by a colonial project that viewed Palestine, in extreme, as a 'land without a people for a people without a land'. But post-1967 in the Occupied Territories this 'frontierity' became particularly acute, with a clearly apartheid-like form. By 1993, when the Oslo peace process began, there were 264,000 Israeli settlers in the West Bank (including East Jerusalem), plus 4,000 in Gaza, living there both in violation of international law and under a completely different legal and institutional regime from the Territories' estimated 2.5 million Palestinians.[15]

In another contrast with the situation in Sudan, hydro-political objectives were central neither to Israel's 1967 capture of the West Bank and Gaza, nor to their subsequent settlement. Not only was the Six Day War no 'water war', as discussed in Chapter 2; equally, Israel's settlement

---

[15] Shafir and Peled, *Being Israeli*, ch. 6; E. Benvenisti, *Legal Dualism: The Absorption of the Occupied Territories into Israel* (Westview, 1990); D. Muir, 'A land without a people for a people without a land', *Middle East Quarterly*, 15:2 (2008), 55–62; B. Kimmerling, *Zionism and Territory: The Socio-Territorial Dimensions of Zionist Politics* (University of California Press, 1983); UN-ESCWA, *Israeli Practices towards the Palestinian People and the Question of Apartheid* (2017); B'Tselem, 'A regime of Jewish supremacy from the Jordan River to the Mediterranean Sea: this is apartheid' (12/01/2021); Foundation for Middle East Peace, 'Comprehensive settlement population 1972–2011' (2012); Palestinian CBS, 'Estimated population in Palestine mid-year by governorate, 1997–2021' (2005).

drive was motivated much more by 'security concerns' (especially establishing a security buffer in the Jordan Valley), by rising religious-nationalist sentiment and by interests in redirecting some of the country's urban growth away from its congested coastal plain, than by anything to do with water (on the latter, consider the words of a 1993 Israeli government advertisement for housing in the Jordan Valley: 'Free yourself from the hysteria that you wake up with every morning. Free yourself of the fear of walking alone in the street.... Free yourself in order to succeed with your wife who will once again smile as you return home .... All this one hour from the center of the country'). That said, Israel's settlement project involved large-scale land appropriation and, as a corollary, the extensive capture of local water resources. Moreover, the occupation of the West Bank was of hydro-political advantage to Israel. A mostly upland region with relatively strong winter rains, the West Bank houses the replenishment zone for the trans-boundary Mountain Aquifer – a groundwater resource which, until Israel's recent desalination investments, provided around a quarter of the country's total water supplies. Israel was already heavily exploiting this Aquifer prior to 1967, via springs and wells below the West Bank. From 1967, however, Israel was also able to use its occupation of the West Bank to restrict and contain rising Palestinian water consumption there, and to ensure that it would remain the Mountain Aquifer's principal user.[16] Occupation, in short, not only opened the gates to local water appropriation; it also created an opportunity for Israel to consolidate an already unequal hydro-political status quo. (It is also worth noting here that, in the process, the West Bank was reimagined as a site of water resource abundance. Pre-1948 the West Bank was not thought of in this way, and hence was not a focus of early Zionist settlement. However, improvements in well technology changed all this, enabling ready access to the territory's rich aquifers: as in South Kordofan and Darfur, technological development opened up a new water frontier).

Many of the consequences are well known. Across the West Bank, the hundreds of Israeli settlements which were established mostly had good, constant water supplies and per capita use well above even the Israeli average, settler swimming pools becoming a potent symbol of this colonial plenty. Settler water supplies were also heavily subsidised. Deep wells were drilled into the Mountain Aquifer to supply the settlements, and from 1981 Israel started integrating them into its national supply

[16] *Ha'aretz* (22/04/1993), in Foundation for Middle East Peace, *Report on Israeli Settlement*, 3:5 (1993), 7; Lowi, *Water and Power*, 185; M. Zeitoun et al., 'Asymmetric abstraction and allocation: the Israeli–Palestinian water pumping record', *Groundwater*, 47:1 (2009), 146–60.

network. In the Jordan Valley, most Palestinian land along with the wells and springs on it was appropriated, in their place coming Israeli settlements with intensive irrigated agriculture and exceptionally high water usage. Elsewhere, Palestinian land and water was widely appropriated too, with the boundaries of settlements typically extending far beyond just their built-up areas. Indeed, to this day springs and wells are repeatedly claimed by armed groups of settlers, with Palestinians being usurped from them or finding access to them denied.[17]

At the same time, a highly restrictive control regime was established to limit West Bank Palestinian water development and demand. A 1968 Military Order declared all local water resources to be public property subject to state control, as within Israel itself. Ownership of all (Palestinian and Israeli settler) water supply systems was subsequently transferred to Israel's parastatal water company, Mekorot. Other Military Orders dictated that all new and existing water installations would require a permit from the Military Government (and, later, what was rebranded the Civil Administration) and that applications could be denied, revoked or amended 'without giving reasons'. Permits were, moreover, routinely denied. From 1967, in a sharp departure from the era of Jordan rule, not a single Palestinian permit was granted for the drilling of wells into the crucial Western Basin of the Mountain Aquifer, with Palestinian exploitation of this basin frozen at 1967 levels. Not one permit was granted for new agricultural wells, either. Irrigation was not permitted after 4 p.m. Meters were installed on existing wells, and abstraction quotas rigorously enforced. 'Illegal' (i.e. non-permitted) wells, pipelines and water storage facilities were, and still are, routinely destroyed by the Israeli military. Palestinians did not benefit from water subsidies either, paying far higher prices for their water supplies. Furthermore, while Palestinian communities did benefit from some Israeli investment in water infrastructure, this investment was decidedly double-edged. While new supply networks were laid, these did not only supply water; they also integrated Palestinian communities, alongside illegal settlements, into Israel's national water supply network. Moreover, far from supplying water equally between settler and Palestinian communities,

---

[17] J. Isaac and J. Selby, 'The Palestinian water crisis: status, projections and potential for resolution', *Natural Resources Forum*, 20:1 (1996), 18–20; Selby, *Water, Power and Politics*, ch. 3; B'Tselem, *Dispossession and Exploitation: Israel's Policy in the Jordan Valley and Northern Dead Sea* (2011), 37; Bimkom, *The Prohibited Zone: Israeli Planning Policy in the Palestinian Villages in Area C* (Planners for Human Rights, 2008); Y. Berger, 'Israeli settlers "upgrade" West Bank springs to usurp Palestinian land', *Ha'aretz* (31/05/2019).

both the material structure of these networks – the size and design of pipes and reservoirs – and the regulations surrounding them worked to ensure that supplies were directed disproportionately, and more reliably, to the settlements. Configured as they were by a racist state project, these supply apparatuses were not merely technologies, but 'techno-political' instruments of colonisation, discrimination and enforced dependency.[18]

The consequence of all this was that, by the early 1990s, Palestinian communities across the West Bank were being plagued by chronic under-supply – while neither Jewish citizens in Israel nor West Bank settlements were experiencing anything equivalent. All Palestinian towns and mains-connected villages were in receipt of intermittent supplies only (with each area of a typical town receiving water for two to three days before having one to two weeks off). Many peripheral and hill-top communities would go several months each summer without piped supplies. Around half of all villages were also not connected. In addition, the West Bank had become increasingly dependent on water supplies from Israel. And overall, according to official Israeli figures, in 1995 Israelis were utilising a staggering 87 per cent of the total yield of the Mountain Aquifer's trans-boundary basins, Palestinians just 13 per cent, while on the crucial Western Basin, the most important 'shared' Israeli–Palestinian groundwater resource, the use ratio was more than 15:1.[19]

The emerging crisis in Gaza was very different, in both timing and form. In comparison to the West Bank, the Gaza region has relatively meagre water resources, with average annual rainfall of around just 300 mm and no perennial surface flows. Yet prior to 1948 the pressure on these resources was quite limited. Historically, Gaza had been known as a verdant oasis, a region of wells, orchards and relief after the long trek across the Sinai Desert ('We have crossed seventy leagues of desert – a very tiring business. Brackish water, often none at all', reported

---

[18] S. Elmusa, *Water Conflict: Economics, Politics, Law and Palestinian–Israeli Water Resources* (Institute for Palestine Studies, 1997), 50, 265–6; Rouyer, *Turning Water*, 48, 53; Jerusalem Media and Communications Centre, *Water: The Red Line* (1994), 46; C. Messerschmid, *The 'Prior Use' Argument: Establishing Benchmarks and Implications of Historic Water Use, 1920–1948*, unpublished report for PLO Negotiation Support Unit (2015), 34; S. Lonergan and D. Brooks, *Watershed: The Role of Fresh Water in the Israeli–Palestinian Conflict* (International Development Research Centre, 1994), 130; Selby interview with Taher Nassereddin, West Bank Water Department (12/04/1998); Al-Haq, *Water for One People Only: Discriminatory Access and 'Water Apartheid' in the OPT* (2013), 59–67; Isaac and Selby, 'The Palestinian water crisis', 18–20; Selby, *Water, Power and Politics*, 83–9.

[19] Selby, *Water, Power and Politics*, 89–90, 173–8; World Bank, *Developing the Occupied Territories: An Investment in Peace. Vol. 5: Infrastructure* (1993), 45; Israel and PLO, *Interim Agreement on the West Bank and Gaza Strip* (1995), Annex II, App. 1, Sch. 10.

Napoleon Bonaparte on the Sinai in 1799; but on Gaza, he observed, 'the climate might be that of Paris' while the 'lemon bushes, olive-groves, and broken ground are just like the scenery of the Languedoc'). In 1948, moreover, the local Palestinian population of Gaza numbered just 70,000 and there were no more than 200 wells abstracting just 10 mcm/y from the shallow alluvial aquifer. Events of 1948–9 changed all this, however. In the space of a year, the local population quadrupled as refugees poured in. The region was enclaved, with the Gaza Strip being established as a 40 km by 9 km ribbon of territory carved out of Mandate Palestine and placed under Egyptian administration. In the process, villages were cut off from their lands and refugees from external employment opportunities. And, altogether, these developments led to an unprecedented rise in groundwater use, setting off the long-term process of groundwater mining and destruction which continues to this day. By 1967, the Strip had an estimated 1,600–1,700 wells and was abstracting an estimated 100–120 mcm/y, a tenfold increase on the 1948 level. With a natural recharge of just 35 mcm/y, Gaza's portion of the Coastal Aquifer was already by this point being seriously mined and thus seeing both declining water table levels and some salinisation (thanks to inflows from both the Mediterranean and, above all, from deep saline aquifers within Israel).[20]

Following Israel's capture of the Strip in 1967, the situation continued to worsen. Within the first two decades after the June War, there were huge increases in local citrus and vegetable cultivation and a more than 50 per cent growth in Gaza's Palestinian-irrigated area, as the territory was incorporated into, and made a dependent periphery of, the dominant Israeli economy. In addition, by the mid-1980s Gaza had become home to a score of Israeli settlements, most of them engaged in intensive irrigated horticulture and benefitting from disproportionate access to the Strip's meagre water resources. Although consistent data is lacking it seems clear, given this, that total water abstraction in Gaza must have further increased post-1967. Whereas in the West Bank the military government imposed strict rules on Palestinian groundwater use so as to ensure that the riches of the Mountain Aquifer would continue to flow towards Israel, in Gaza no equivalent hydro-political interest applied and

---

[20] Napoleon letter to Desaix (27/02/1799), in J. M. Thompson, *Letters of Napoleon* (Basil Blackwell, 1934), 58–9; UNCCP, *Final Report of the United Nations Economic Survey Mission for the Middle East* (1949), 19; Elmusa, *Water Conflict*, 93–4; C. Messerschmid, *Water in Gaza: Problems and Prospects* (Ibrahim Abu-Lughod Institute of International Studies, 2011), 2; World Bank, *Developing the Occupied Territories: Infrastructure*, 54; A. Vengosh et al., 'Sources of salinity and boron in the Gaza Strip: natural contaminant flow in the southern Mediterranean coastal aquifer', *Water Resources Research*, 41:1 (2005), 4, 18.

there was thus no equivalent permit or quota regime. Only from the mid-1980s did this change, as it became evident that over-abstraction and salinisation levels were becoming critical and significantly affecting citrus yields. Thus in 1984 metres were installed and quotas imposed on Palestinian agricultural wells, and from 1987 onwards all new Palestinian well licences were refused. As in the West Bank, Palestinian water demand in Gaza was henceforth capped and suppressed, with water use declining as a result.[21]

Paralleling this, Gaza also faced deepening wastewater problems. Israel had made only the most limited investments in wastewater collection and treatment in the Occupied Territories, but whereas in the West Bank the main consequence of this was foul-smelling *wadis* and streams, in Gaza wastewater simply infiltrated into the territory's shallow aquifer, polluting the very water resource on which its population depended. Indeed, within the context of severe over-abstraction and declining groundwater levels, the pollution of Gaza's aquifer was a conscious Israeli strategy: wastewater was not allowed to be channelled to the sea, as ordinarily happens elsewhere; instead all of Gaza's collected wastewater was stored in internal sewage 'lagoons' which had boreholes drilled into them to facilitate percolation into the ground. As internal World Bank documentation correctly noted, Gaza thus faced an acute 'water quantity–quality dilemma', wastewater returns being 'the reason why depletion of Gaza's groundwater' was 'only happening very slowly', but also 'why quality deterioration' was 'happening relatively quickly'.[22]

By the early to mid-1990s, as a result, the overall water situation in Gaza was as follows (we consider more recent developments in Chapter 7). Most of its groundwater was brackish or saline, with chloride levels exceeding 250 mg/l (the World Health Organization, WHO, standard for fresh water) in almost all areas and exceeding 1,000 mg/l in some. Nitrate levels were also well above accepted WHO levels, thanks to wastewater pollution. Most of Gaza's groundwater was not fit for drinking. And there were also problems of water availability. By the World Bank's conservative estimates, water use for domestic purposes was just 101 l/c/d, higher than in the West Bank (85 l/c/d) but still only around a third of the level in Israel (280 l/c/d). The disparity in agricultural water use was even larger. And average Gaza settler water consumption was

---

[21] D. Kahan, *Agriculture and Water Resources in the West Bank and Gaza (1967–1987)* (West Bank Data Project, 1987), 94, 130, 143–5; S. Roy, *The Gaza Strip: The Political Economy of De-Development* (Institute for Palestine Studies, 1995); Elmusa, *Water Conflict*, 93–6; World Bank, *Developing the Occupied Territories: Infrastructure*, 53.

[22] Messerschmid, *Water in Gaza*, 5–8; World Bank, *The West Bank and Gaza: Water Sector Review, Vol. I: Main Report* (internal report, 1997), 7.

more than seven times that of the average Gaza Palestinian. There was little political or violent water-related conflict, much less so than in the West Bank. However, so tangible and omnipresent were Gaza's water problems that leading Israeli journalist Amira Hass could name her powerful account of everyday life in the territory after them. And into the new millennium, so degraded have Gaza's groundwater resources become that the UN, and many others, have repeatedly warned that the territory could become 'unusable as early as 2016, with the damage irreversible by 2020'. While such claims are in truth without scientific basis – illustrative, once again, of Jared Diamond–style eco-collapse rhetoric – they are nonetheless indicative of the scale of Gaza's water crisis.[23]

How should this situation be explained? In most accounts, Gaza's water crisis is depicted as the inexorable product of limited natural resource endowments combined with rapid population growth and rising water demand, accentuated by problems of poor and politicised governance; it is presented, in sum, in standard Malthusian fashion, as an essentially internally generated environmental scarcity crisis (as well as evidence of the supposed inability of Palestinians to properly manage their water resources). Yet this is misleading on at least two levels. First, the population pressures on Gaza's water resources are much less a function of natural growth than, as indicated above, of displacement and enclavement – of the mass displacement of Palestinians to Gaza during 1948–9 (which will be considered more fully in the next chapter) and of the inability of most Gazans to escape from what has effectively been turned into an open air prison. In the language of the UK government's 2011 foresight report on environmental migration, Gazans are a paradigmatically 'trapped population' whose chronic water vulnerabilities stem, at root, from politically-induced immobility. Second, although Gaza's groundwater resources are much more limited than those of the West Bank, internal resource scarcity is not its principal supply-side problem. Consider the fact that most cities today obtain their water supplies from well beyond their municipal boundaries. Consider too that the Coastal Aquifer around Tel Aviv was suffering from serious over-abstraction and salinisation as early as the 1930s, well before Gaza – and

---

[23] World Bank, *Developing the Occupied Territories: Infrastructure*, 49, 55, 67; Elmusa, *Water Conflict*, 122–3; B. Shomar et al., 'Elevated nitrate levels in the groundwater of the Gaza Strip: distribution and sources', *Science of the Total Environment*, 398:1–3 (2008), 164–74; B. Shomar, 'Groundwater of the Gaza Strip: is it drinkable?', *Environmental Geology*, 50:5 (2006), 743–51; World Bank, *Developing the Occupied Territories: An Investment in Peace. Vol. 1: Overview* (1993), 10; A. Hass, *Drinking the Sea at Gaza: Days and Nights in a Land Under Siege* (Henry Holt, 1996); UNSCO, *Gaza in 2020: A Liveable Place?* (2012), 11.

that this problem was only ameliorated through the large-scale transfer of water from the upper Jordan River. And consider, lastly, that Israel's highly integrated national water supply system, discussed in the previous chapter, largely bypasses Gaza, sending water in bulk to Beer Sheva and Negev farms but providing only a few million cubic metres each year to Palestinians in the Strip. Viewed thus, Gaza's core water supply problem is not so much local resource scarcity but that, unlike every other equivalent built-up area on the planet – Gaza essentially now being a city, and not a particularly large one at that – it is all but cut off from the lands around it and denied a meaningful hinterland (see Figure 6.1).[24] In these respects, Gaza's water crisis is neither an internal nor a Malthusian one, but essentially a product of politics at the frontier.

Israel's contrasting approaches towards the West Bank and Gaza are also instructive here. Whereas Israel has extensive religious, settlement and external defence interests in the West Bank, Gaza has always been primarily viewed instead as a waste ground of unwanted Palestinian bodies, a territory which, as Yitzhak Rabin expressed it, Israeli leaders fantasised might 'sink into the sea'.[25] Reflecting this, the West Bank was heavily settled and in most respects economically and infrastructurally incorporated into Israeli territory, while Gaza was not; the West Bank was turned into a colonisation frontier, but Gaza into a frontier enclave. Patterns of water infrastructure development and water insecurity reflected this contrast. The West Bank, as discussed, was heavily if asymmetrically integrated into Israel's national water supply network. Gaza, by contrast, was not; instead it was circumnavigated by Israel's supply lines to the Negev and left to the mercy of its own limited water resources and the encroaching sea. None of this was dictated by patterns of resource availability, still less by scarcity. Indeed it is notable that while the West Bank is a land of relative water plenty but acute water conflict – of political hostility over water supply inequalities and regular state violence against Palestinian supply infrastructures – Gaza is a land of relative water scarcity but no equivalent hydro-political discord or violence. As in Sudan, it is local resource abundance rather than scarcity that is most closely correlated with water-related conflict.

---

[24] J. Schwarz, 'Water resources in Judea, Samaria, and the Gaza Strip', in J. D. Elazer (ed.), *Judea, Samaria and Gaza* (American Enterprise Institute, 1982), 95–100; E. Weinthal et al., 'The water crisis in the Gaza Strip: prospects for resolution', *Ground Water*, 43:5 (2005), 653–60; UK Government Office for Science, *Foresight: Migration and Global Environmental Change, Final Project Report* (2011); Messerschmid, *Water in Gaza*; A. Hass, 'The most logical solution to the Gaza water crisis is the most political one', *Ha'aretz* (16/10/2018).

[25] D. Landau, 'Rabin expresses his frustration with Palestinian stance in talks', *Jewish Telegraphic Agency Daily News Bulletin* (04/09/1992).

Water Apartheid in the West Bank and Gaza 189

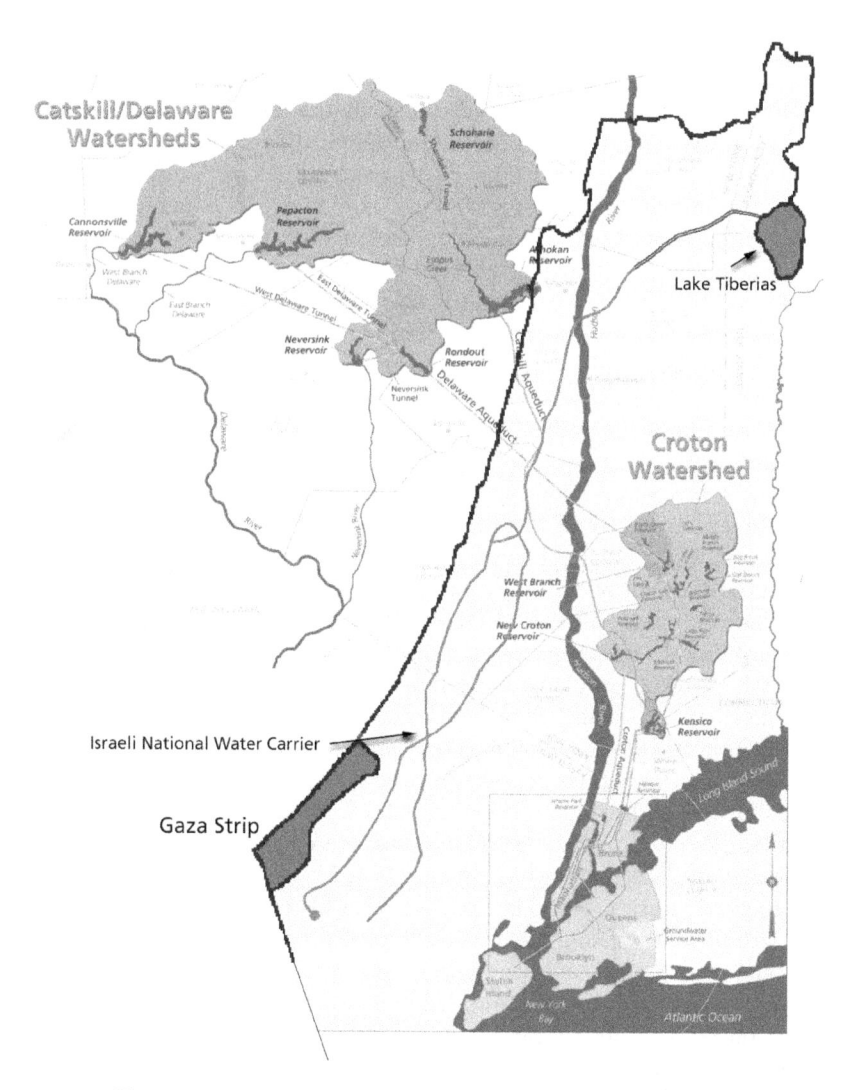

Figure 6.1 Gaza is Manhattan. This image, produced by Clemens Messerschmid, overlays Israel's and New York City's water supply systems, to scale. Like most cities New York is dependent on a hinterland for its water supply. The Gaza Strip, by contrast, is denied one – despite being not much bigger than Manhattan and having a far smaller population.

The Rise and Fall of Hasakah

Let us now return to Syria.[26] We have already seen, in the previous chapter, that over the four decades from the 1960s Syria's Ba'athist regime pursued a project of rapid state-led hydraulic and agricultural development. Moreover, we have also seen, while considering the Syria drought–conflict thesis, that Syria was experiencing a deepening agrarian crisis from the early 2000s onwards, well before the extreme drought of 2006/7–2008/9. This structural crisis, as discussed, was at a national level rooted in the decisive turn towards the market economy inaugurated by Bashar al-Assad following his accession to the presidency in 2000. In line with World Bank and International Monetary Fund (IMF) advice, a rural economic system, which for decades had been defined by state farms, agricultural input subsidies, food price controls and heavy investment in hydraulic infrastructures, was rapidly dismantled, leading to a nationwide decline in agricultural employment and a mass exodus to the peri-urban slum areas around Syria's cities. An economically unsustainable agrarian expansion was quickly followed by crisis, involution and out-migration.

Yet the above provides only a partial explanation of Syria's pre–civil war agrarian and rural out-migration troubles, for alongside these national-level dynamics were important regional ones, most crucially in that province where these problems were most extreme: the north-eastern-most province of Hasakah. Some context is required. Environmentally, Hasakah ranges from rolling, fertile plains along the border with Turkey, where annual precipitation can be over 600 mm, to steppe and desert lands in the south which receive rainfall of 200 mm or less. It is crossed north to south by what historically has been Syria's largest Euphrates tributary, the Khabour, which in turn is fed by a mass of perennial and seasonal tributaries; plus Hasakah has also historically possessed extensive groundwater resources, especially in the north. Prior to the 1930s, the area was barely cultivated let alone enclosed, its rural economy dominated instead by semi-nomadic Bedouin pastoralists working on open rangelands; in Bernard Lewis' formulation, Hasakah was part of Syria's 'zone of nomadism'. But from the 1930s onwards, and especially from the 1960s, it became Syria's major pioneering zone, the home of both the Ba'athist state's agricultural modernisation project and its nascent oil industry. Large-scale irrigation began in the 1950s, centred on the floodplains of the Khabour and its tributaries, before expanding significantly from the 1970s as the Syrian government, acting on advice from the FAO and the US Department of Agriculture, embraced the systematic

---

[26] This section summarises Selby, 'Climate change and the Syrian civil war, part II'.

exploitation of the region's groundwater resources. Open range lands were demarcated, privatised and appropriated, while the Bedouin pastoralists who had previously farmed them were pushed ever deeper into the *badiya* (desert) or isolated in marginal areas, progressively sedentarised and impoverished. In their place, people moved in or were resettled from elsewhere in the country, such that Hasakah came to witness exceptionally high in-migration and population growth. It was these processes of horizontal agricultural expansion underpinned by surface and especially groundwater irrigation, combined with land enclosure, displacement and large-scale in-migration and settlement, which led to Hasakah becoming Syria's 'breadbasket' region – accounting, by the 1990s, for up to half of the country's total wheat as well as cotton production.[27]

The ecological, economic and social consequences of these frontier dynamics were profound. For one, the transformation of Hasakah into the country's 'breadbasket' was premised on, and only made possible by, unsustainable levels of groundwater abstraction. By the turn of the millennium, 78 per cent of the irrigated land area in the Khabour basin was receiving its water from wells, with total irrigation water use equivalent to more than 300 per cent of the basin's safe yield. As early as the late 1990s, groundwater levels were declining throughout Hasakah at a rate of several metres per year. Of course, Hasakah was not alone in facing such problems: by 2001, most basins across Syria were in water deficit; the country's annual water consumption exceeded sustainable yield by 20 per cent; and irrigation water use alone amounted to 99 per cent of the country's sustainable water yield as well as 85–90 per cent of total water use – with most of this coming from groundwater. That said, no other region of the country was home to such extreme groundwater dependence or over-abstraction. Well before the 2006/7–2008/9 drought, Hasakah's water resources were already being mined to a degree that, even by Syrian standards, was extraordinary.[28]

Hasakah's surface water resources were also in steep decline. The Khabour River, which had historically been the lifeblood of the region's

---

[27] B. Lewis, *Nomads and Settlers in Syria and Jordan, 1800–1980* (Cambridge University Press, 1987), 4; Ababsa, 'Agrarian counter-reform', 85; J. Kolars and W. Mitchell, *The Euphrates River and the Southeast Anatolia Development Project* (Southern Illinois University Press, 1991), 144; Hole and Zaitchik, 'Policies, plans, practice'; D. Chatty, 'The Bedouin in contemporary Syria: persistence of tribal authority and control', *Middle East Journal*, 64:1 (2010), 29–49; Khawaja, *Internal Migration*, 25; Selby, 'Climate change and the Syrian civil war, part II', 263.

[28] C. Varela-Ortega and J. Sagardoy, 'Irrigation water policies in Syria: current developments and future options', in Firillo and Vercueil, *Syrian Agriculture*, 340, 343; D. Orešić and G. Bahnan, 'Water resources in agriculture in north-eastern Syria (governorate Al Hasakah)', *Hrvatski Geografski Glasnik*, 68:1 (2006), 87–9 (in Croatian); World Bank, *Syrian Arab Republic Irrigation*, 11, 13.

agricultural economy, ceased being a perennial river in the late 1990s and became a seasonal stream, dry at the regional capital of Hasakah for six months of the year. The main source spring of the Khabour River, at Ras al-Ayn, which had hitherto discharged around 1.5 bcm/y – a flow volume which made it one of the world's great springs, the so-called 'great karst spring of Mesopotamia' – had, by 2002–10, declined to just a tenth of this level. Both agricultural production and settlement were inevitably affected. Between 1990 and 2000, as the Khabour dried, the cultivated area within the upper Khabour, north of Hasakah city, declined by more than 40 per cent, and within the lower Khabour by nearly 75 per cent. Even prior to 2000, settlements were being abandoned both within the Khabour floodplain and in areas of particularly excessive groundwater irrigation. As these processes unfolded, production shifted instead to off-river and increasingly marginal plots, including areas of the steppe with average rainfall below 200 mm where groundwater dependency was even higher and agricultural production more precarious. And cotton and wheat production, which had risen steeply during the 1980s and 1990s, peaked around 2000 before then going into secular decline. Although this pattern was not unique to Hasakah, nowhere else in Syria did it take such an extreme form. As is typical of frontier and late developing regions, the excesses and contradictions of development were especially acute there. In only a few decades, Hasakah went through the full cycle of boom and bust – from open rangeland to national breadbasket before collapsing into what Watts, in another context, has called a 'recessional frontier'.[29]

While the proximate cause of this pattern was the discovery and exploitation of groundwater resources followed by their rapid depletion, three other sets of frontier dynamics were also at work; these are key to explaining both the region's overall rise and fall, and its water problems in particular. To start with, both because of its peripheral location and its unique political economy, Hasakah's agrarian transformation was not accompanied by any equivalent industrial let alone service sector

[29] D. Orešić and G. Bahnan, 'River regime changes in the upper part of the Khabour catchment area in north-eastern Syria in the second half of the XXth century – part I: annual runoff', *Hrvatski Geografski Glasnik*, 67:1 (2005), 56 (in Croatian); D. Orešić and G. Bahnan, 'River regime changes in the upper part of the Khabour catchment area in north-eastern Syria in the second half of the XXth century – part II: mean monthly runoff', *Hrvatski Geografski Glasnik*, 67:2 (2005), 84 (in Croatian); Syria CBS, 'Length of rivers within the Syrian land and their flow rates' (2010); UN-ESCWA, *Inventory of Shared Water*, 91; D. J. Burdon and C. Safadi, 'Ras-el-Ain: the great karst spring of Mesopotamia: an hydrogeological study', *Journal of Hydrology*, 1:1 (1963), 58–95; Hole and Zaitchik, 'Policies, plans, practice', 141–8; Hole, 'Drivers of unsustainable land use', 6–7; Selby, 'Climate change and the Syrian civil war, part II', 262–4; M. J. Watts, 'Frontiers: authority, precarity and insurgency at the edge of the state', *World Development*, 101 (2018), 482.

investment, leaving the province dominated by primary agricultural (plus fossil fuel) production. Hasakah and neighbouring parts of the north-east were thus the poorest in Syria: in 2003–4, according to the UNDP, 58 per cent of Syria's poor resided there and 18 per cent of the region's rural population were living in extreme poverty. Moreover, during the early 2000s, as Bashar al-Assad's liberalisation programme unfolded, poverty levels in the region deteriorated, despite improving across most of the country. It is not difficult to see why this was. A region that had benefitted disproportionately, if unevenly and distortedly, from a national agrarian project would inevitably also be disproportionately affected when the core underpinnings of this boom – subsidised fuel and fertilisers, price controls on agricultural products, state farms and locally abundant water resources – were suddenly removed. It should be no surprise, given this, that while farming employment declined right across Syria from 2000 onwards it declined particularly sharply in Hasakah: by 2008 its official unemployment rate was 28 per cent, near double that of any other province in the country.[30] Hasakah's frontier water crisis was in these senses a product of its distinctively frontier economy.

Part of the reason for this, second, lay in the province's ethnic composition: majority or near-majority Kurdish, and home to the largest concentration of non-Arabs in the country. Although Syria's Kurds had historically not been subject to anything like the same level of extreme state violence and repression as in neighbouring Iraq or Turkey – at least not prior to the civil war – the Syrian state and its Arab nationalist leaders had long waged a quiet war against them. At extreme, Syria's Kurds were viewed, in the rightly infamous words of Muhamed Talab Hilal, head of security in Hasakah during the 1960s, as 'a people who have no history, civilisation, language, or ethnic origin'; who are 'dirt' that requires 'purification' by 'the Arab conscience'; who are 'a malignant tumour on the side of the Arab nation and must be removed'.[31] And this shaped Hasakah's frontier dynamics in two ways.

On the one hand, from the 1970s onwards the Syrian state pursued a programme of Arabisation aimed at expanding and consolidating Hasakah's Arab population. The most well-known instance of this was during 1973–6 when, in line with proposals made by Talab Hilal, more than 20,000 Arab villagers were transferred from land flooded during the creation of Lake Assad to forty-two model farming settlements in Kurdish areas of northern Hasakah, the so-called 'Arab Belt'. Like the other state-led colonisation schemes examined in this chapter, these Arab

---

[30] El Laithy and Abu-Ismael, *Poverty in Syria*, 1, 3, 27; Aita, *Labour Market Policies*, 32.
[31] M. Gunter, *Out of Nowhere: The Kurds of Syria in Peace and War* (Hurst, 2014), 22–3.

settlements benefitted from wide-ranging state subsidy and support, and involved the widespread expropriation of land and attendant water resources. More broadly, the many other Syrian Arabs moving into Hasakah from the 1970s onwards also benefitted from easy access to government credit and subsidies, to the extent that Syria's supposedly nationwide system of agrarian support effectively functioned as an instrument of Arabisation. As one telling example of this, the 2009 UN and Syrian government drought relief programme in Hasakah focused entirely on one of the provinces' Arab districts, despite the fact that Kurdish communities were also affected by the drought and indeed were probably more vulnerable to it. Hasakah became Syria's pre-eminent frontier region and then experienced such a steep agrarian collapse, not only because of its untapped land and water resources but – just as within the adjacent Kurdish-dominated region of southeast Turkey – because agricultural development facilitated internal colonisation and the extension of state control over a heavily ethnic minority-populated territory.[32]

Alongside this project of Arabisation, the Syrian state also pursued efforts to exclude, control and weaken Hasakah's indigenous Kurdish population. The 1962 Hasakah census, undertaken, spuriously, in just one day and in Hasakah province alone, created a class of several hundred thousand 'non-citizen Kurds' who, shorn of their citizenship rights, could not own or rent land, work in the public sphere or receive state loans or benefits; by 2009, there were an estimated 250,000–300,000 such 'stateless Kurds' in Syria, mostly in Hasakah. This denial of political rights and citizenship was accompanied by widespread land expropriation, first during the 1960s – both as a direct consequence of the 1962 census and through Syria's 1963 agrarian reform programme – and later during the creation of the Arab Belt; reflecting this, an unusually high proportion of land was expropriated in Hasakah and the Jazira. In addition, from 1964 the whole of Hasakah was defined, unusually among Syrian provinces, as a 'border area', such that the Ministry of Interior and other departments were granted multiple vetoes over land use development. These arrangements were selectively enforced to restrict Kurdish development, and were further tightened through Presidential Decree 49 of 2008, which once again was applied to all of Hasakah (and Qunaytirah, bordering the Israeli-occupied Golan Heights) only – resulting in it being widely denounced by Kurdish rights activists as 'a continuation of the Arab

---

[32] Meyer, 'Rural development', 249–59; HRW, *Syria: The Silenced Kurds* (1996); UN-OCHA, *Syria Drought Response Plan 2009–10, Mid-Term Review*, 4; De Schutter, *Report of the Special Rapporteur*, 12; 'The forgotten people: Kurds in Syria'; *Kurdish Aspect* (09/11/2010).

Belt policy' and 'a decree of ethnic cleansing and demographic change'. Such systemic discrimination must have been an important factor in Hasakah's pre–civil war socio-ecological crisis.[33]

Far from being merely internal issues these frontier dynamics also had international dimensions, rooted in regime concerns about the transnational Kurdish question and its hold over a valuable if peripheral province. Thus the Arab Belt was intended not just to help Arabise Hasakah's population, but also to create a cordon sanitaire along the border with Turkey which would separate Hasakah's Kurds from the much more numerous and politically radical Kurdish population of south-east Anatolia, it being for this reason that it took the form of a 15 km-deep band of villages extending 280 km along the Turkish border. Similarly, the Syrian government justified its denial of citizenship to 'non-citizen Kurds' on the grounds that, following its 1963 agrarian reforms, there had been an increase in illegal immigration by Turkish Kurds hoping to benefit from land redistribution.[34] The Syrian state's exclusion and dispossession of Hasakah's Kurds was rooted in insecurities that were international and geopolitical, as much as internal.

Last, Hasakah's borderland location also contributed to its mounting water crisis in a more direct way, since the surface and groundwater resources on which Hasakah is naturally dependent are not internal but cross-border, being mostly recharged over the Taurus Mountains. An estimated 83 per cent of the total flow of the Khabour River naturally originates in Turkey, such that the latter's large-scale development of these resources since the late 1980s has had a major bearing on Hasakah's water supplies. It is impossible to say where the balance of responsibility lies for the drying up of Ras al-Ayn and the Khabour River and the general collapse of Hasakah's groundwater levels, since this would require coordinated monitoring and modelling on both sides of the border – and no such work has ever been undertaken. Yet irrigation development on the Turkish side of the border was so intense between 2001 and 2015 that there was a secular increase in vegetative cover there, by contrast with a marked decrease within Hasakah itself.[35] What seems clear, therefore, is that Turkey must bear at least some share of responsibility for the rapid degradation of Hasakah's water

---

[33] HRW, *Group Denial: Repression of Kurdish Political and Cultural Rights in Syria* (2009), 10; De Schutter, *Report of the Special Rapporteur*, 14; H. Allsopp, *The Kurds of Syria: Political Parties and Identity in the Middle East* (I.B. Tauris, 2015), 25–7, 149–55, 160; Meyer, 'Rural development', 251; Tejel, *Syria's Kurds*, 60; Kurdwatch, *Decree 49: Dispossession of the Kurdish Population?* (European Center for Kurdish Studies, 2010).
[34] Meyer, 'Rural development', 258; HRW, *Syria*, App. A.
[35] Kolars and Mitchell, *Euphrates River*, 191, 222; L. Eklund and D. Thompson, 'Differences in resource management affects drought vulnerability across the borders between Iraq, Syria and Turkey', *Ecology and Society*, 22:4 (2017), 9.

resources, perhaps even the lion's share. As in Gaza, the politics of the border has been constitutive of groundwater crisis.

The combination of deep poverty, major oilfields, abundant waters, wide open spaces for enclosure and colonisation, distance from Damascus, a substantial Kurdish population and proximity to both Turkey and Iraq all conspired to make Hasakah a uniquely sensitive as well as opportunity-laden frontier region, and modern Syria's pre-eminent site of internal colonisation, ethnic discrimination and unsustainable development. The region's pre–civil war groundwater crisis and its extreme vulnerability to the 2006/7–2008/9 drought – in short, its exceptional levels of water resource degradation and water insecurity – were the inexorable consequences. As in South Kordofan, Darfur and the West Bank and Gaza, in Hasakah too frontier abundance proved to be double-edged. Yet, as within our other cases, this relative resource abundance was not the fundamental cause of what subsequently ensued. Rather, Hasakah's pre–civil war water crisis was, like the broader military and insecurity crises to follow, in large measure a legacy of dynamics set in motion by the Sykes–Picot Agreement and Treaty of Lausanne eighty years previously.

The Lake Chad Basin: Nigeria's Outstanding Internal Frontier

More than our other cases, the frontier character of the Lake Chad region derives, at one level, from brute facts of geography: its cross-border location; its huge distance from three of the four national capitals (Niamey is 1,200 km away from Lake Chad, Abuja and Yaoundé each around 1,000 km away); its Sahelian ecology jutting out into the Sahara; the complex wetland social ecology of Lake Chad itself; and the long history of attenuated state authority and local resistance which has, almost inevitably, followed. And yet, as we have sought to illustrate, contemporary frontiers are above all products of uneven development and state-building within the context of global capitalist intensification and transformation; 'frontierity' today is a function of history and political economy much more than physical geography. In what follows we illustrate this once again, focusing mainly on the Nigerian portion of the Lake Chad basin, the states of Borno, Adamawa and Yobe which have been at the epicentre of the 'Lake Chad crisis'.

We must start with the colonial era. Under British rule from 1861 onwards, northern and southern Nigeria were administered largely separately, with distinctive systems of law, land tenure and governance, and Native Administrations structured around regional 'tribal' majorities – in northern Nigeria, the Hausa-Fulani. Indeed, Lord Lugard's system of indirect rule was pioneered in Nigeria (and Uganda), before being rolled out elsewhere including, as discussed in Chapter 4, within Sudan. Unlike Sudan, however, colonial development in Nigeria focused on the south, while northern development was actively restricted with a view to preventing the erosion of traditional authority and the rise of threats to British rule. As a result, on independence in 1960 northerners accounted, for instance, for just 2 per cent of senior civil service posts and the north had under 6 per cent of the country's secondary schools. In addition, both under British rule and afterwards the northeast was treated as peripheral even within Nigeria's north. Following independence the north-east remained politically under-represented, economically neglected and conspicuously poor, with particularly high poverty rates, low educational access and more. Indicatively, only in 1964 was Nigeria's railway network extended to Maiduguri; not until the late 1970s was a paved road laid between Maiduguri and the town of Baga, near Lake Chad.[36]

In a departure from this pattern of colonial and early post-colonial marginalisation, however, from the 1970s north-east Nigeria became a site of intense state-led and internationally supported agrarian development. Nationally, agricultural expansion and intensification became prioritised as never before, partly in response to the Biafran war and famine of 1967–70 and the Sahelian droughts of the early 1970s, and partly simply in line with policies being adopted across the post-colonial world, including, as we have seen, in Syria and Sudan. Successive initiatives and national development plans – 'Operation Feed the Nation' launched in 1976 by the military government of Olusegun Obasanjo; Shehu Shagari's 'Green Revolution' programme which replaced it – prioritised irrigation development and increased wheat and rice production, in particular. A system of River Basin Development Authorities was established, including, among the first of these, the Chad Basin Development Authority. And the north-east specifically was identified as a prime region for agricultural expansion and intensification, for reasons which should, by now, be familiar: its assumed abundance of land; its abundance of under-exploited water resources; its relatively sparse population; and the benefits of developing

---

[36] A. R. Mustapha, *Ethnic Structure, Inequality and Governance of the Public Sector in Nigeria* (UN Research Institute for Social Development, 2006); B. Archibong, 'Historical origins of persistent inequality in Nigeria', *Oxford Development Studies*, 46:3 (2018), 325–47; Mamdani, *Citizen and Subject*, 7, 104, 113–14; G. Magrin and M.-A. Pérouse de Montclos (eds.), *Crisis and Development: The Lake Chad Region and Boko Haram* (Agence Française de Développement, 2018), 86; G. Magrin and G. M. Ngaressem, 'Le lac Tchad et les échanges: un pôle agricole exportateur', in Lemoalle and Magrin, *Développement du Lac Tchad*, 565, 570.

and integrating what successive national governments viewed as a strategically important but long resistant borderland region.[37]

Hydraulics were central to this mission of developing the north-east. Under the South Chad Irrigation Project (SCIP), launched in 1972 as 'the largest and perhaps the most ambitious' irrigation project in Nigeria, water was to be pumped from Lake Chad and distributed through canals to irrigate over 100,000 hectares of land for wheat, rice and cotton production. Under the Baga Polder project, launched at the same time, a 32 km barrier was constructed around part of the lake to capture seasonal flooding, with the aim of irrigating 20,000 hectares of land. In addition, and more significantly, dams were constructed along all the rivers feeding Lake Chad, together with state-owned irrigation schemes. In total since the 1970s, Nigeria has constructed twenty-seven dams along the Komadougou-Yobe River and its tributaries, including the 1,345 mcm-capacity Tiga Dam, the 1,140 mcm-capacity Hadejia Dam and the 930 mcm-capacity Chellawa Gorge Dam; while, in parallel, both Chad and Cameroon have built major dams on the Chari-Logone. In addition, a series of World Bank supported Agricultural Development Projects (ADPs) were launched to assist the 'traditional' agricultural sector – these initially focusing on rain-fed agriculture but then, from the early 1980s, on small-scale groundwater irrigation, providing farmers with subsidies and loans for drilling tube wells and purchasing diesel pumps.[38]

Many of these initiatives fell way short. Lake Chad's contraction during the 1970s and early 1980s left many of the SCIP's distribution canals stranded well away from the water's edge such that, by 1983–4, only around a tenth of the promised area was being cultivated. By the mid-1990s, the Baga Polder project had only about 1,000 hectares under irrigated cultivation. More broadly, recurring problems of water distribution and salinisation, combined with reduced state support within the context of declining oil revenues and structural adjustment programmes, meant that many of Nigeria's large-scale hydraulic and agricultural

---

[37] H. O. Sano, *The Political Economy of Food in Nigeria 1960–1982* (Scandinavian Institute of African Studies, 1983), 24–30; K. Kimmage, 'The evolution of the "wheat trap": the Nigerian wheat boom', *Africa: Journal of the International African Institute*, 61:4 (1991), 471–501; A. Kolawole, 'RBRDAs and vulnerability to hunger in Nigeria: the case of the South Chad Irrigation Project', *Food Policy*, 13:4 (1988), 389–96.

[38] Kolawole, 'RBRDAs', 392; M. Bertoncin and A. Pase, 'Interpreting mega-development projects as territorial traps: the case of irrigation schemes on the shores of Lake Chad', *Geographica Helvetica*, 72:2 (2017), 247–9; GIZ, *Joint Environmental Audit on the Drying up of Lake Chad* (2015), 154–6; Sano, *Political Economy*, 42; K. Kimmage, 'Small-scale irrigation initiatives in Nigeria: the problems of equity and sustainability', *Applied Geography*, 11:1 (1991), 5–20.

development projects failed to realise their potential. Nonetheless, the dam schemes along with small-scale groundwater irrigation quickly transformed the north-east into Nigeria's 'breadbasket' region, focused on wheat, rice, maize and tomato production. Between 1976 and 1995 alone, Borno's cultivated land area increased from 2.9 to 3.8 million hectares, around 51 per cent of its total land area, while in the decade to 2010 it increased by a further 380,000 hectares.[39]

The environmental consequences have, unsurprisingly, been profound. Since the 1960s the Chari-Logone River, the main source for Lake Chad's southern pool, has seen its average flow drop by at least 50 per cent. The Komadougou-Yobe River, feeding the northern pool, has reportedly declined by even more. Those studies that have analysed this have all identified local human activities – that is, surface and groundwater withdrawals – rather than drought as the main cause of this low river flow and the lake's minimal recovery since the 1970s–1980s. In particular, although data is scant on this, Nigeria's development of the Komadougou-Yobe may largely explain the non-recovery of the northern pool since the 1980s. The construction of dams has also disrupted downstream seasonal flooding, with attendant ecological as well as social consequences. Across the north-east, the clearance and levelling of land to make way for irrigated agriculture has led to soil erosion and damage to soil structure and soil fertility. And abstraction for irrigation has been identified as a major cause of widespread groundwater depletion in north-east Nigeria, groundwater levels across Borno having declined by several metres per year since the 1960s.[40]

These environmental consequences aside, agrarian development has also been accompanied by sweeping changes to the political economy of Nigeria's north-east. As in Sudan, the new state-led agrarian development drive coincided with, and was in part facilitated by, transformations

---

[39] W. M. Adams, 'Large scale irrigation in northern Nigeria: performance and ideology', *Transactions of the Institute of British Geographers*, 16:3 (1991), 287–300; Bertoncin and Pase, 'Interpreting mega-development'; UNEP, Lake Chad Basin: GIWA Regional Assessment *43* (2004), 49, 62; World Bank, *Assessment of Vegetation and Land Use Changes in Nigeria between 1976/78 and 1993/95* (1998), 8/24, 8/107; A. O. Arowolo et al., 'Assessing changes in the value of ecosystem services in response to land-use/land-cover dynamics in Nigeria', *Science of the Total Environment*, 636 (2018), 597–609.
[40] Lemoalle, 'Le fonctionnement hydrologique du lac Tchad', in Lemoalle and Magrin, *Développement du Lac Tchad*, 23–5; GIZ, *Joint Environmental Audit*, 45; Mahmood and Jia, 'Assessment of hydro-climatic trends'; Zhu et al., 'Relative contribution'; T. Wallace, 'Agricultural projects and land in northern Nigeria', *Review of African Political Economy*, 7:17 (1980), 65–6; Kimmage, 'Evolution of the "wheat trap"', 492–5; UNEP, *Lake Chad*, 51; S. Adamu et al., 'Groundwater depletion in the upper aquifer of the Chad formation, Chad Basin, north-eastern Nigeria', *Nigerian Journal of Technology*, 39:2 (2020), 621–31.

in local authority structures – the dissolution of the Native Authority system in 1968 and the reorganisation of the federal system in the 1970s through the creation of new states and the establishment of Local Government Areas (LGAs) – which resulted in a 'virtual collapse' of community-level governance structures. Property relations were also transformed; the 1978 Land Use Act nationalising communal lands and assigning their management, formerly under traditional authorities, to LGAs and state governors. In turn, the main beneficiaries of the large-scale agricultural projects during this period – those benefitting most from allocations of land, water, loans and subsidies – were an elite group of absentee military officers, high-ranking civil servants and wealthy urban business people. Rapidly rising wheat prices plus the ready availability of subsidies resulted in 'a wheat-induced land scramble in the northern states of Nigeria', led by wealthy 'overnight farmers'. Large swathes of land were cleared, levelled, enclosed and then cultivated, intruding on land previously used for dry season grazing and contributing to an increase in land-related conflicts. Moreover, unable to afford water, seeds, machinery, fertilisers or labour costs, many smaller landholders rented or sold off their land, such that land ownership became increasingly concentrated. Although the ADPs ostensibly targeted small-scale producers with plots of just a hectare or two, in practice the result was the growth of cultivated plots of up to 300 hectares.[41]

One knock-on consequence was extensive migration, resettlement and displacement. The SCIP involved the relocation of around 50,000 people living in the project area – with land enclosed and people resettled on new plots as tenant farmers – plus plans to bring in another 50,000 people from other areas. Around 13,000 people were displaced for the construction of the Tiga Dam, while several thousand more were displaced to make way for canals, pilot farms and other irrigation project infrastructure – many to areas far from water and floodplain land and without adequate compensation given the rapidly rising land prices. Many of the newly landless poor migrated to regional urban centres. Meanwhile, others migrated to new agricultural frontiers, including land opened up by the recession of Lake Chad – an area of rich soils, good water availability and freedom from irrigation scheme restrictions. Indeed, during the decades prior to the eruption of the Lake Chad crisis there was net in-migration into the lake

---

[41] A. R. Mustapha, 'Understanding Boko Haram', in A. R. Mustapha (ed.), *Sects and Social Disorder: Muslim Identities and Conflict in Northern Nigeria* (James Currey, 2014), 178; Magrin and Pérouse de Montclos, *Crisis and Development*, 88–95; Sano, *Political Economy*, 34–6; A. C. Okolie, 'Oil rents, international loans and agrarian policies in Nigeria, 1970–1992', *Review of African Political Economy*, 22:64 (1995), 200; Kimmage, 'Small-scale irrigation', 9–12.

region, at least within Nigeria. While the droughts of the 1970s–1980s certainly contributed to this migration, overall it was – as in Darfur – more a consequence of frontier-style grabbing of land and water resources than of a generalised condition of scarcity.[42]

None of which is to suggest, of course, that either the emergence of Boko Haram in the mid-1990s as a sectarian political–religious movement critical of the Nigerian state, or its subsequent escalation and militarisation following the extrajudicial killing of its leader Mohammed Yusuf, and still less the security and humanitarian crises which followed, were essentially caused by the contradictions of agrarian development or environmental change. And yet it is clear that frontier dynamics, including those touched on above, were key to north-east Nigeria's and the Lake Chad region's early-2000s descents into violence and insecurity. By the early 2000s, north-east Nigeria had become a land of elite enrichment, home to levels of unemployment and poverty higher than anywhere else in the country. Moreover, Boko Haram's recruits would largely derive from rural areas and small towns. Studies of the crisis' socio-economic roots consistently point to such factors, noting the simultaneous marginalisation, exploitation and relative decline of the region, and the pervasive economic precarity there, as crucial to understanding the rise of Boko Haram. Like Hasakah in Syria's north-east, north-east Nigeria had become, in Watts' terms, an abandoned 'recessional frontier'.[43]

Resistance and Resilience

Long as this chapter already is, there is one further issue to discuss before concluding. For, from Lake Chad to Syria's north-eastern borderlands, our analysis of the political ecology of frontiers has so far focused on their opening up, colonisation and development by the state – that is, on the power and agency of the state in relation to frontier spaces and populations, and on the insecurities, violence and environmental degradation which so often follow. Where local populations have been discussed, this has essentially been as victims of these state- and elite-led processes. Yet, to leave things here would be misleading for, however unequal the power relations are between colonising core and expropriated periphery, the

---

[42] A. Kolawole, 'Farm tenancy on the South Chad Irrigation Project, Nigeria', *Land Use Policy*, 5:4 (1988), 438, 444; Wallace, 'Agricultural projects', 61–5; F. Réounodji et al., 'Histoire du peuplement et logiques de mobilité', in Lemoalle and Magrin, *Développement du Lac Tchad*, 155–60; Magrin and Pérouse de Montclos, *Crisis and Development*, 35–8.

[43] Mustapha, 'Understanding Boko Haram', 171–6; Anugwom, *Boko Haram Insurgence*, 96–103; Magrin and Pérouse de Montclos, *Crisis and Development*, 130–1; Watts, 'Frontiers', 482–5.

latter is never just a victim and is never without agency. Frontiers are spaces of particularly acute state-induced socio-ecological crises, but they are also, because of this, spaces of heightened agency, adaptation, resistance and resilience.

Noting this is important for two inter-linked reasons. It is important, on the one hand, because while frontier agency does feature right across mainstream nationalist, neo-liberal, Malthusian and colonial discourse on development and the environment, this is typically in problematic ways. Thus local populations are widely portrayed as obstacles to progressive development, as beholden to outdated traditions and as inefficient water users; they are often criticised for their 'non-payment culture', for their 'illegal' abstractions and for over-breeding; and they are typically depicted, in sum, as authors of their own environmental insecurity. Such pejorative readings of frontier agency need contesting. Yet, on the other hand, poor and marginalised communities are all too rarely represented as possessing political as opposed to social agency vis-à-vis their environments, that is, as engaging in forms of environmental politics or environmentalism. Environmentalism itself is still widely thought of as a predominantly Northern middle-class concern, reflecting 'post-materialist' values and economic security. Eco-determinist thought includes no environmentalists, as already discussed in Chapter 2. And the goal of 'protecting nature' has historically often been allied to elite, state and colonial ambitions – as we see worldwide in militarised conservation practices and, within our cases, in the central place of the environment in Israeli society, where the protection and ostensible repair of nature has served as ideological cover for the expropriation of land. Within these contexts, 'frontier environmentalism' – and, more broadly, what Ramachandra Guha and Joan Martinez-Alier label 'empty belly environmentalism' and 'the environmentalism of the poor' – are even today often hidden from view.[44]

Across our cases, such frontier or more broadly subaltern socio-ecological activism has taken a range of more or less explicitly political forms. In South Kordofan, the Nuba Mountains General Union – founded in the 1960s by Nuba intellectuals to expand their national political representation, promote pan-Africanism and oppose the Sudanese regime's taxation policies – has opposed the appropriation of

---

[44] H. Gvirtzman, The Truth behind the Palestinian Water Libels, BESA Center Paper 238 (Begin-Sadat Center for Strategic Studies, 2014); R. Inglehart, 'Post-materialism in an environment of insecurity', *American Political Science Review*, 75:4 (2014), 880–900; R. Duffy et al., 'Why we must question the militarisation of conservation', *Biological Conservation*, 232 (2019), 66–73; Guha and Martinez-Alier, *Varieties of Environmentalism*; Nixon, *Slow Violence*.

their water-rich land by non-Nuba and non-Moro merchants and called for the eradication of 'feudalistic land policies and relations of production'. On the Nile, an array of indigenous rights, women's rights and dam-specific groups and committees mobilised to contest the Bashir regime's post-2000 dam building, as discussed in Chapter 5. And in the Occupied Palestinian Territories, the first intifada against Israel's occupation – which involved extensive grassroots mobilisation and local institution-building under the banners of Palestinian nationalism and self-reliance – led to the establishment of a raft of popular committees and NGOs to protect and support Palestinian agriculture and land and water resources, and advocate for Palestinian water rights.[45] Across these diverse sites women have often been at the forefront of resistance, reflecting the hyper-masculine character, and gendered consequences, of both mega-dam building and frontier colonisation. As these examples suggest, subaltern environmentalism is inevitably political, typically one element in a broader project of defending threatened subjectivities, identities, communities and livelihoods against state development ambitions, and militarised (and masculine) state power.

This is even more evidently the case in Syria where, from 2012, an autonomous Kurdish region – the region of Rojava – was established in the north-east of the country, committed both to defending Kurdish rights and to an explicitly eco-socialist political agenda. Inspired by the writings of US anarchist Murray Bookchin and their adaptation by Kurdistan Workers' Party founder Abdullah Öcalan, Rojava's revolution was built on three pillars: decentralised 'Democratic Confederalism', feminism and social ecology. The latter involved a commitment to moving away from the wheat monoculture and associated environmental degradation and exploitation to which north-east Syria had long been subjected by the Ba'athist regime, and in its stead developing a model of sustainable, localised and diversified cooperative agriculture. Indeed, this agenda was no accident, being a political response to the region's longstanding – and simultaneously ethnic and environmental – frontier domination by the Syrian state. Under Ba'ath Party rule, it 'would have been impossible to assemble three sewing machines for a textile workshop', explains Remziye Mihemed, finance minister of Rojava's Jazira canton, 'because a day or two later, regime functionaries would storm in and shut it down'. 'Tree planting had to be approved by three ministries, which made it all but impossible' and resulted in Hasakah looking 'like a single

---

[45] M. A. M. Salih, 'Generation and migration: identity crisis and political change among the Moro of the Nuba Mountains', *GeoJournal*, 25:1 (1991), 54–5; Kadouf, 'Marginalization and resistance'; Schmidinger, 'Spatial control'; Zeitoun et al., 'A "justice" reading'; J. R. Nassar and R. Heacock (eds.), *Intifada: Palestine at the Crossroads* (Praeger, 1990).

huge wheatfield', observes one commentary. 'Rojava could feed two or even three times its current population yet still 60 percent of Syrians living below the poverty threshold are Kurds from Rojava', notes Ahmed Yousef, chairman of Afrin University. Whereas the Syrian state was happy to blame drought for the region's agrarian crisis, as discussed in Chapter 3, a truer picture of its causes can be found in such statements from Rojava. The paradox, of course, is that the existence of an autonomous Rojava remains dependent on US support – and that large-scale agriculture, along with oil production, remain strategic assets for it within this context.[46]

Less explicitly political, but political nonetheless, are everyday practices of frontier resilience. A note of explanation here. Within contemporary critical scholarship, 'resilience' is regularly characterised as a discursive invention of neo-liberal governance that has apolitical and individualising consequences. Yet as Caitlin Ryan has persuasively argued, within contexts of endemic state violence and dispossession, everyday practices of adaptation, improvisation and getting by can be thoroughly political – to the extent that the distinction between political 'resistance' and apolitical 'resilience' becomes erased. In the Palestinian context, most notably, practices of coping with, circumnavigating and quietly defying Israel's occupation, while maintaining dignity in the face of its ritual humiliations and remaining stubbornly on the land, have been elevated to the level of collective national strategy: the strategy of steadfastness, or *sumud*. Practices relating to water provide a case in point. Thus in the West Bank households and communities respond to regular water supply cuts through a kaleidoscope of supply- and demand-side means: by using diesel generators and rubber tubing to fill rooftop water tanks; by collecting water from nearby springs; by purchasing it on informal tanker markets; by collecting groundwater for irrigation in large cisterns dug into steep-sloping land; by reusing grey water; by connecting to settler supply lines and irrigating their crops by night; through municipal rotation of water supplies to different parts of town; by showering at the houses of those friends or relatives who happen to have water; by leaving cleaning and laundry until the water returns; and much else

---

[46] M. Bookchin, *The Ecology of Freedom: The Emergence and Dissolution of Hierarchy* (AK Press, 1982); A. Öcalan, *The Political Thought of Abdullah Öcalan: Kurdistan, Woman's Revolution and Democratic Confederalism* (Pluto, 2017); Internationalist Commune of Rojava, *Make Rojava Green Again* (Dog Section Press, 2018), 67–71; M. Knapp et al., *Revolution in Rojava: Democratic Autonomy and Women's Liberation in Syrian Kurdistan* (Pluto, 2016), 320, 322; M. Bassiki, 'Race for wheat heats up between Damascus, opposition', *Al-Monitor* (07/04/2020).

besides. In Gaza, similarly, adaptations to the territory's chronic water quality situation have included the rise of private water treatment and tanker markets and household-level desalination: almost all Gazans rely on local desalination for drinking purposes. The point here is not to romanticise such practices: coping is a burden as well as anything positive and its weight is far from equally distributed: women, illustratively, tend to be the primary practitioners of *sumud*. Yet cognisance of people's immense 'adaptive capacities' is crucial to understanding the political ecology of the frontier – to understanding, for instance, why the despoilation of Gaza's groundwater resources has not made it 'unliveable' or an 'environmental refugee' crisis waiting to happen.[47] Frontiers are sites of extreme insecurity but they are also, as a corollary, ones of particularly profound resilience and resistance.

The New Frontier Lands of Climate Change

The sorts of desert, dryland and semi-arid frontiers examined in this chapter are a recurring obsession within eco-collapse and environmental conflict narratives. The Sahara Desert, we are told, is voraciously expanding. Everywhere, desertification is thought to be unfolding apace. The Sahel – the Sahara's 'border' or 'shore' – is a particular concern, imagined within UN organisations and European stabilisation units and development ministries as a region full of climate change–induced strife and potential climate refugees. The fears of collapse discussed in Chapter 2 find their main geographical corollary just here: at the semi-arid frontier.[48]

---

[47] J. Joseph, 'Resilience as embedded neoliberalism: a governmentality approach', *Resilience*, 1:1 (2013), 38–52; C. Ryan, 'Everyday resilience as resistance: Palestinian women practicing sumud', *International Political Sociology*, 9:4 (2015), 299–315; P. Bourbeau and C. Ryan, 'Resilience, resistance, infrapolitics and enmeshment', *European Journal of International Relations*, 24:1 (2018), 221–39; R. Shehadeh, *The Third Way* (Quartet, 1982); Selby, *Water, Power and Politics*, ch. 8; J. Selby, 'Governance and resistance in Palestine: simulations, confrontations, sumoud', in F. Cochrane et al. (eds.), *Global Governance, Conflict and Resistance* (Palgrave, 2003), 118–34; Hass, *Drinking the Sea*, 59–60; Oxfam GB, 'Water Markets in Gaza: An Emergency Market and Mapping Analysis (EMMA) Survey on Private and Public Supply Markets in Gaza' (2013); J. Peteet, *Gender in Crisis: Women and the Palestinian Resistance Movement* (Columbia University Press, 1991); UNSCO, Gaza in 2020; G. Bromberg, 'When Gazan refugees pour into Tel Aviv', *Times of Israel* (06/09/2015).
[48] C. Werrell and F. Femia, 'AFRICOM commander on climate change: Sahel receding almost a mile per year', Center for Climate and Security blog (18/03/2018), https://climateandsecurity.org/2018/03/africom-commander-on-climate-change-sahel-receding-almost-a-mile-per-year/; UNEP, *Livelihood Security, Climate Change, Migration and Conflict in the Sahel* (2011); J. Ribot et al., 'Climate of anxiety in the Sahel: emigration in xenophobic times', *Public Culture*, 32:1 (2020), 45–75.

That such fears are typically without empirical basis has already been amply demonstrated in previous chapters. Beyond this point, though, what is evident is that eco-collapse and environmental conflict narratives fundamentally misread the nature of frontier lands today. While such narratives essentially read frontier violence as arising from resource scarcities and, in turn, social contraction and retreat, the evidence in this chapter – and what we know of resource frontiers more broadly – suggests very differently. From South Kordofan, Darfur and Lake Chad to Israel-Palestine and north-east Syria, it is real or imagined abundances of land and associated water resources which have been the more fundamental invitations to conflict. In each of these contexts, technological or infrastructural changes combined with a state-led developmental project – some combination of pipelines, rail-lines, roads, tractors and tubewells, together a nationalist ideology and resources of state – have turned hitherto 'under-developed' lands and waters into highly valued resources and commodities. Across these cases, also, frontiers have consistently been marked by economic dynamism, in-migration and settlement, widespread illegality, the expropriation and colonisation of land and resources, the dispossession of local populations and extreme levels of political contestation and violent conflict. The defining feature of contemporary frontier lands, in short, is not scarcity-driven recession, but their expansionary and conflict-laden incorporation into both circuits of state power and logics of global capital.

This is not to suggest, of course, that frontier lands are immune to recession or ecological crisis. From the American wildwood to late twentieth-century Hasakah, the development of barely regulated and contested frontier regions has always involved local environmental destruction, with this in turn leading to either out-migration or the 'opening up' of yet more, and more ecologically precarious, frontier territory. Moreover, because frontier development is typically facilitated and made possible by state projects and far-away interests, such regions are acutely vulnerable to political and economic changes, whether these be new strategies of population control or shifts in national subsidy regimes or international commodity markets. Recession, environmental degradation and with them exploitation and conflict are all endemic features of the frontier. Crucially, however, these are not consequences of natural scarcity but of the reverse: the 'water frontiers' examined in this chapter were all, at least once, regions of relative water abundance.

This does not mean, we should stress, that it is the natural distribution of water resources or water-related interests that are the principal drivers of conflict at the frontier. And yet, judging by our cases, water is much more central to the dynamics of contemporary frontiers than is often recognised – including within research on resource frontiers and water politics. Classically, it was soil exhaustion which was the main limit factor on frontier agricultural production, with the clearing of forests and exploitation of 'virgin soils' being quickly followed by their rapid depletion, declining yields and onward migration. Paucities of fuel and labour and were also sometimes critical, though the one was typically mitigated by deforestation, the other by slavery or indentured servitude. Contemporary semi-arid frontier zones present a very different picture, however. Artificial fertilisers now enable multi-cropping and monocultures even in the thinnest of soils. Mechanisation and cheap diesel and electricity have hugely reduced demand for wood fuel and labour.[49] And the upshot is that, today, water scarcity has become the main limit factor on agricultural expansion in many semi-arid frontier regions while local water abundances – of groundwater resources in particular – have become its sine qua non. In many frontier regions water is effectively the new soil: the attribute that most gives value to land and which most makes land worth commodifying, appropriating, developing and fighting over.

What, though, of the future, especially within the context of accelerating climate change? Over the coming decades the world's dryland frontier regions will inevitably experience significant climatic changes with equally inevitable (though also place-specific) environmental, economic and political consequences: nothing in the foregoing is meant to dispute this. However, if the past is any guide these climatic transformations will not be the central cause of frontier conflicts and insecurities since, within a global capitalist system, the latter are essentially rooted in projects of expansion, incorporation, colonisation and resource exploitation and capture. So long as our global order remains capitalist, new abundances and new projects of development and state expansion will inevitably be accompanied by new frontier rushes and frontier conflicts. Struggles to control land and its resources will intersect with longstanding or newly constructed divides between core and marginalised populations and, within these populations, along lines of class, race, gender and more. Unsustainable exploitation of local resources – the mining of soils, groundwater and so on – will repeatedly induce recession, out-migration and the further opening up of new frontiers. And the need to adapt to and mitigate climate change will furnish powerful new

---

[49] Turner, *Frontier*, 21–2; J. W. Moore, 'Sugar and the expansion of the early modern world-economy: commodity frontiers, ecological transformation, and industrialization', *Review*, 23:3 (2000), 409–33; B. Clark and J. B. Foster, 'Ecological imperialism and the global metabolic rift: unequal exchange and the guano/nitrates trade', *International Journal of Comparative Sociology*, 50:3–4 (2009), 311–34.

motivations and legitimations for frontier development, in some cases compounding pre-existing problems of unsustainable exploitation and social marginalisation. Semi-arid 'water frontiers' will thus in all likelihood continue to be sites of extreme environmental and socio-political contradictions – if anything, increasingly so. Yet at the same time climate change will open up entirely new resource frontiers far away from the divided environments considered in this book, especially in Greenland, the Arctic Ocean and the northern latitudes of Eurasia and North America. Although beyond the scope of the present analysis, a frontier optic surely suggests that, as the ice retreats in these northern regions, the developmental and, in turn, political and security consequences are likely to be huge. 'Frontiers', as Lord Curzon observed in 1907, are 'the razor's edge on which hang suspended the modern issues of war and peace' and 'the chief anxiety of nearly every Foreign Office in the civilized world'. Under conditions of accelerating climate change, something similar may come to apply once again.[50]

---

[50] M. Nuttall, 'Imagining and governing the Greenlandic resource frontier', *Polar Journal*, 2:1 (2012), 113–24; K. Dodds and M. Nuttall, *The Scramble for the Poles: The Geopolitics of the Arctic and Antarctic* (Polity, 2016); G. Curzon, 'Text of the 1907 Romanes lecture on the subject of frontiers', www.dur.ac.uk/resources/ibru/resources/links/curzon.pdf.

7 War

Where is the Bloody Business?

That war remains both all too common and devastating in its consequences is unfortunately all too clear. Each year, at present, there are at least fifty state-based armed conflicts and sixty-five non-state-based ones globally. Forced displacement is at record levels. Global military expenditures continue to rise. And with this the environmental and resource impacts of war and the preparations for it are inexorably rising too.[1] Quite apart from the devastating direct impacts of warfare – the loss of lives and livelihoods, the razing of cities, mass displacement and more – there are its profound longer-term legacies, the imprints that it can continue to leave on polities, economies, identities and environments decades or even centuries later. War's importance as an aspect of modern socio-ecological existence is both all too obvious, and shows few signs of abating.

Yet despite this, many of the intellectual resources available for understanding war and its consequences, including its consequences for water (in)security, are in crucial respects quite limited, often being organised around a set of questionable exclusions. Thus most research on war tends to focus on those periods and places where fighting takes place, in abstraction from longer-term processes, and broader structures and geographies, of war preparation. Today's wars are usually categorised and analysed as 'civil wars' even in cases where many of their key participants, causes and consequences are international. Within contemporary conflict research, 'war' is also typically defined in relation to battle death numbers – for instance, as occurring when there are more than 1,000 battle-related deaths in any given year – even though there are often strong continuities between periods of war and periods of supposed 'non-war',

---

[1] T. Pettersson and M. Öberg, 'Organised violence, 1989–2019', *Journal of Peace Research*, 57:4 (2020), 597–613; UNHCR, *Global Trends: Forced Displacement in 2020* (2021), 6; D. L da Silva et al., *Trends in World Military Expenditure, 2020* (SIPRI, 2021); G. E. Machlis and T. Hanson, 'Warfare ecology', *BioScience*, 58:8 (2008), 729–36.

and although direct killing may not be the main objective, or best indicator, of it. Moreover, for eminently understandable reasons, the vast bulk of research and commentary on war concentrates much more on its direct local impacts and its destructiveness than on its productive and positive dimensions or broader societal consequences. While this is obviously so with respect to conflict reporting, it also applies, for instance, to much post–Cold War research on 'new wars' which, in line with a framework first developed by Mary Kaldor, assumes that contemporary warfare is inherently chaotic, disorganised and destructive, by contrast with the rule-bound and productive 'old wars' of high modernity. Moreover, it also applies, albeit in a different way, to liberal interpretations of economic and social development. Within such readings, war is typically viewed as an aberration and as involving a reversal of normal developmental processes – what Paul Collier called 'development in reverse'.[2] Right across such accounts, the place of war and war preparation within social change and capitalist development feature only at the margins, or are neglected altogether.

Most discussions of water, war and development reflect one or more of these shortcomings. Thus, on the one hand, work on the impacts of war on water security typically focuses largely, even exclusively, on the direct, negative and location-specific impacts of military activity, especially the destruction of water-related infrastructures. And, on the other, accounts of water development typically pay scant regard to the roles of war and violence therein. More broadly, it is striking just how little research there is on the impacts of war on water resources, water supplies and water-related insecurities. Unlike the literature on the contribution of water scarcities to war, one can almost count on the one hand the recent scientific studies of the reverse causal relationship, the impacts of war on water. Most research and research funding on environmental security, it would seem, is more comfortable analysing war as an imagined future threat than in considering how lives, livelihoods and environments are already being brutalised and transformed by it. Within this body of research, as well as within most studies of water development, the actuality of the bloody business is for the most part ignored and concealed.

Consider, for example, two recent accounts of Israeli water development: Seth Siegel's widely publicised 2017 book *Let There Be Water: Israel's Solution for a Water-Starved World* and a 2019 report from the Tony Blair Institute for Global Change, 'How Israel Became a World

---

[2] M. Kaldor, *New and Old Wars: Organized Violence in a Global Era* (Polity, 1999); P. Collier, 'Civil war as development in reverse', in P. Collier et al. (eds.), *Breaking the Conflict Trap: Civil War and Development Policy* (Oxford University Press, 2003), 13–32.

Leader in Agriculture and Water: Insights for Today's Developing Countries'. As their titles suggest, these two texts have much in common. They both chart Israel's supposed journey from water scarcity to abundance, the one exploring Israel's rise to becoming a 'water superpower', and the other cataloguing the factors behind 'Israel's agricultural miracle' (many of which relate to water). They both identify a range of technical, economic and institutional reasons for Israel's demonstrable successes in water and agriculture. And they both extol the Israeli experience as an important source of lessons for the rest of the world, and for developing states in particular (indicatively, the Blair Institute report was launched at the 2019 conference of the Alliance for a Green Revolution in Africa, the Rockefeller and Gates Foundation–funded initiative for advancing the continent's 'doubly green' agricultural transformation). As Tony Blair puts it in the Forward to his report, 'Israel has much to offer others from its experience of "making the desert bloom" – building a thriving agriculture sector under conditions of considerable adversity'.[3]

The problem is that, whether this is true or not, both the Blair Institute report and Siegel's text are silent on key aspects of Israeli development. Neither includes more than the briefest mention of Palestinians (indeed, Palestinians appear not once in the Blair Institute report, either as part of Israel's agricultural development or as subjects who might have something to say about it: only Israeli experts and officials were interviewed); the operative assumption here seems to be that Palestinians are somehow external to Israeli history and society. Neither text even alludes to the roles of war and violence within Israel's hydro-politics. There is no mention let alone analysis of how resource capture through war has underpinned Israeli hydraulic or agricultural development. There is no consideration of whether population displacement was a condition of possibility for this. There is no examination of how Israeli–Palestinian inequalities in access to land and water are maintained by military force. And there is no analysis of how water supplies or agricultural production have been affected by military activity (other than Siegel's observation that 'Palestinian claims' about this are 'rebutted by Israel'). Indeed, with irony upon irony, in the extreme case of the Blair Institute report we encounter not only one of the high priests of Western liberal interventionism recommending that post-colonial Africa draw inspiration from a European settler-colonial state; we also encounter an organisation committed to 'putting the smallholder farmer at the center of [Africa's]

---

[3] Siegel, *Let There Be Water*, 172; D. Abraham et al., *How Israel Became a World Leader in Agriculture and Water: Insights for Today's Developing Countries* (Tony Blair Institute for Global Change, 2019), 7; G. Conway, *The Doubly Green Revolution: Food for All in the 21st Century* (Penguin, 1997).

growing economy' being advised to follow the example of a polity that achieved its objectives via the ethnic cleansing of its own indigenous smallholder population; and, whether out of ignorance or disregard, we find not even a hint of this.[4] Instead, in both this and Siegel's text the reader is offered, in addition to strong doses of pro-Israeli *Hasbara*, contradiction- and violence-free Whig histories of water development.

Thankfully, however, there do exist resources for thinking more critically about war and its consequences. Here, we build upon this work – drawing especially upon contributions from historical sociology, critical geopolitics, critical war studies and political ecology – to offer what we hope is a more rounded assessment of the impacts of war on water security. 'War', as we use the term here, refers not just to periods when military-on-military combat crosses some arbitrary battle death threshold but instead, and more broadly, to the organised and large-scale use of force for political purposes – this including widespread and often genocidal actions against civilian populations, the military targeting of those spaces, infrastructures and resources through which fighting and living are sustained, and what Raphael Lemkin called efforts to permanently 'cripple' groups and populations. In turn, we understand war as having wide-ranging impacts, some intended, others unintended; some direct, others indirect; some short-term, others long-term; and some local and others distant, ranging from the places where bombs fall to the metropolitan laboratories, factories and ministries where they are planned, produced and procured. Moreover, we understand war as not only destructive but also productive in its consequences, repeatedly serving as the violent midwife to social and political change. Crucially, though, these consequences are usually highly asymmetrical, inflicting death and destruction on some while being highly productive for others – war being in this sense a particularly extreme means of division and redistribution. Far from being exceptional features of some wars, or a product of 'new wars' specifically, we view all colonial and modern warfare as essentially of this form.[5]

---

[4] Abraham et al., *How Israel Became a World Leader*, 48; Siegel, *Let There Be Water*, 181; Alliance for a Green Revolution in Africa, *Corporate Profile* (2018), 3.

[5] C. Tilly, 'War making and state making as organized crime', in P. B. Evans et al. (eds.), *Bringing the State Back In* (Cambridge University Press, 1985), 169–91; C. Cramer, *Civil War Is Not a Stupid Thing: Accounting for Violence in Developing Countries* (Hurst, 2006); T. Barkawi and S. Brighton, 'Powers of war: fighting, knowledge, and critique', *International Political Sociology*, 5:2 (2001), 126–43; D. Gregory, 'War and peace', *Transactions of the Institute of British Geographers*, 35:2 (2010), 154–86; Peluso and Vandergeest, 'Political ecologies'; R. Lemkin, 'Genocide as a crime under international law', *American Journal of International Law*, 41:1 (1947), 147.

From these starting points, this chapter maps out and explores what we view as the four main ways in which contemporary war matters for patterns of water (in)security: through infrastructure destruction; through population displacement; through the expropriation of resources and infrastructures; and through war's profound if mostly indirect ramifications for state-building and development. In exploring these various impacts and consequences, the chapter draws on evidence from all five of our divided environments. We focus above all, given the availability of evidence, on the recent and ongoing wars across our cases: South Sudan since 2013, Syria since 2011, Lake Chad since 2009, Darfur 2003–5, and Israel's 2008–9, 2012, 2014, 2021 wars on, and post-2007 blockade of, Gaza. We also consider some key historical conflagrations, most notably the wars of partition in Israel–Palestine (1948–9), Cyprus (1974) and Sudan (1983–2005); and we also touch on the impacts of global wars, specifically the World Wars and Cold War, on our cases. War, the chapter argues, is simultaneously transformative and contradictory in its impacts on water security and vulnerabilities – and is likely to remain no less so in an era of climate disruption.

Infrastructure Destruction and its Consequences

If, as Peter Sloterdijk argues, the 'twentieth century will be remembered as the period whose decisive idea consisted in targeting not the body of the enemy, but his [sic] environment' – this beginning with the first use of chlorine gas in 1915, which targeted the air of the enemy combatant – then twenty-first-century warfare will perhaps by extension be known, above all, for its destruction of vital infrastructures. For, as societies have become more and more dependent upon infrastructures – those giant material systems of circulation and provision which underpin abstract space-time and enable the constant production and consumption demanded by capitalism (as discussed in Chapter 5) – so have these societies also become more and more vulnerable to infrastructural disruption. What Stephen Graham has called 'infrastructural warfare' has been the inevitable result, especially within urban spaces.[6]

The infrastructural destruction visited on Syria since 2011 is a clear case in point. Across the country entire neighbourhoods have been levelled, with the Syrian regime, opposition forces, the United States and Russia all conspiring in violence. Water towers, treatment plants, pumping stations, supply pipelines and sewage systems have all been

---

[6] P. Sloterdijk, 'Airquakes', *Environment and Planning D: Society and Space*, 27:1 (2009), 43; S. Graham, *Cities under Siege: The New Military Urbanism* (Verso, 2010).

affected, on a scale which almost beggars belief. By 2017, according to the World Bank, nearly two-thirds of Syria's water treatment plants, over half of its pumping stations and wells, a quarter of its sewage treatment plants and a third of water towers and tanks had been either damaged or destroyed. In some places, the damage has been even more extensive still. In the Kurdish-majority city of Kobani, according to the Bank, over 70 per cent of water supply and sanitation infrastructure was non-operational in 2017; in Aleppo, Hama, Homs and Idlib around 60 per cent of all water and sanitation assets had been damaged. Moreover, individual targeted airstrikes have repeatedly had dire consequences for local water supply. A May 2014 attack on the al-Khafsah water pumping station in Aleppo – one of the most important in Syria, producing an average of 18 million litres of drinking water daily – cut off water for nearly three million people, with government and opposition forces each blaming the other for the attack. In December 2016, the Syrian air force deliberately bombed water sources north-west of Damascus, affecting water supply for 5.5 million people for more than a month – an act deemed a war crime by the UN's Independent Commission of Inquiry. During its mid-2017 operation in Raqqa, US forces directly targeted communal water points. And, in October 2019, the Turkish armed forces shelled the Aluk water station in north-east Syria, cutting off water supplies to nearly 500,000 people for over a month. As numerous reports have concluded, the guiding rationale behind these and the many other such instances of infrastructural warfare in Syria has been clear: to empty opposition-held areas of their civilian populations, to isolate combatants and to prevent the return of civilians by rendering bombed areas uninhabitable.[7]

Israel's repeated bombardments of Gaza have been differently motivated but similarly disastrous for water infrastructures and supplies. Consider Israel's Operation Cast Lead of 2008–9. This individual operation, as the UN and others extensively documented, saw the destruction of an estimated 74 wells, 5,700 water tanks and 20,000 metres of water

---

[7] World Bank, *The Toll of War: The Economic and Social Consequences of the Conflict in Syria* (2017), 29–30; World Bank, *Syria Damage Assessment of Selected Cities: Aleppo, Hama, Idlib – Phase III* (2017); World Bank, *The Mobility of Displaced Syrians: An Economic and Social Analysis* (2020), 137–42; N. Shamout, 'Syria faces an imminent food and water crisis', Chatham House (24/06/2014); UNHRC, *Human Rights Abuses and International Humanitarian Law Violations in the Syrian Arab Republic, 21 July 2016–28 February 2017* (2017); Amnesty International, *War of Annihilation: Devastating Toll on Civilians, Raqqa – Syria* (2018), 44; UNHRC, *Report of the Independent International Commission of Inquiry on the Syrian Arab Republic* (2020), 10; Syria Institute and PAX, *No Return to Homs: A Case Study on Demographic Engineering in Syria* (2017); J. L. Sowers et al., 'Targeting environmental infrastructures, international law, and civilians in the new Middle East wars', *Security Dialogue*, 48:5 (2017), 410–30.

and wastewater pipes; the apparently 'deliberate and premediated' bombing of part of Gaza City's wastewater treatment plant which in turn led to the outflow of 200,000 cm of sewage into surrounding farmland; the complete destruction of a well complex supplying 25,000 people in Jabalia, including the killing of its operator, which the UN judged a violation of customary international law and possibly a war crime; and much else besides. The UN's Goldstone Report concluded that, in this operation, there was 'a deliberate and systematic policy on the part of the Israeli armed forces to target ... water installations' with 'the specific purpose of denying sustenance' to the local population. More broadly, the repeated attacks on Gaza since 2000 – combined, from 2007, with the effects of Israel's blockade, as detailed below – have had devastating, cumulative impacts on Gaza's water and wastewater infrastructures.[8]

While infrastructure destruction affects urban areas in particular, it is not limited to them; irrigation infrastructures, in particular, are routinely targeted too. During the June 1967 war, Israeli forces engaged in intentional destruction of Palestinian pumps drawing water from the Jordan River. During Sudan's second civil war, government and SPLM/A forces systematically destroyed wells and hand pumps, poisoned water sources and dumped grain into wells to rot during their attacks on southern villages. During the 2003–5 war in Darfur, government forces and associated Janjaweed engaged in similarly widespread destruction of rural water infrastructures, physically destroying pumps and poisoning wells by throwing chemicals, oil, dead animals and bodies into them; during one attack on a village in West Darfur, all of its wells were poisoned with DDT. Since 2010, both Boko Haram and the Nigerian military have deliberately damaged and destroyed water points and pipes and polluted and mined water sources; by 2017, according to UNICEF, three-quarters of water and sanitation infrastructure in conflict-affected areas in northeast Nigeria had been damaged or destroyed. Since 2011, among countless such instances across rural Syria, several districts in the Orontes basin have seen over 40 per cent of their agricultural wells plugged and over half of all well pumps looted or sabotaged. And since the onset of the South Sudanese civil war in 2013, community water points have been extensively destroyed alongside the broader plundering and gutting of homes, businesses, schools and hospitals. In one illustrative Jonglei state village

---

[8] Al Mezan Center for Human Rights, *The Impact of the Israeli Offensive on the Right to Water in the Gaza Strip* (2009), 6; UNRWA, *Updated Quick Response Plan for Gaza: An Assessment of Needs Six Months after the War* (2009), 9; UNHRC, *Report of the United Nations Fact-Finding Mission on the Gaza Conflict* (2009), 22, 26, 206–10, 217, 266; E. Weinthal et al., 'Targeting infrastructure and livelihoods in the West Bank and Gaza', *International Affairs*, 95:2 (2019), 319–40.

alone, 100 water pumps were bulldozed by a neighbouring ethnic militia early in the civil war. By 2016, an estimated 40 per cent of water and sanitation facilities in conflict-affected states within South Sudan had been destroyed.[9] From the Jordan Valley to Jonglei, the pattern is depressingly clear.

Paralleling this destruction of water infrastructures is the extensive targeting and restriction of electricity systems and energy supplies, with knock-on water security consequences. In Bentiu, Bor, Malakal and elsewhere in South Sudan, the civil war has seen government and opposition forces alike looting generators and solar panels needed to power water pumping stations and water supply systems. During July 2016, in Juba, government troops even looted World Food Programme and FAO compounds, stealing and destroying generators, fuel and other equipment. In 2006, Israel bombed Gaza's only power plant, rendering it inoperable and in the process halving the territory's electricity supply with devastating consequences for water supplies and wastewater treatment. In 2014, it was bombed again, to similar effect. And, in 2008, at the height of Israel's siege of the Strip, electricity cuts at one point left 40 per cent of Gaza's population without running water and led to such a breakdown of the territory's wastewater system that 30 million litres of sewage were spilling into the sea each day. In Syria, meanwhile, power plants, transmission stations, high voltage power lines and more have repeatedly been targeted by the regime, opposition forces, Russia and the United States alike. Some major power plants have been completely destroyed, while others have been damaged beyond repair. A 2018 assessment found that less than 10 per cent of power sector assets in fifteen of the country's major cities were fully functional, and under a third were barely functioning at all. Electricity generation in Syria has in turn plummeted: between 2010 and 2015 alone it decreased by nearly two-thirds. And, altogether, this destruction had widespread consequences for the

---

[9] UN-ECOSOC, *Report of the Secretary-General Prepared in Pursuance of General Assembly Decision 39/442* (1985), 202; HRW, *Civilian Devastation: Abuses by All Parties in the War in Southern Sudan* (1994); HRW, *Darfur in Flames: Atrocities in Western Sudan* (2004); J. Hagan and J. Kaiser, 'The displaced and dispossessed of Darfur: explaining the sources of a continuing state-led genocide', *British Journal of Sociology*, 62:1 (2011), 1–25; UNICEF, '27 Million people lack safe water in countries facing or at risk of famine' (29/03/2017); Amnesty International, 'Nigeria: military razes villages as Boko Haram attacks escalate' (14/02/2020); R. Jaubert et al., *Syria: The Impact of the Conflict on Population Displacement, Water and Agriculture in the Orontes River Basin* (Swiss Agency for Development and Cooperation, 2014); HRW, *"They Burned It All": Destruction of Villages, Killings, and Sexual Violence in Unity State, South Sudan* (2015); B. Trew, 'South Sudan, where a water crisis is leading to child kidnappings and rape', *Independent* (04/04/2019); UNICEF South Sudan, *Program Component Strategy Note: Water, Sanitation and Hygiene* (2016), 3.

country's water sector: July 2016 attacks on a transmission station near Aleppo, for instance, left nearly two million people without access to water supplies.[10]

From Syria to South Sudan war has also led to fuel shortages and fuel price inflation, again with consequences for water security. In Syria, where fuel oil prices increased tenfold and diesel prices fivefold after 2011, this translated into higher water prices, limiting the use of groundwater pumps and irrigation across the country. By 2016, according to FAO assessments, 20 per cent of Syrian farming households had lost access to irrigation systems entirely, while an additional 40 per cent of households could access irrigation but used less water because of fuel prices and availability. In Juba, sharply increasing diesel prices between 2013 and 2016 pushed up the price of water from filling stations and tankers – the cost of a drum of water near doubled during this period – leading water trucks to restrict deliveries to areas closer to filling stations and reduce supplies to poorer areas on the outskirts of the city. From Syria to South Sudan and elsewhere, war has also disrupted access to everything from chemicals for water treatment to spare parts and equipment for the repair of generators, pipelines and major infrastructures. Indeed, in Gaza this has become a semi-permanent state of affairs, with the Israeli and international blockade having had dire consequences for equipment, materials and fuel imports, in turn affecting everything from new infrastructure works to network maintenance, sewage treatment, leakage and losses.[11]

The repercussions for sanitation services and water supply are typically profound. In Syria, roughly 90 per cent of households had access to mains networks prior to 2011; by 2017, however, just 36 per cent of households could rely on public networks as their main source of water. Between

---

[10] UNICEF South Sudan, *Program Component Strategy*; Amnesty International, *"We Did Not Believe We Would Survive": Killings, Rape and Looting in Juba* (2016), 21; B'Tselem, *Act of Vengeance: Israel's Bombing of the Gaza Power Plant and Its Effects* (2006); HRW, 'Gaza: widespread impact of power plant attack' (03/08/2014); UN-OCHA, *Gaza Closure: Situation Report* (18–24/01/2008); World Bank, *Toll of War*, 30–3; World Bank, *Mobility of Displaced Syrians*, 155–6; Sowers et al., 'Targeting environmental infrastructures', 419.

[11] World Bank, *Toll of War*, 69; FAO and WFP, *FAO/WFP Crop and Food Security Assessment Mission to the Syrian Arab Republic* (2017), 34; FAO, *Counting the Cost: Agriculture in Syria after Six Years of Crisis* (2017), 12; K. King, *Deteriorating Economic Situation and its Effect on Safe and Adequate Water Supply in Juba, South Sudan* (Oxfam, 2015), 11; Oxfam, *A City Exposed: Clean Water Runs Dry in Juba* (2015); B. Mosello et al., *Improving WASH Service Delivery in Protracted Crises: The Case of South Sudan* (ODI, 2016), 12–15; M. Zeitoun et al., 'Urban water ecology: a study of water supply in Basrah', *International Journal of Urban and Regional Research*, 41:6 (2017), 904–25; Centre on Housing Rights and Evictions, *Hostage to Politics: The Impact of Sanctions and the Blockade on the Human Right to Water and Sanitation in Gaza* (2008).

2011 and 2017 water network coverage at least halved in the majority of Syrian provinces; in Aleppo it declined from being near universal prior to 2011 to just 30 per cent in 2017. Even in government-controlled urban areas, around half of households – and the majority of female-headed households – would periodically go without piped water supplies for several days at a time. In Idlib and Aleppo, a quarter of households reported not having enough water to meet household needs. Wastewater services have also collapsed: whereas an estimated 65 per cent of people were served by a functioning wastewater treatment plant before the war, by 2019 this figure had dropped to less than 10 per cent, with at least 70 per cent of sewage remaining untreated. In South Sudan, whereas an estimated 70 per cent of the country's population had access to improved drinking water sources in 2010, by 2018 this had dropped to a third. And in Gaza, Israel's blockade and repeated attacks have compounded already severe water and sanitation problems. During the blockade's first year, water use declined from 95 l/c/d to less than 60 l/c/d. During that same year, increased pressures on Gaza's wastewater system arising from Israel's attacks and blockade led to the collapse of an earth embankment forming one of the territory's sewage lagoons, partially submerging the Bedouin village of Umm al-Naser with five fatalities. Then, during Operation Cast Lead, an estimated one-third of the population lost complete access to running water. Such war and blockade impacts on Gaza's water and wastewater infrastructures have been linked to an array of health problems, including the territory's high rates of blue baby syndrome and diarrhoea.[12]

The war-induced destruction or decay of water infrastructures also typically has various knock-on environmental, economic and social consequences. Infrastructural collapse is often followed by increased local groundwater abstraction, as households and neighbourhoods take water supply matters into their own hands: thus in Syria illegal well drilling mushroomed after 2010 – during the regime's siege of eastern Ghouta, for

---

[12] World Bank, *Mobility of Displaced Syrians*, 136–43; S. Doocy and E. Lyles, 'Humanitarian needs among displaced and female-headed households in government-controlled areas of Syria', *American Journal of Public Health*, 107:6 (2017), 950–9; UNHCR et al., *Shelter and NFI Assessment: Syria* (2017); UN-OCHA, *2019 Humanitarian Needs Overview: Syrian Arab Republic* (2019), 25, 81; South Sudan Ministry of Health and National Bureau of Statistics, *The Sudan Household Health Survey 2010* (2011), 42–6; REACH, *South Sudan Water, Sanitation and Hygiene Brief: July/August and November/December 2018* (2019); UN-OCHA, *The Humanitarian Monitor: Occupied Palestinian Territory*, 24 (2008), 19; M. Zeitoun, 'Complicity when the dam breaks', *New York Times* (06/08/2007); Al Mezan Center for Human Rights, *The Impact of the Israeli Offensive*, 6; UNHRC, *Report on the Gaza Conflict*, 268; UNEP, *Environmental Assessment of the Gaza Strip following the Escalation of Hostilities in December 2008–January 2009* (2009), 57.

example, in which access to mains water was cut, civilians dug some 600 wells and installed manual water pumps to supply the area – while there was a similar upsurge in well drilling in Gaza in 2006, in the wake of Israel's (initial) bombing of the Strip's only power plant. Simultaneous with this, private water vending typically proliferates, with household spending on water increasing steeply. In Syria, tanker prices rose to $10–12 per cm or more, such that the average household reliant on tankers was spending 10 per cent of their income on water supplies, and some reportedly more than half. In Juba, just a third of households were reliant on water tankers in 2008–9, but by 2016 this had risen to over three-quarters; by 2017 households were spending, on average, 30 per cent of their income on water, six times higher than the internationally recognised burden threshold of 5 per cent. Indeed, a World Bank study of Juba found that, in 2016, the poorest households were spending 120 per cent of their income on water supplies. Last (for now), war conditions also typically result in households spending a much increased portion of their time fetching and managing water, in an extreme version of the practices of resilience discussed in the previous chapter. And while this may be interpreted, in part, as evidence of adaptative capacities and social agency, it is also clear that it has regressive distributive consequences – not least in that prolonged violence, which is mostly instigated and waged by men, can end up resulting, among many other things, in poor women and girls walking for hours in search of water.[13]

The Antinomies of Displacement

Mass wartime population displacements are common to all of our cases: the numbers speak for themselves. Around 751,000 displaced during the war of 1948–9, a full 88 per cent of the Palestinian population of what would henceforth become Israel. Several hundred thousand more displaced in 1967. Over 5 million Palestinians who decades later still have refugee status. Around 160,000 displaced during Turkey's 1974 invasion of northern Cyprus, amounting to nearly 90 per cent of the area's Greek Cypriot population (and with the remainder to depart following a subsequent transfer agreement). An estimated 4.5 million south Sudanese displaced during

---

[13] UNHRC, *Report on the Syrian Arab Republic* (2018), 20; Messerschmid, *Water in Gaza*, 9; UNHCR et al., *Shelter and NFI Assessment*, 77; World Bank, *Syria Damage Assessment*, 41; UN-OCHA, *2019 Humanitarian Needs Overview*, 81–2; JICA, *Juba Urban Water*, 2.4; World Bank, *Opportunities for Improving Urban Service Delivery in South Sudan: A Tale of Two Cities. PART III: Synthesis Report* (2017), 46–51; M. Matoso, 'Water tankers and bicycles: a lifeline in South Sudan', Oxfam (29/08/2017); Sphere Association, *The Sphere Handbook: Humanitarian Charter and Minimum Standards in Humanitarian Response*, 4th ed. (2018), 106; Trew, 'South Sudan'.

the second Sudanese civil war. Over a third of South Sudan's population displaced since 2013, nearly 1.7 million internally and 2.3 million internationally to become refugees in Uganda, Sudan and Ethiopia. Around 2 million displaced during the 2003–5 war in Darfur. Nearly 4.5 million displaced since 2009 in the Lake Chad basin, the vast majority within north-eastern Nigeria. And a staggering 12 million Syrians, over half of the country's population, displaced since 2011 – around half displaced internally and half internationally, the majority to become refugees in Turkey, Lebanon and Jordan. Brute displacement numbers aside, across these cases war has also created entirely new population and social geographies, with huge flows of people relocating from the countryside to camps and urban fringes – and in the process being transformed into an informal urban precariat and in some cases urban revolutionaries – as well as between regions and across borders. In most cases, moreover, there has also been little or no refugee return, making displacement not just protracted, but semi-permanent. This is no accident: especially within modern state-building wars, displacing populations and preventing return is a recurring strategic objective.[14]

The multiple insecurities that typically result are all too obvious, in relation to water and sanitation included. Within Syria, almost a third of internally displaced households receive unsafe drinking water, over a third are unable to access sufficient water to satisfy basic needs and more than half lack access to functioning bathing facilities. In Lebanon, no more than a third of Syrian refugees have access to the water network, with most relying instead on open wells and reservoirs, standpipes, illegal network tapping, tankers and/or bottled water. In Jordan, nearly all refugee households in camps rely on tankers or other private sources for both general use and drinking, and as many as half of those in informal settlements cannot access sanitation infrastructure. And in Turkey, more than two-thirds of Syrian refugee households in rural areas, mostly living in informal settlements, lack access to public water, sanitation and electricity services. In Sudan, Khartoum's rapid expansion during the second civil war – by 2004 an estimated 40 per cent of the city's residents, up to

---

[14] UNCCP, *Final Report*, 22; European Commission of Human Rights, *Application Nos. 6780/74 and 6950/75, Cyprus against Turkey: Report of the Commission, Vol. 1* (1976), 6; UNHCR, *UNHCR Global Appeal 2006: South Sudan* (2005), 132; IOM, *Displacement Tracking Matrix: South Sudan – 2019 at a Glance* (2020); UNHCR, *Regional Overview of the South Sudanese Refugee Population* (31/05/2020); UN, *Darfur Humanitarian Profiles No. 11 and 12* (01/03/2005); IOM, *Within and Beyond Borders: Tracking Displacement in the Lake Chad Basin* (2019); UN-OCHA, *2019 Humanitarian Needs Overview*, 18; R. Sayigh, *The Palestinians: From Peasants to Revolutionaries* (Zed, 1979); A. G. Lichtenheld, 'Explaining population displacement strategies in civil wars: a cross-national analysis', *International Organization*, 74:2 (2020), 253–94.

two million people, were internally displaced, mainly from the south – created multiple water insecurities, with most of the city's peripheries relying on water supplied by tanker, bicycle or donkey cart, most areas being unconnected to the sewage system, and seasonal flooding combined with poor sanitation regularly leading to outbreaks of waterborne diseases. Juba's rapid expansion as a result of the North–South war and subsequent refugee return was such that by 2008 only about 10 per cent of its population had access to piped water supplies. Elsewhere in South Sudan, Protection of Civilian sites around UN bases, which by 2017 were housing around 230,000 IDPs, have experienced serious pressures on water supplies and sanitation and, as a result, demonstrations. In IDP camps in Adamawa, Borno and Yobe states in Nigeria, there were just five latrines per 5,000–6,000 people in 2014, leading to open defecation and contamination of well water. Even West Bank refugee camps, which are in most cases now around seventy years old, typically have far less reliable water supplies than neighbouring Palestinian communities.[15]

Direct water security impacts aside, displacement also typically has a series of acute indirect consequences, foremost among them those associated with the spread of waterborne diseases. After many years free from polio and cholera, Syria has experienced repeated outbreaks of both since the start of the civil war associated with transmission through contaminated water. In western Darfur, a large Hepatitis E outbreak during 2004 saw over 2,600 cases and 45 deaths in one IDP camp alone, with reliance on untreated river water for drinking supplies identified as a major factor. In South Sudan, cholera epidemics were declared every year between 2014 and 2017 (after four years with no confirmed cases), with nearly 30,000 cases reported and 650 deaths. Large IDP settlements have been most heavily affected by these outbreaks, particularly in Juba and around the

---

[15] REACH, *WASH Overview: Camps and Sites – HNO 2020* (2019); UNHCR et al., *VASyR 2019: Vulnerability Assessment of Syrian Refugees in Lebanon* (2019), 55–6; REACH and UNICEF, *Syrian Refugees Staying in Informal Tented Settlements in Jordan: Multi-Sector Assessment Report* (2014), 2; Médecins du Monde Turkey, *Multi-Sectoral Needs Assessment of Syrian Refugees in Turkey* (2019), 14, 21; M. A. M. Assal, Displaced Persons in Khartoum: Current Realities and Post-War Scenarios (Population Council, 2004), 14–19; S. Pantuliano et al., City Limits: Urbanisation and Vulnerability in Sudan – Khartoum Case Study (Humanitarian Policy Group, 2011), 29–30; S. Pantuliano et al., The Long Road Home: Opportunities and Obstacles to the Reintegration of IDPs and Refugees Returning to Southern Sudan and the Three Areas – Report of Phase II (Humanitarian Policy Group, 2008), 23; J. Munive, 'Resilience in displacement and the protection of civilians in South Sudan', *Journal of Refugee Studies*, 34:2 (2019),1879–99 ; ACAPS, *Nigeria: Boko Haram Insurgency* (20/01/2014), 2, 6; A. De Donato, 'Politiques de l'eau dans la construction d'un État-nation Palestinien: le cas du camp de réfugiés de Deisha et du district de Bethléem (West Bank)', *Journal des Anthropologues*, 132-3 (2013), 169–95.

Sudd, where communities rely on swamp waters for drinking and cooking (during the 2014 outbreak, for example, IDP settlements in Juba, Torit and Wau Shilluk accounted for almost 80 per cent of the country's 6,500 cholera cases). And around Lake Chad, there have been two major cholera outbreaks since 2010. In 2010–11, 127,000 cholera cases and over 4,600 deaths were reported across the basin, the largest outbreak in the region in two decades. In 2018, a further cholera outbreak saw nearly 40,000 reported cases and 845 deaths in Cameroon, Niger and Nigeria. And in 2017, a Hepatitis E outbreak saw nearly 4,000 cases and more than 50 deaths in the Lake Chad border regions of Chad, Niger and Nigeria.[16]

Perhaps most strikingly, the negative impacts of war on water supplies and sanitation provision can be detected right through to data on mortality. In Darfur, for instance, where according to the Center for Research on the Epidemiology of Disasters (CRED) there were an estimated 131,000 'excess deaths' during the 2003–5 war, 70 per cent of these were the result not of violence directly but disease and malnutrition; according to the WHO, diarrhoea arising from poor sanitation was 'the main cause of death' during the Darfur war. Moreover, evidence suggests that this pattern became even more striking in subsequent years, as the fighting in Darfur abated: from the start of the war through to 2008, a full 80 per cent of the 300,000 excess deaths in Darfur are estimated to have been due to factors other than direct violence. Similarly in South Sudan, where there were an estimated 383,000 excess war-related deaths between 2013 and 2018, an estimated 190,000 were the direct result of violence while almost 200,000 deaths resulted from other factors – food insecurity most clearly, but probably also increased transmission of infectious diseases arising from poor water and sanitation.[17]

Displacement also alters and increases water demand in host areas, in turn placing additional pressures on local water resources, groundwater

---

[16] Center for Operational Analysis and Research, *Annex: Public Health Responses to Syria's Past Epidemics* (2021); J.-P. Guthmann et al., 'A large outbreak of Hepatitis E among a displaced population in Darfur, Sudan, 2004: the role of water treatment methods', *Clinical Infectious Diseases*, 42:12 (2006), 1685–91; J. Dunoyer et al., In-Depth Cholera Epidemiological Report for South Sudan (Prospective Cooperation and UNICEF, 2018); UNICEF et al., *Lake Chad Basin Cholera Outbreak: Key Messages – September 2018* (2018); C. Roby, 'Hepatitis E outbreak sparks concern in Lake Chad Basin', *Devex* (10/08/2017).

[17] WHO, *Retrospective Mortality Survey among the Internally Displaced Population, Greater Darfur, Sudan, August 2004* (2004), iv; D. Guha-Sapir and O. Degomme, *Darfur: Counting the Deaths – Mortality Estimates from Multiple Survey Data* (CRED, 2005); D. Guha-Sapir and O. Degomme, *Darfur: Counting the Deaths (2) – What Are the Trends?* (CRED, 2005); D. Degomme and O. Guha-Sapir, 'Patterns of mortality rates in Darfur conflict', *Lancet*, 375:9711 (2010), 294–300; F. Checchi et al., Estimates of Crisis-Attributable Mortality in South Sudan, December 2013-April 2018: A Statistical Analysis (London School of Hygiene and Tropical Medicine, 2018).

in particular. Thus in Darfur, IDP camps have been associated with local deforestation and bush clearance for cultivation, as well as more intense cultivation arising from the ready availability of cheap labour, all increasing local water demands. In Syria, conversely, the establishment and growth of IDP and refugee camps has swallowed up agricultural land and orchards, displacing agricultural water use. In the coastal provinces of Lattakia and Tartous, internal displacement has resulted in a 50 per cent population increase and, alongside this, increased groundwater abstraction and saltwater intrusion. In Lebanon and Jordan, the arrival of hundreds of thousands of Syrian refugees has pushed up overall water demand – increasing it by over 20 per cent in Jordan, according to the national government, and by as much as 40 per cent in some northern areas with large refugee populations. Zaatari camp in northern Jordan – the largest single Syrian refugee camp and one of the largest in the world – has been associated with a severe decline in groundwater levels in the Amman Zarqa basin. Bidibidi refugee camp – at the time of writing the second largest in the world, housing a quarter of all South Sudanese refugees in Uganda – has also seen water levels dropping and many boreholes drying up.[18] It also bears emphasising that the strains on both Israel's and Gaza's waters are in large measure historical legacies of displacement, in the one case because of the Holocaust, in the other 1948–9. Contrary to Malthusian caricature, in Gaza, Israel, Jordan and other spaces of refuge it is war, not natural growth, which most underpins the extreme population pressures on local water resources.

Yet the displacement–resources relationship is not defined by this for, converse to what happens in spaces of refuge, spaces of net displacement and out-migration typically see reduced economic activity, reduced water usage and, in turn, a degree of environmental recovery. Contemporary Syria offers particularly striking evidence of this. Between 2011 and 2016

---

[18] M. Hagenlocher et al., 'Integrated assessment of the environmental impact of an IDP camp in Sudan based on very high resolution multi-temporal satellite imagery', *Remote Sensing of Environment*, 126 (2012), 27–38; O. Kranz et al., 'Assessment of environmental changes induced by internally displaced person (IDP) camps in the Darfur region, Sudan, based on multitemporal MODIS data', *International Journal of Remote Sensing*, 36:1 (2015), 190–210; Harvard Humanitarian Initiative, *Displacement and Destruction: Analysis of Idlib, Syria 2017–2020* (2020); G. Faour and A. Fayad, 'Water environment in the coastal basins of Syria: assessing the impacts of the war', *Environmental Processes*, 1:4 (2014), 533–52; Jordan Ministry of Planning and International Cooperation, *The Jordan Response Plan for the Syria Crisis 2017–2019* (2016); H. H. Jaafar et al., 'Refugees, water balance, and water stress: lessons learned from Lebanon', *Ambio*, 49 (2020), 1179–93; M. A. Al Wreikat and A. A. Al Kharabsheh, 'Impact of over-pumping on groundwater resources sustainability at Amman Zarqa basin, Jordan: a case study of arid areas affected by Syrian refugees crisis', *Environmental Earth Sciences*, 79:1 (2020), 19; R. Ssekandi, 'Ugandan experts urge action on impact of South Sudan refugees on environment', *Xinhua* (11/06/2017).

Syria's rural population dropped by nearly a third, which in combination with the infrastructural destruction already discussed, as well as input shortages, price rises and the destruction of orchards and crops, led to significant declines in agricultural activity, including levels of irrigation. The total area within Syria under irrigated cultivation declined by nearly three-quarters between 2011 and 2017–18, from about 1.5 million to just over 400,000 hectares, as did the area cultivated with cotton. Countrywide, cattle numbers declined by 40 per cent; in Idlib province, they reportedly declined by 90 per cent. In many areas of the country, large-scale irrigation schemes saw similarly dramatic declines. In the Al Eis irrigation scheme near Aleppo, for instance, cultivated land reportedly decreased by roughly 60 per cent between 2011 and 2016. The Orontes basin, a key agricultural area which saw particularly high levels of displacement in the early years of the war, there was a halving of the area under irrigation between 2010 and 2014, and a more than 70 per cent decline in crop production. The Yarmouk basin in southern Syria likewise saw an estimated 50 per cent decline in irrigated land area. And the consequence, in many areas of the country, has been increased river flows and water table levels. The volume of water received by Jordan through the Yarmouk River, for instance, is estimated to have increased by 350 per cent since the start of the civil war.[19]

In South Sudan, similarly, agricultural production has declined significantly since 2013. In 2014 alone, the harvested area dropped by more than half in Upper Nile state and more than two-thirds in Jonglei – both epicentres of fighting and displacement during the first year of the civil war. Harvested areas have also declined significantly in Central Equatoria (a 60 per cent drop between 2013 and 2017), Western Equatoria and Western Bahr el Ghazal (both 40 per cent). Alongside mass population displacement, livestock and seasonal migration routes have been disrupted, resulting in the loss of regular water and grazing sources, disease outbreaks and, in some areas, herd size reductions of 40–60 per cent. More broadly, it is evident that repeated war and mass displacement in South Sudan – not only since 2013 but just as crucially during the 1983–2005 North–South civil war – have been major long-term retardants of population expansion, resource exploitation and economic development

---

[19] FAO, *Counting the Cost*; World Bank, *Toll of War*, 56–60; FAO and WFP, *FAO/WFP Crop and Food Security Assessment Mission to the Syrian Arab Republic* (2019); Jaubert et al., *Syria*, 24; H. H. Jaafar et al., 'Impact of the Syrian conflict on irrigated agriculture in the Orontes Basin', *International Journal of Water Resources Development*, 31:3 (2015), 436–49; M. F. Müller et al., 'Impact of the Syrian refugee crisis on land use and transboundary freshwater resources', *Proceedings of the National Academy of Sciences*, 113:52 (2016), 14932–7.

across the country and are key to understanding the exceptionally low level of stress on the country's water resources detailed in Chapter 2. Studies suggest, for instance, that forest cover in many areas of South Sudan increased significantly during the second civil war as a result of conflict-driven displacement.[20]

Although the evidence here is only suggestive, a similar pattern may have been at work around Lake Chad. For, as part of their counter-insurgency operations against Boko Haram, agricultural production around the lake was intentionally restricted, particularly after states of emergency were declared in Nigeria in 2013 and in Niger, Cameroon and Chad in 2015. With the general aims of limiting Boko Haram's tax revenues as well as its access to materials and places to hide, a range of extreme measures were introduced, among them the clearing of crops, restrictions on crop planting (in north-eastern Nigeria the government banned crops higher than a metre), restrictions on access to fertilisers (which might be used for making improvised explosive devices), restrictions on the movement of crops and livestock, the closure of markets and cross-border trade and the creation of vast militarised 'no-go' zones. The production and, in turn, livelihood consequences were dire. In Nigeria's Borno state, food staple production dropped dramatically between 2010 and 2015 – millet by 55 per cent, sorghum by 82 per cent, rice by 67 per cent – and by 2017 the state's wheat production, which had previously accounted for a quarter of the country's total, had reportedly stopped altogether. In the Far North region of Cameroon, food staple production reportedly dropped by more than 50 per cent between 2012 and 2014 and by another two-thirds between 2014 and 2017. In Niger's Diffa region, production of the main high-value cash crop, red peppers, was banned – such that cultivation immediately declined by about 75 per cent on previous years. And in Chad, in the polder areas of the Lac region, cropped areas had declined by up to 25 per cent by 2016 due to insecurity and the abandonment of farms. It is impossible to say with any precision how this affected irrigation levels. However, what we do know is that, by 2009, there were at least 20,000 boreholes across the Lake Chad basin – a large number of which would subsequently have been abandoned. And what we also know is that, after 2012, there appears

---

[20] FAO and WFP, *FAO/WFP Crop and Food Security Assessment Mission to South Sudan* (2018), 30; D. Onyango et al., Contribution of Livestock to the South Sudan Economy (IGAD Centre for Pastoral Areas and Livestock Development, 2015), 10; V. Gorsevski et al., 'Analysis of the impacts of armed conflict on the Eastern Afromontane forest region on the South Sudan–Uganda border using multitemporal Landsat imagery', *Remote Sensing of Environment*, 118:15 (2012), 10–20; Q. Duan and M. Tan, 'Spatial and temporal variations of forest cover in developing countries', *Sustainability*, 11 (2019), 1517.

to have been a notable uptick in the lake's total and especially dry season volumes, which does not correlate with changes in annual rainfall; and that the lake's northern pool, fed by the Komadougou-Yobe River that flows through Boko Haram-affected areas of north-east Nigeria, expanded particularly anomalously at the same time.[21] We can only speculate here on whether these changes may have been caused by military restrictions on agriculture and war-induced displacement. But in our view this seems quite possible, even likely. A case which is widely interpreted within climate security discourse as evidence of how environmental change contributes to violent conflict may, in fact, illustrate the exact reverse.

Displacement is thus a thoroughly contradictory phenomena as far as water resources and insecurities are concerned. From Jordan to Lake Chad, mass displacement typically places immense pressure on water infrastructures and resources within host areas, leading to deep water insecurities for immigrant and host communities alike. Yet, at the same time, regions suffering mass out-migration typically see plummeting economic activity, reduced irrigation and with it a degree of environmental recovery. Displacement can thus fundamentally reshape the geographies of water consumption, vulnerability and repair. But that is not all. For, by way of a further contradiction, the prevention of migration – the effective confinement and 'trapping' of populations in areas under deep environmental stress – can further exacerbate water resource depletion and water-related vulnerabilities.[22] Gaza's terrible water situation, most obviously, is simultaneously a function of war-induced displacement, going back to 1948–9, and of the fact that its ghettoised population has effectively nowhere to go. Such are the socio-environmental antinomies of mass displacement.

Expropriation

If destruction and displacement are two hallmarks of modern warfare then a third is large-scale expropriation, whether this be of land, resources or infrastructures. Of course, and as previous chapters have illustrated,

---

[21] Magrin and Pérouse de Montclos, *Crisis and Development*, 158–64; ICG, *Cameroon's Far North: Reconstruction amid Ongoing Conflict* (2017), 4; ICG, *Instruments of Pain (IV): The Food Crisis in North East Nigeria* (2017), 3; 'How Boko Haram brought hunger to northern Cameroon', *New Humanitarian* (28/01/2015); C. Sissons and C. Lappartient, *Smoked Fish and Dried Red Pepper Income Market Systems, Diffa Region, Eastern Niger* (Oxfam, 2016), 27–36; UN-OCHA, 'Niger: small farmers are paying the price for the lingering insecurity in Diffa' (03/10/2017); FEWS NET, *Chad Food Security Outlook: Deterioration in the Food Security Situation for Most Agropastoral Sahelian Chad* (2016), 2; LCBC, *Lake Chad Sustainable Water Management: Project Activities Report No. 3* (2010); Vivekananda et al., *Shoring Up Stability*, 42–3; Pham-Duc et al., 'The Lake Chad', figs. 2, 3; Mahmood and Jia, 'Assessment of hydro-climatic trends'.

[22] UK Government Office for Science, *Foresight*.

Expropriation is not a feature of war alone: dam building, frontier settlement and the use of discriminatory policy and legal mechanisms can all function as means of dispossession. War, however, surpasses all these others in its capacity to transform patterns of possession over water resources and supplies. This is not to suggest, we must emphasise, that those wars where this has applied have been either principally or in large measure motivated by hydro-political concerns: as discussed in Chapter 2, water is rarely a central reason for or cause of armed conflict. Yet what happens during war typically exceeds initial intentions; water may not be the primary objective of war but, whether for tactical reasons or by dint of accident or circumstance, it is often purloined along the way.

Two basic forms of expropriation may be distinguished – one merely 'political', involving the capture of water resources or infrastructures by one state or quasi-state actor from another, and the other 'social', involving shifts in different social groups' control of, and access to, water resources and infrastructures. The establishment of Islamic State (IS) over large swathes of Syrian (and Iraqi) territory between 2012 and 2019 illustrates the former, for although IS's rise was not motivated by hydro-politics, the capture and control of the Tishrin, Tabqa and Ba'ath dams on the Euphrates River (and other dams along the Euphrates and Tigris in Iraq) was evidently 'a central pillar' of its state-building strategy. Thus IS described these dams as key targets in its 'great Caliphate project'; captured the three Syrian dams during early military operations in 2012–13; and fought lengthy battles to retain control of them, including a near two month-long battle in 2017 over control of the Tabqa Dam. Islamic State's motivations appear to have been twofold. Control of water was, on the one hand, crucial to IS's economic and fiscal strategies. Large volumes of water were needed for the extraction and processing of the crude oil on which IS revenues depended. Water and electricity charges provided direct revenue streams. Control of the Tabqa Dam enabled IS to temporarily increase electricity supplies to Raqqa and other IS-held towns as and when needed. And control of Lake Assad also meant that agricultural production in areas under IS control – which, at their height, accounted for over a third of Syria's total cropped area – could be sustained and in some areas increased, reducing food import dependency while generating tax revenues (as much as 30 per cent of which is estimated to have derived from agriculture). At the same time, control of key dams enabled IS to both use and threaten to use 'water as a weapon' – as is often emphasised in discussions of the subject. Islamic State repeatedly cut water (and electricity) supplies to towns under Syrian government control, as well as to IS-occupied areas that were deemed to require disciplining. In February 2017, Lake Assad was used to flood villages east of Aleppo under the control of the advancing Syrian army. Islamic State threatened to retaliate against attacks by destroying the Tabqa Dam, which would have released 11 mcm of water and caused untold destruction. The Tabqa Dam later served as an IS refuge given US reluctance to bomb it – which apparently slowed the US-led offensive against IS during 2016–17. And Turkey likewise employed water as a weapon against north-east Syria's Kurdish community following its invasion of the region in late 2019.[23]

By contrast with such essentially 'political' capture, others among our divided environments illustrate far-reaching war-enabled social changes in water access and control. Thus the 2003–5 war in Darfur saw the extensive capture of Fur, Masalit and Zaghawa land by groups linked to pro-government militias, especially in areas close to major *wadis* (livestock herds were also looted 'on an almost industrial scale', with non-Arab populations in Darfur losing an estimated 50–90 per cent of their livestock to the combined forces of theft and disease). In South Kordofan, the second civil war saw nearly 4 million hectares of land being expropriated for rain-fed cultivation, as discussed in Chapter 6. And in Syria, mass displacement has enabled the Syrian regime and the country's business, political and security elites to seize huge swathes of land and property, including property made valuable by water resources and infrastructures.[24] Across such diverse instances, what we repeatedly see is the capture of water as a corollary of land expropriation and population

---

[23] T. von Lossow, 'The rebirth of water as a weapon: IS in Syria and Iraq', *International Spectator*, 51 (2016), 82, 87–8, 93–4; M. Daoudy, 'Water weaponization in the Syrian conflict: strategies of domination and cooperation', *International Affairs*, 96: 5 (2020), 1356, 1360, 1362; 'Islamic State seizes Iraq's largest dam', *Al Jazeera* (07/08/2014); E. Francis, 'U.S.-backed Syria militias say Tabqa, dam captured from Islamic State', *Reuters* (10/05/2017); I. Mazlum, 'ISIS as an actor controlling water resources in Syria and Iraq', in Ö. Z. Okta et al. (eds.), *Violent Non-state Actors and the Syrian Civil War: The ISIS and YPG Cases* (Springer, 2018), 115–20; L. Eklund et al., 'How conflict affects land use: agricultural activity in areas seized by the Islamic State', *Environmental Research Letters*, 12:5 (2017), 7; S. Almukhtar, 'Life under the Islamic State: fines, taxes and punishments', *New York Times* (29/05/2016); M. D. King, 'The weaponization of water in Syria and Iraq', *Washington Quarterly*, 38 (2015), 162; P. H. Gleick, 'Water as a weapon and casualty of armed conflict: a review of recent water-related violence in Iraq, Syria, and Yemen', *Wiley Interdisciplinary Reviews: Water*, 6:4 (2019), 9; 'One million affected by Turkish water cutoff from Alouk station in Syria's Hasakah: UNICEF', *North Press Agency* (16/07/2021).

[24] O. Olsson, 'After Janjaweed? Socioeconomic impacts of the conflict in Darfur', *World Bank Economic Review*, 24:3 (2010), 386–411; HRW, *Darfur Destroyed: Ethnic Cleansing by Government and Militia Forces in Western Sudan* (2004), 31; Young et al., *Darfur*, xi; Suliman, 'Civil war', 105; H. Baumann (ed.), *Reclaiming Home: The Struggle for Socially Just Housing, Land and Property Rights in Syria, Iraq and Libya* (Friedrich-Ebert-Stiftung, 2019).

displacement: not only land–water nexuses in operation, but displacement–expropriation nexuses too.

In this sense northern Cyprus and Israel–Palestine present mixed cases, combining war-induced shifts in political control of water resources together with mass displacement and ensuing changes in social access. Prior to 1974, Greek Cypriots had been the major landowners across Cyprus, holding around 70 per cent of all privately owned land including most of the water-rich agricultural areas of Morphou (Güzelyurt, in Turkish) and the Mesaoria. With the ethnic cleansing of Greek Cypriots and Turkey's occupation of northern Cyprus, however, there was large-scale redistribution of Greek Cypriot property, with some of it being redistributed to Turkish Cypriot refugees from the south, some of it being leased to settlers from mainland Turkey, and the remainder coming under the control of Turkish Cypriot parastatal institutions. The Cypfruvex company, most notably, was founded in 1974 by a conglomerate of state banks and other institutions, whereupon it took over 2,300 hectares of Greek Cypriot land plus processing facilities to become the largest citrus exporter in northern Cyprus. Turkish military bases, with their heavy water resource use, were mostly built on expropriated lands too. The displacement of Greek Cypriots and redistribution of their land was thus at the same time an instance of large-scale water resource capture.[25]

That Israel–Palestine illustrates the same in even sharper form may surprise some readers given that Israel was already extensively exploiting the West Bank Mountain Aquifer prior to the June 1967 war, as already discussed in Chapter 6. Where it does apply, however, is the war of 1948–9 – the outstanding event in the history of the Israeli–Palestinian conflict, and worth discussing at length. Prior to 1948, around 91 per cent of cultivated land in Mandate Palestine was Palestinian-owned, with Palestinians accounting for an estimated 78 per cent of the territory's crop value (not including citrus) and the majority of its irrigated land. In the mid-1940s, Mandate Palestine's Palestinian and Jewish communities were each consuming an estimated 350 mcm/y of water, mostly for agriculture, with Palestinians accounting for around two-thirds of the usage from springs and Jews for the equivalent usage from wells.[26] In other words, despite the Zionist movement's progressive colonisation of Palestine during the Mandate period – a process underpinned by its far superior political,

---

[25] A. Gürel and K. Özersay, 'Cyprus and the politics of property', *Mediterranean Politics*, 11:3 (2006), 354; T. Ekici, *The Political and Economic History of North Cyprus: A Discordant Polity* (Palgrave, 2019), 44, 80, 195.

[26] Anglo-American Committee of Inquiry, *A Survey of Palestine, Vol. 1* (1946), 323, 327, 339; Messerschmid, *The "Prior Use" Argument*, 2.

economic and technological capabilities – as this period drew to a close the two communities were roughly equal users of Palestine's water resources.

The war of 1948–9 changed this situation irrevocably. The culmination of the settler-colonial encounter discussed in Chapter 4, this war – a 'War of Independence' for Israelis, *al-Naqba* ('the Catastrophe') for Palestinians – decisively reshaped not only Palestine's borders and demography, but also its landscape and patterns of water access and use. Palestine's social geography was transformed almost overnight. In 1947, there had been around 500 Palestinian villages within what would become Israel; of these, 418 were abandoned during the war and then immediately destroyed in order to prevent the refugees' return and create space for Jewish settlement (the refugees were almost without exception prevented from coming back – 'this is our policy: that they are not returning', stated Israeli Foreign Minister Moshe Sharett as early as June 1948). Mass expulsions from a dozen urban centres also took place, including the cities of Jaffa, Jerusalem and Haifa and exclusively Arab regional centres such as Beer Sheva. In total, according to the UN Conciliation Commission for Palestine (UNCCP), Israel acquired over 1.6 million hectares of land from Palestinian refugees – more than 80 per cent of its total area – including a cultivable area of nearly 0.5 million hectares.[27]

Onto this ethnically cleansed land – an ideal 'blank slate' for development – Jewish settlements, agriculture and JNF forests were planted in their place. Between 1948 and 1953 alone, at least 350 new Jewish settlements were established on refugee property (out of a total of 370 established during this period), most of them for Mizrahi Jews arriving from the Middle East and North Africa. Most abandoned Arab farmland became Jewish-farmed: as early as mid-1949, two-thirds of all land sown with grain in Israel was formerly Arab property. To a far greater degree than in post-partition Cyprus, the vast majority of signs of the former population and their livelihoods were erased from Israel's landscape: native villages were levelled, modes of farming changed and a new toponymy was imposed by an official government naming committee. As Ben-Gurion observed in justification, 'not destroying abandoned villages would be contrary to the policy of development and revitalisation of wasteland'.[28]

---

[27] W. Khalidi et al., *All That Remains: The Palestinian Villages Occupied and Depopulated by Israel in 1948* (Institute for Palestine Studies, 1992); B. Morris, *The Birth of the Palestinian Refugee Problem Revisited* (Cambridge University Press, 2003), 319; M. R. Fischbach, *Records of Dispossession: Palestinian Refugee Property and the Arab–Israeli Conflict* (Columbia University Press, 2003), 120–1.

[28] D. Peretz, *Israel and the Palestine Arabs* (Middle East Institute, 1958), 143; A. George, '"Making the desert bloom": a myth examined', *Journal of Palestine Studies*, 8:2 (1979),

Expropriation 231

These changes inevitably included radical discontinuities in patterns of water control and use; indeed, Palestinians' loss of water was, in proportional terms, even greater than their loss of territory or population. The 1949 armistice lines furnished Israel with over 92 per cent of the recorded well and spring flows of Mandate Palestine (compared with 78 per cent of its territorial area), as well as access to, and thus the opportunity to exploit, the upper Jordan River.[29] The wells and springs of the 726,000 Palestinians who had fled were simply expropriated alongside the rest of their property. Moreover, the areas that they fled to or remained in – the Gaza Strip, the West Bank and the upper Galilee, above all – were in most regards water-poor. Israel came to occupy Mandate Palestine's water-rich north and fertile coastal plain, as well as its thinly populated desert south – areas of practical colonisation, plus a zone of Zionist fantasy – while Palestinians were variously displaced to, or allowed to remain on, its residual arid, semi-arid and upland margins.

The fate of the Huleh Valley illustrates these dynamics in microcosm. This 177 km^2 expanse of lake and marshland on the upper Jordan River had long been of interest to the Zionist movement, partly for its agricultural potential, and partly its strategic location bordering Syria and Lebanon. Hence the first Jewish settlement was established there as early as 1882; a concession for the drainage of the central Huleh was purchased in 1934; and from the late 1930s, as the prospect of partition loomed, Huleh became a key focus of Zionist land acquisition and settlement. Yet for various reasons the Huleh concession was never implemented, and hence prior to the war the Valley had a mixed population, comprising over 15,000 Palestinians in 37 long-established villages and 24 Jewish settlements with a population of around 4,700 and owning a third of the Valley's land. During April–May 1948, however, all but one of the Palestinian villages in the Valley were emptied and subsequently destroyed. This was no accident: the fertile valleys of northern Palestine all witnessed particularly heavy village destruction during the war, as the Israeli state and its agricultural settlers sought to take possession of the richest Arab land. By the end of May Jewish settlers were already harvesting Arab fields, where winter crops had ripened. Although some villagers, exceptionally, were allowed to return following the armistice with Syria, once drainage began in the 1950s they were again expelled. Emptied of its original population, the Valley's lake and marshes were then quickly transformed into arable land. The Jordan River was canalised, the

---

98; M. Benvenisti, *Sacred Landscape: The Buried History of the Holy Land* (University of California Press, 2000), 11–54, 164, 168.
[29] Messerschmid, *The 'Prior Use' Argument*.

landscape modernised, all signs of the Valley's native inhabitants were extinguished and, perhaps most importantly for Israel, its sovereign possession of a strategically important region was consolidated.[30]

None of the above will be counter-intuitive to those familiar with post-nationalist accounts of the Israeli–Palestinian conflict. Yet Israeli water officials and commentators routinely claim, ignoring 1948–9 and focusing on 1967 only, that Israel has 'prior' or 'historical use' rights over Palestine's water resources – as if the *al-Naqba* had never occurred or those Palestinians displaced during it had never been resource users. Moreover, most academic studies of Israeli–Palestinian and Jordan basin water politics (including past work by one of the co-authors) completely ignore the signal role of the 1948–9 war in upending established patterns of water control and use.[31] In truth, however, 1948–9 was the outstanding moment of violent water resource capture in modern Israeli–Palestinian history – one whose consequences remain essentially in place more than seventy years later.

Institution-Building and Development

As if this were not enough, war also has profound if highly uneven consequences for institution-building and development, including in relation to water. The negative institutional consequences cut across most of our cases and are not difficult to comprehend: the loss of skilled staff, whether through death, displacement or middle class out-migration; associated losses of institutional memory; those shortages of fuel, material, equipment and capital already discussed; reduced access to areas that are contested, insecure or beyond government control; logistical and system-management difficulties arising from this, as well as from increased local autonomy and the proliferation of community and household water management practices (including those usually considered 'illegal'); the understandable prioritisation of humanitarian and crisis-related activity over more standard institutional work; and sometimes

---

[30] Morris, *Birth of the Palestinian Refugee Problem*, xi, xiv, 132, 248–52, 344–7; Government of Palestine, *Village Statistics 1945: A Classification of Land and Area Ownership in Palestine* (1945), 69–71; Khalidi, *All That Remains*, 427–511; W. P. N. Tyler, 'The Huleh concession and Jewish settlement of the Huleh Valley, 1934–48', *Middle Eastern Studies*, 30:4 (1994), 853, 849; Benvenisti, *Sacred Landscape*, 129–31, 170; D. Rabinowitz and S. Khawalde, 'Demilitarized, then dispossessed: the Kirad Bedouins of the Hula Valley in the context of Syrian–Israeli relations', *International Journal of Middle East Studies*, 32:4 (2000), 511–30.

[31] H. Gvirtzman, *The Israeli–Palestinian Water Conflict: An Israeli Perspective*, Working Paper No. 94 (Begin-Sadat Center for Strategic Studies, 2012), 26–7; Selby, *Water, Power and Politics*.

too, problems arising from the increased involvement of international actors, including international organisations, aid agencies and NGOs, within war-affected water sectors. Thus, after 2011, hydro-electric production at Syria's Tabqa Dam plummeted not so much because of physical damage but because only a few hundred qualified staff remained of the 2,500 needed to operate the dam at full capacity. Since 2013 South Sudan has seen, among many other things, local government water officials being forced out of rebel-held areas, intensified competition among water sector institutions, and water development financing being suspended as donors divert resources to emergency water, sanitation and hygiene (WASH) responses. During the 2001–4 intifada, Israeli military and settler violence in the West Bank, plus obstructions on movement, meant that Palestinian water officials largely stopped monitoring wells and springs. And on a different temporal scale, Turkey's invasion and occupation of northern Cyprus, combined with its predominantly military-led approach there, bequeathed the territory a weak and ill-funded central Water Works Department squeezed between strong municipal and individual farmer control over water resources and the large water transfer projects of Turkey's state water works, the DSI. These and many other such examples all bear out, on a governance level, Collier's characterisation of war as 'development in reverse'.[32]

The same, moreover, applies to the economic and social impacts of war. Many of these have already been illustrated above: the widespread destruction of water and sanitation infrastructures; the reductions in irrigation, agricultural and overall economic activity; the increased labour of household water management borne by women and girls; and increased mortality as a result of inadequate water and sanitation and resurgent infectious diseases. Beyond this, however, war can have long-term distorting and deterrent effects upon development. Thus Turkey's invasion of Cyprus not only partitioned its water resources and infrastructures but also prevented the realisation of plans for island-wide integrated water management, including for a water conveyance from the northern Troodos similar to the one in the south. The North–South civil war in Sudan not only saw strikes on the Jonglei Canal but the project's long-term abandonment. And joint Syrian–Turkish construction of a 'Friendship Dam' on the Orontes River – a foundation stone for which was laid by the two

---

[32] World Bank, *Mobility of Displaced Syrians*, 155–7; International Energy Agency, 'Electricity generation by source', www.iea.org/data-and-statistics/data-browser; Mosello et al., *Improving WASH*; Selby discussions with Palestinian water officials (various dates); Hoffmann, 'From small streams', 273; Collier, 'Civil war'.

countries' prime ministers in February 2011 – has been halted since the start of the Syrian conflict that same year.[33]

Alongside this and other such evidence of war impeding water development, it must also be recognised, however, that war has also often done the reverse – that is, been a productive and generative force for the mobilisation and management of water resources. This should come as no surprise. From technological innovation to gender relations and regional development – from the origins of the Internet, to the extension of democratic franchises, to the rise of East Asian 'developmental states' – war and war preparation have often been powerful stimuli to socioeconomic change, and the water sector is no exception. Thus the TVA, as noted in Chapter 5, was in large measure a by-product of US war industry needs during the First and Second World Wars. Canada's high level of hydropower development, similarly, can be traced back to the Second World War, when wartime production demanded – and simultaneously legitimated – the mass damning of the country's rivers to serve the requirements of nation and war. Both the Green Revolution and, with Sneddon, the 'concrete revolution' of worldwide dam building that accompanied it were explicitly conceived and pursued as part of US strategy for winning the Cold War, the TVA becoming so well-known globally for this very reason. Syria's Tabqa Dam was built and funded by the Soviet Union for similar reasons, made possible by the huge upsurge in aid – in geopolitical rents – following the 1967 war with Israel, as the Soviet Union sought to shore up its regional client following its devastating military defeat. And the Republic of Cyprus' water development programme was accelerated in the wake of Turkey's 1974 invasion, as part of a series of emergency plans for refugee integration and economic reconstruction and modernisation.[34] As all these examples testify, the mass mobilisation of people and resources during war often demands and enables the no less mass mobilisation – and with it, exploitation and destruction – of nature.

---

[33] G. Socratous, 'Integrated water resources planning in Cyprus', in P. Koundouri (ed.), *Water Resources Allocation: Policy and Socioeconomic Issues in Cyprus* (Springer, 2011), 119; Collins, *Waters of the Nile*, 387–401; W. Scheumann and O. Al-Shmaly, 'The Turkish–Syrian Friendship Dam on the Orontes River: benefits for all?', in A. Kibaroglu and R. Jaubert (eds.), *Water Resources Management in the Lower Asi-Orontes River Basin: Issues and Opportunities* (Graduate Institute of International and Development Studies and MEF University, 2016), 125–37.

[34] M. Woo-Cummings (ed.), *The Developmental State* (Cornell University Press, 1999); M. Evenden, *Allied Power: Mobilizing Hydro-Electricity during Canada's Second World War* (Toronto University Press, 2015); N. Cullather, *The Hungry World: America's Cold War Battle against Poverty in Asia* (Harvard University Press, 2010); Sneddon, *Concrete Revolution*; V. Perthes, *The Political Economy of Syria under Asad* (I.B. Tauris, 1995), 32; R. King, 'Land consolidation in Cyprus', *Geography*, 65:4 (1980), 323.

Israel provides a final illustration of the generative character of war. We have already seen, in Chapter 4, that early Zionism's settler-colonial encounter with the Palestinians was not between pre-set identities, still less institutions; rather, this encounter was a constitutive force, productive of the two distinct nationalisms as well as of many of the defining 'internal' features of Yishuv society and post-independence Israeli politics, most notably their strongly centralised and quasi-socialist institutions. In this and other respects, the Israeli–Palestinian conflict has from the outset been societally productive and immensely destructive in equal measure. Yet if this holds as a general rule, it applies most clearly to the war of 1948–9, the single most important event in Israeli–Palestinian history. Within a decade of the war, Israel had formulated a national Water Law which is globally unique in defining all water resources as public, state property, and under which all surface, sub-surface and supra-surface water resources are public property. Now, this hyper-centralised system of water law has often been represented as a natural outgrowth of Zionist principles – Zionism illustrates 'the power of ideas', remember – and praised as enabling the rational, sustainable management of the country's water resources; it is lauded by both of the texts discussed at the outset of this chapter, for instance, and characterised by leading Israel geographer Arnon Soffer as 'one of the reasons why we are able to be a villa in the surrounding jungle'. Yet it is clear that this system would not have been possible without the 1948–9 war and its mass displacement of the Palestinians. Within what became Israel, as in most places, water rights had traditionally been tied to property ownership – and without mass displacement established ownership patterns would inevitably have prevented such thoroughgoing state control over water resources. Even today, as Julie Trottier has shown, this dynamic can be observed in the West Bank, where the Palestinian Authority (PA) and Israeli military government alike have struggled to extend control over communal and farmer-owned wells and springs. In Cyprus, similarly, post-independence hydraulic development was pursued without any parallel transformation of property relations, such that water sector development has in most areas had a deeply clientelist quality – with both investment choices and the benefits therefrom being largely dictated by landowners and the Water Development Department having only limited regulatory powers (as indicated by the fact that almost half of all wells operate either without permits or above permitted levels).[35] In Israel, by

---

[35] Kaplan, 'The revenge of geography', 98; Siegel, *Let There Be Water*, 16; Abraham et al., *How Israel Became a World Leader*, 27; A. Soffer in Siegel, *Let There Be Water*, 19; J. Trottier, *Hydropolitics in the West Bank and Gaza Strip* (Palestinian Academic Society for the Study of International Affairs, 1999), ch. 3; Zoumides et al., 'Quantifying the

contrast, the mass displacement of Palestinians during 1948–9 created an ideal blank slate for development, water development included, furnishing the conditions of possibility for the establishment of a uniquely centralised water governance regime. While this water regime has been shaped by ideas, no doubt, it was only via war and ethnic cleansing that these ideas could be realised.

War and Climate Change Vulnerability

This chapter has sought to explore the impacts of war on water security and insecurities, analysing four different ways in which war matters for water: through infrastructure destruction; through the mass displacement of populations; through the capture and expropriation of water resources and infrastructures; and through war's consequences for state-building and development. The evidence to this effect is, to our minds, compelling. Not only does war have direct and immediately transformative impacts on water resources, infrastructures and patterns of water security and vulnerability, with impacts which are mediated by racial, class, gender and other hierarchies. Just as crucially, war's consequences may endure for decades or more. They can range, however uncomfortable it may be to acknowledge this, from the deeply destructive to the profoundly productive, bringing death and destruction to some but water security to others – war in this sense being the greatest of all redistributors, including in relation to water. Moreover, given that war is, by Clausewitz's famous aphorism, 'a continuation of politics by other means', it follows that what this chapter has once again demonstrated is that 'the political', broadly construed, lies at the heart of water-related crises and insecurities.[36] Indeed, there exists much more direct and extensive evidence on the impacts of war on water than on the reverse, water resource availability as a cause of or contributor to war. The sad truth is that contemporary patterns of water security and insecurity are impossible to understand except with heavy and sustained reference to organised political violence and its legacies.

One would barely guess this, however, from the weight of scholarly research on the subject, for not only is there very little research on the impacts of war on water security, as noted at the outset of this chapter; beyond this, there is currently very little research on war's implications for climate change, for ecology or for socio-environmental vulnerabilities

---

poorly known role', 2504, 2511; I. Iacovides, 'Water management: administration and legislation', in Koundouri, *Water Resources Allocation*, 52.

[36] C. von Clausewitz, *On War* (Penguin, 1968), 119.

more broadly. Early work on environmental security did attend to the ecological impacts of war: Arthur Westing, for instance, combined a questionable Malthusianism with a deep personal commitment to exposing the environmental horrors associated with contemporary militarism. Yet today the situation is sharply otherwise. For whatever reasons – among them Western state-led funding priorities and practical problems of data access and availability – mainstream environmental, climate and water security research focuses today almost entirely on the contribution of environmental factors to violent conflict, largely ignoring the reverse causal question of the contribution of violent conflict to environmental destruction and change. Most critical research has been little better, dominated as it is by an approach, securitisation theory, which, in its preoccupation with the discursive construction of security threats, has directed attention away from the materiality of war and war preparation as well as their consequences. Whig histories of water development largely ignore war too, as already discussed. Indeed, only those with political ecology inclinations have consistently highlighted and pursued research agendas on the environmental impacts of militarisation and war.[37] This chapter's heavy reliance on humanitarian and development reports is but one minor consequence.

While for both these and more generic reasons, there are thus huge uncertainties associated with how climate change and related vulnerabilities are, and will be, affected by war, the foregoing analysis does provide some pointers. It illustrates, first, something which is often noted but rarely explored in depth, namely, that the destruction and displacement wrought by armed conflict – the destruction of infrastructures, the despoilation of resources, the decay of economies and institutions, mass dislocation into refugee camps and peri-urban areas – can both significantly deepen and create wholly new environment-related vulnerabilities and, by extension, can and will increase vulnerabilities to climate change. Stated simply, droughts, floods, extreme temperatures, rising sea levels and so on are, and will be, all the more difficult for communities and households to adapt to in contexts where the resources, supply networks and livelihood systems on which adaptative capacities are founded have been expropriated, degraded or destroyed through war. This much is fairly obvious and

---

[37] A. H. Westing, *Arthur H. Westing: Pioneer on the Environmental Impact of War* (Springer, 2013); A. Stavrianakis and J. Selby (eds.), *Militarism and International Relations: Political Economy, Security, Theory* (Routledge, 2012), 9–10; Peluso and Vandergeest, 'Political ecologies'; Sowers et al., 'Targeting environmental infrastructures'; Zeitoun et al., 'Urban water ecology'; Duffy et al., 'Why we must question the militarisation of conservation'.

in line with both mainstream adaptation thinking and characterisations of war as 'development in reverse'.[38]

Yet second, the foregoing suggests that the opposite also hold true, namely, that because war is productive as well as destructive, it follows that war and war preparation may increase some groups' economic, infrastructural and institutional capacities – and by extension reduce their vulnerabilities to climate change. Paradigmatically, it is impossible to understand Israel's successes as a water manager adequately – or how its officials can view the country as largely immune to whatever serious droughts climate change may throw up – without registering the historical contributions of land and water expropriation therein, or the many indirect ways in which war and militarism have shaped its economic, diplomatic and institutional capacities, and in turn control over water. Moreover, even where war and militarism are not particularly productive, they may still function and be pursued as means of preserving status quo interests and privileges, with significant implications for how climate vulnerabilities are distributed. Whether mass international migration becomes a viable climate change adaptation strategy, for instance, will in large measure depend upon Northern military and associated responses – upon whether the current war on migrants from the global South, with its border walls, gunboats, surveillance systems and anti-migrant rhetoric, is further ramped up or not. Equally, whether geoengineering is pursued as a global adaptation strategy may ultimately pivot on its functionality to US military interests.[39] War and militarism are, and will in all likelihood continue to be, decisive in the global sharing out of adaptive capacities and vulnerabilities to climate change.

This all suggests, third, that at least within the sort of 'conflict-affected contexts' analysed in this book, climate change and war are very far from being equivalent causes of human vulnerability, and are unlikely to become so anytime soon. Northern donors and think tanks often refer today to climate and conflict as 'intersecting' causes of vulnerability and fragility, as if they are somehow on a par. Donors insist on climate-smart financing and climate adaptation policies, even as the missiles rain down. They obsess, rather obscenely, with imagined future 'climate refugees', even as the number of actually conflict-displaced marches ever higher. It

---

[38] H. Buhaug and N. von Uexkull, 'Vicious circles: violence, vulnerability, and climate change', *Annual Review of Environment and Resources*, 46:1 (2021), 545–68; M. R. Muzamil et al., 'An extreme climatic event and systematic vulnerabilities in the face of conflict: insights from the Taliban insurgency in Swat, Pakistan', *Regional Environmental Change*, 21:1 (2021), 1–13; Collier, 'Civil war'.

[39] K. Surprise, 'Geopolitical ecology of solar engineering: from a "logic of multilateralism" to logics of militarization', *Journal of Political Ecology*, 27:1 (2020), 213–35.

is regularly reported, to give but one specific example, that many more people are displaced each year by weather disasters than by conflict. Yet all of this is profoundly misleading. Most weather shock displacement is temporary, much and perhaps most of it comprising pre-emptive evacuations and bearing no resemblance to the long-term expulsions suffered by Sudanese, Syrian, Cypriot let alone Palestinian refugees. The evidence on climate impacts is typically somewhere between indirect and baseless, whereas that on war impacts is mostly direct, copious and beyond serious doubt. In the West Bank and Gaza, as even UNDP has recognised, climate change's impacts on livelihoods pale in comparison with those of the Israeli occupation. Indeed, as Michael Mason has observed of his time working on UNDP's West Bank and Gaza climate adaptation strategy during 2008–9, just months after the 2008–9 war, 'climate change was irrelevant in this flattened landscape', the work striking him, on occasion, as 'surreal, even absurd'.[40] We concur: in contexts of prolonged violent political conflict, it is war, war preparation and their enduring legacies which are the primary structural causes of vulnerability, including in relation to climate. And while we cannot know the future of war, we see little reason to doubt that this dynamic will continue. Within such contexts and perhaps globally, the future of climate change vulnerabilities will in large measure depend upon the future of war.

---

[40] B. van Bronkhorst and F. Bousquet, 'Tackling the intersecting challenges of climate change, fragility and conflict', World Bank blog (27/01/2021), https://blogs.worldbank.org/dev4peace/tackling-intersecting-challenges-climate-change-fragility-and-conflict; IDMC, *Global Report on Internal Displacement 2021: Internal Displacement in a Changing Climate* (2021); Selby and Daoust, *Rapid Evidence Assessment*; UNDP, *Climate Adaptation Strategy and Programme of Action for the Palestinian Authority* (2010); M. Mason et al., 'Compounding vulnerability: impacts of climate change on Palestinians in Gaza and the West Bank', *Journal of Palestine Studies*, 41:3 (2012), 39; M. Mason, 'Climate change and conflict in the Middle East', *International Journal of Middle East Studies*, 51:4 (2019), 626.

8 Peace

The Promises of Environmental Peacebuilding

Ever since the early 1990s, one standard response to eco-crisis and environmental security narratives has been to insist that the environment can be a source of cooperation as much as of conflict, of amity rather than just enmity. Political leaders from Soviet President Mikhail Gorbachev to UN Secretary-Generals Kofi Annan, Ban Ki-moon and António Guterres have all argued thus. Numerous peacebuilding and development organisations have too. Ken Conca, perhaps the best-known scholar of environmental peacebuilding, submits that there is 'a good deductive foundation' deriving especially from constructivist and liberal functionalist IR theories 'for the claim that environmental cooperation can promote and enhance peace'. And an ever-expanding body of literature has tended to agree.[1]

With the possible exception of biodiversity conservation, no issue has been more central to these promises of peacemaking via nature than trans-boundary water management. This is partly because, unlike on many other issues – and in contrast also with the meagre evidence of inter-state 'water wars' – there exists extensive bilateral and regional cooperation over trans-boundary water resources, some of it deeply institutionalised. Yet for advocates of hydro-political peacebuilding, the case for water typically extends well beyond this simple fact. Thus it is often suggested that water is 'a resource whose characteristics tend to induce cooperation and incite violence only in the exception' and, at the same time, that technical or political cooperation over trans-boundary

---

[1] M. Gorbachev, 'The unity of water', *Project Syndicate* (07/07/2014); UN, 'World's water problems can be "catalyst for cooperation" says Secretary-General in message on World Water Day' (26/02/2002); Ban Ki-moon, 'Remarks at High-Level Interactive Dialogue on Water Cooperation' (03/22/2013); UN Security Council, *7959th Meeting* (S/PV.7959) (06/06/2017), 2–3; K. Conca, 'Environmental cooperation and international peace', in P. F. Diehl and N. P. Gleditsch (eds.), *Environmental Conflict: An Anthology* (Routledge, 2001), 230, 245.

water resources, as well as post-conflict reconstruction of water infrastructures and services, can be instrumental in the nurturing of wider peace.[2] Indeed water, it is often posited, is especially and perhaps uniquely well-suited to facilitating water-to-peace transitions; at the extreme, it is sometimes even claimed that the natural affinity between water and peace is so profound as to prevent descents into bloodshed and war.

This chapter, it will be no surprise to learn, offers a quite different assessment. Building upon extant critiques of environmental peacebuilding, especially by political ecologists and critical water politics researchers, it argues that water does not have innate cooperation-inducing properties and is not particularly important or instrumental in nurturing peace.[3] It shows that, within contemporary peace processes, water peacebuilding and cooperation are essentially 'dependent variables', structured and shaped by broader political and economic imperatives rather than vice versa. And it demonstrates that because peace processes are themselves often deeply problematic – in extreme reproducing or even radicalising pre-peacemaking divisions and attendant patterns of conflict, appropriation and inequality – so the same applies to water-related peacebuilding and cooperation specifically. Water under contemporary global capitalism, we argue in this chapter, is less a catalyst to peace than a vivid pointer to the latter's many, recurring limitations. The rhetoric of hydropolitical 'peace' and 'cooperation' is little more than that.

Our arguments to this effect are framed around the idea of 'peace processes': protracted processes of peacemaking which cut across many different socio-political domains and include everything from unofficial track II negotiations through to 'post-conflict' reconstruction and development programmes funded by international donors. In adopting this focus, we depart both from the dominant tendency within contemporary critical peace scholarship, which generally takes 'peacebuilding' – that is, post-conflict reconstruction – as its central object of analysis before interrogating it as an inherently liberal or neo-liberal project, as well as from most recent work on water and peace, which typically considers the issues of 'water cooperation' (or 'diplomacy') and water peacebuilding separately. Taking the notion of 'peace processes' as our starting point, by

---

[2] Wolf, 'Conflict and cooperation', 251.
[3] M. Zeitoun and J. Warner, 'Hydro-hegemony: a framework for analysis of transboundary water conflicts', *Water Policy*, 8:5 (2006), 435–60; C. Sneddon and C. Fox, 'Rethinking transboundary waters: a critical hydropolitics of the Mekong basin', *Political Geography*, 25 (2006), 181–202; K. Aggestam and A. Sundell-Eklund, 'Situating water in peacebuilding: revisiting the Middle East peace process', *Water International*, 39:1 (2014), 10–22; R. Duffy, 'Peace parks: the paradox of globalisation', *Geopolitics*, 6:2 (2001), 1–26.

contrast, we seek to analyse post-conflict reconstruction and development efforts and peace negotiations and agreements in tandem; to examine how the former are variously shaped and misshaped by the latter; and to draw attention to the specificities of peacemaking in different political contexts – something which is all too often absent from depoliticised celebrations of 'water for peace' and in equally generic denunciations of the 'liberal peacebuilding' project.[4] In this, the Israeli–Palestinian, Cyprus and Sudan–South Sudan peace processes provide our primary evidence base. We begin by interrogating key liberal environmental peacebuilding claims before considering these three peace processes in greater detail; and we conclude, as elsewhere, by considering the implications of our analysis for a world that is rapidly warming.

Is Peace Really Blue?

The idea that water has innate, even unique, cooperation-inducing characteristics comes in a number of forms. Sometimes it is suggested that the undoubted biological importance of water generates a natural tendency towards cooperation, especially under conditions of scarcity: that 'water is too vital a resource to be put at risk by war'; that water scarcity 'concentrates the minds of decision-makers'; that water induces cooperation 'because it is essential to life and so highly charged'; and that '[w]hen states go to war it is usually over something far less important than water'. Others suggest that the borderless character of the hydrological cycle – the fact that, in nature, 'water knows no borders' – and the dependence of so many states on trans-boundary watercourses, makes water cooperation so widespread.[5] Within post-conflict reconstruction planning, water sector investment is often advocated on the grounds of the tangible, material benefits that it can undoubtedly bring to communities devastated by war. And still others have argued, inspired by liberal functionalist reasoning, that it is the relatively technical character of water issues which makes them amenable to building peace and cooperation.

---

[4] This approach builds on J. Selby, 'The political economy of peace processes', in M. Pugh et al. (eds.), *Whose Peace? Critical Perspectives on the Political Economy of Peacebuilding* (Palgrave, 2008), 11–29; J. Selby, 'The myth of liberal peace-building', *Conflict, Security and Development*, 13:1 (2013), 57–86.

[5] M. Dolatyar and T. Gray, 'The politics of water scarcity in the Middle East', *Environmental Politics*, 9:3 (2000), 67; F. W. Frey and T. Naff, 'Water: an emerging issue in the Middle East?', *Annals of the American Academy of Political and Social Science*, 482 (1985), 67; UNDP, *Beyond Scarcity: Power, Poverty and the Global Water Crisis* (1996), 20; E. Akçalı and M. Antonsich, '"Nature knows no boundaries": a critical reading of UNDP environmental peacemaking in Cyprus', *Annals of the Association of American Geographers*, 99:5 (2009), 940–7.

The more rhetorical of these claims can be dealt with quite briefly (the others will be addressed in the next section). It is, first of all, nonsensical to suggest that biology or preciousness are barriers to war: water resources and supplies are regularly put at risk during war, as amply documented in the previous chapter; other critical resources and services, from oilfields to harbours and hospitals, are targeted during war precisely because of their strategic, economic or symbolic importance; and, in any case, under late twentieth and early twenty-first-century global capitalism water is not life, as argued in Chapter 2. Equally, it is hard to see why the borderless character of atmospheric, river and groundwater flows should ipso facto generate cooperation more than conflict. By contrast with standard environmental 'commons' situations – in which actors are equally or, in Ulrich Beck's terms, 'democratically' dependent on a shared resource and where there thus exist strong mutual incentives for cooperation – on trans-boundary watercourses there are always powerful asymmetries of one form or another between upstream and downstream riparians. Moreover, in practice water today is often not borderless at all, as the Gaza–Israel and Syria–Turkey borders vividly and visibly demonstrate. Trans-boundary watercourses do, no doubt, invite, and maybe even demand, cross-border interactions. The multi-purpose character of surface water flows – their use across transport, agriculture, energy production, washing, waste disposal, drinking, leisure, livelihoods and more – creates ideal opportunities for 'issue linkage' and the creation of systems of reciprocity, as beloved by neo-liberal institutionalists (it creates opportunities, for example, for arrangements in which an upstream state might use water primarily for electricity generation, while also storing water to be used by downstream states for irrigation).[6] However, issue linkage can be a cause or feature of conflict as much as of cooperation, and there is no in principle reason why multi-functionality should lead to one of these outcomes more than the other.

The most egregious contemporary example of misplaced liberal water cooperation thinking is provided by the 'Blue Peace' initiative, developed by the Mumbai-based Strategic Foresight Group (SFG). Funded by the Swiss and Swedish international development agencies, this initiative has attracted a great deal of international attention, culminating in a 2016 address to the UN Security Council and convenorship of a self-styled 'Global High-Level Panel on Water and Peace'. Blue Peace's central 'findings' are twofold: that 'any two countries engaged in active water

---

[6] Beck, *Risk Society*, 36; C. W. Sadoff and D. Grey, 'Beyond the river: the benefits of cooperation on international rivers', *Water Policy*, 4:5 (2002), 389–403; S. Dinar, *International Water Treaties: Negotiation and Cooperation along Transboundary Rivers* (Routledge, 2008).

cooperation do not go to war for any reason whatsoever' (with 'war' being understood as 'armed hostilities of any kind or direct or indirect involvement in cross-border terrorism or a risk of any such events'); and that water cooperation is the independent, explanatory variable behind this correlation – 'most of the countries in the Middle East', Blue Peace claim, 'are at risk of war *because* they have avoided regional cooperation in [sic] water'.[7] However, these assertions are so conspicuously flawed that it is difficult to know where to begin. Active water cooperation, as defined by Blue Peace, did not prevent Sudan's attempted assassination of Egyptian President Hosni Mubarak in 1995, nor the outbreak of the second Palestinian intifada in 2001. What Blue Peace calls a 'new spirit of cooperation' on the Euphrates from 2008 onwards did not stop Syria's descent into civil war or Turkey's active military support of the Syrian opposition. Membership of the Lake Chad Basin Commission has not prevented (considering conflicts since 2010 only) the Libyan civil war, the late stages of the Chadian civil war, numerous armed wars and armed conflicts in Sudan, armed conflict in the Central African Republic or armed conflict with Boko Haram, Islamic State and/or Al-Qaeda in the Islamic Maghreb in Algeria, Cameroon, Niger and Nigeria. Similarly, membership of the Southern African Development Community has not prevented (again, since 2010 only) armed conflict with Cabindan secessionists in Angola and with Renamo in Mozambique, or the continuation of war and armed conflict in the Democratic Republic of Congo. Myanmar and Ukraine, not to mention the United States, France and the UK, are all categorised by Blue Peace as *not facing* the 'risk of war', when all have either experienced or participated in war in recent years. There are other problems too, but these few illustrations should suffice. In truth, Blue Peace's water cooperation–war equation is even less rigorous than Thomas Friedman's infamous Golden Arches Theory of Conflict Resolution, which claimed that any two countries with a Macdonald's franchise would not go to war with one another (while Friedman's theory was falsified almost as soon as it was advanced, it was at least rooted in a semi-plausible causal account of the impact of economic development on conflict; the Blue Peace equation, by contrast, provides no account at all of how and why water cooperation should have such determining impacts).[8] It is a sad commentary on the Blue Peace

---

[7] SFG, *Water Cooperation for a Secure World: Focus on the Middle East* (2013), 2–4, 120 (italics added); UN Security Council, *7818th Meeting* (S/PV.7818) (22/11/2016); Geneva Water Hub, *Matter of Survival*.
[8] SFG, *Water Cooperation*, 5, 20–1, 34–5; SFG, *Water Cooperation Quotient 2017* (2017); T. Friedman, *The Lexus and the Olive Tree* (HarperCollins, 1999), 195.

initiative's funders, and on the state of environmental security policy discourse more broadly, that it has achieved such prominence.

Contrary to those who contend that it is so vital as to inspire peace, water has generally not been a central issue or source of dispute in recent peace processes, as all three of the cases examined in this chapter confirm. Water received only the briefest of mentions in the 1993 Oslo Agreement between Israel and the PLO. It was relegated to Article 40 of Annex III of the 1995 Oslo II Agreement which followed (while there was considerable disagreement over the terms of this article, ultimately water was not deemed of sufficient importance by the PLO leadership for them to hold out – with consequences which are made clear below). Water has not usually been understood as a core Israeli–Palestinian final status issue either, but instead as one of any number of 'other issues of common interest' to be resolved in addition to the core questions of 'Jerusalem, refugees, settlements, security arrangements, borders, relations and cooperation with other neighbors'. Similarly, the 2005 Naivasha Agreement between Sudan and the SPLM/A barely mentioned water issues, despite being 260 pages long. Accounts of the negotiations underpinning this agreement suggest that the Nile barely featured at all. In turn, water issues did not receive any great attention during the 2005-11 interim phase of the Sudanese peace process, and have not featured prominently in the negotiations and agreements aimed at ending the post-independence violence in South Sudan. And water does not feature at all within the main articles of the 2004 Annan Plan for Cyprus.[9] Altogether, the record of post–Cold War peace negotiations and agreements lends very little support to Blue Peace–style claims about the political and peacemaking importance of water.

Indeed, the centrality of water issues within peacemaking initiatives may, if anything, be in long-term decline. In the mid-1950s, US Secretary of State John Foster Dulles characterised regional water development as one of the United States' top three global peacemaking priorities, while water was viewed by the United States, as well as Israel, as one of the five core issues to be addressed in any final Arab–Israeli peace settlement. It is not difficult to see why. Water was accorded such importance by the United States because of its broader Cold War–informed philosophy of

---

[9] Israel and PLO, *Declaration of Principles on Interim Self-Government Arrangements* (1993), art. 7 (4), annex III (1), art. 5 (3); Israel and PLO, *Interim Agreement*, app. 1, annex III, art. 40; *The Comprehensive Peace Agreement between the Government of the Republic of the Sudan and the Sudan People's Liberation Movement/Sudan People's Liberation Army* (2005), ch. 2, part 5 (A33); *Revitalised Agreement on the Resolution of the Conflict in the Republic of South Sudan (R-ARCSS)* (2018), ch. IV (4.8.3); J. Young, *The Fate of Sudan: The Origins and Consequences of a Flawed Peace Process* (Zed, 2012); *The Comprehensive Settlement of the Cyprus Problem* (2004), main arts.

stabilisation through agrarian development, to which we return in the next section. Moreover, within the Middle East specifically, water was viewed by the United States and Israel as providing a way 'to liquidate the Palestinian refugee problem', as one analyst put it, through mass refugee resettlement on the rich but water-poor agricultural lands of the Jordan Valley. United States policy was to develop the Jordan Valley to house hundreds of thousands of refugees and, as the National Security Council noted, to 'link the development of the Jordan Valley to Arab refugee resettlement to the maximum extent practicable'.[10] A half century later, by contrast, no one envisages Palestinian refugees becoming *fellahin* again; and no one imagines that agrarian development is the key to US hegemony or international order. It should be no surprise, given this, that the place of water within either the Israeli–Palestinian or any other peace process is not what it once was.

Liberal Functionalism

In contrast to the view that it is water's vital importance which makes it such a wellspring of cooperation, others suggest, conversely, that it is the relatively technical character of water issues which makes them so well-suited to nurturing peace. This view is rather more substantial and deserving of fuller consideration. Intellectually, its origins lie in liberal functionalist – or more precisely neo-functionalist – approaches to IR, above all in the work of Ernst Haas on the European Coal and Steel Community (ECSC) and the roots of European integration. In Haas' assessment, the formation of the ECSC in 1951 created new bureaucratic and economic interests solicitous of further integration, sparking the process which eventually led to the establishment of the European Economic Community and, later, EU. For neo-functionalists, bilateral or multilateral regional cooperation on welfare, technical and economic issues can help in building new social and supra-national interests, and in turn mutual trust and understanding, which can then 'spill over' into other areas, incrementally softening long-entrenched political hostilities and forging ever-deeper cross-border relations.[11]

Such thinking is often reflected in peacemaking discourse and practice. During the 1990s, it was extensively invoked in relation to the Middle East in particular, with then Israeli Foreign Minister Shimon Peres calling

---

[10] G. G. Stevens, *Jordan River Partition* (Hoover Institution, Stanford University 1965), 17; 'Statement of Policy by the US National Security Council' (23/07/1954); Lowi, *Water and Power*, 81–3, 100, 206.

[11] E. B Haas, *The Uniting of Europe: Political, Social and Economic Forces 1950–1957* (Stanford University Press, 1958).

for a 'new Middle East' of peace through regional development and integration, and the entire multilateral negotiation track being designed around it. Moreover, trans-boundary water development is often depicted in these terms. 'Just as coal and steel laid the foundation for the creation of the European Union, so can oil, gas, water and any other natural resource pave the way for reunifying Cyprus', commented a Turkish Cypriot official as construction began on Turkey's water pipeline to northern Cyprus. 'If regional trade agreements over resources in Europe could lead to [its] eventual unification' then perhaps the 'creation of a water–energy trade community' could 'kick-start a process of larger regional cooperation, and perhaps even regional unification, in the Levant region', suggest EcoPeace, an Israeli–Jordanian-Palestinian (but in practice Israeli-led) peacebuilding NGO. Or as Aaron Wolf, perhaps the leading scholar of water cooperation, maintains, water can 'help induce ever-increasing cooperation ... between otherwise hostile riparians, in essence "leading" peace talks.'[12]

Unfortunately, both history and logic suggest otherwise. Consider, first, the wastewater treatment plant in Cyprus' divided capital, Nicosia (in Turkish, Lefkoşa), which has often been lauded as 'a perfect example of bi-communal cooperation' and characterised in neo-functionalist terms. Inaugurated in 1980, less than a decade after Nicosia's partition, and modernised from 2010 to produce high-quality effluent for irrigation, this plant was designed – and is maintained – through long-term collaboration between Greek Cypriot and Turkish Cypriot engineers; to this day it serves both sides of the city, providing a perfect example of how enduring technical cooperation can flourish even under circumstances of geopolitical conflict and division. Yet to this rosy picture a series of qualifiers are required. To start with, far from being a consequence of technical collaboration alone, the treatment plant was from the beginning also a political project, conceived and promoted by Lellos Demetriades and Mustafa Akinci, the city's two municipal leaders, as part of their shared bicommunal political agenda. In turn, there exists no evidence, so far as we are aware, of 'spill over' from wastewater to other functional spheres. Nicosia's wastewater system remains in key respects divided, with the two networks that it serves remaining separate and subject to different legislation. Moreover, the 'bicommunal' design and operation of the plant has only been possible thanks to various unofficial and

---

[12] S. Peres, *The New Middle East* (Holt, 1993); J. Peters, *Pathways to Peace: The Multilateral Arab–Israeli Peace Talks* (Royal Institute for International Affairs, 1996); Associated Press, 'Cyprus water plan: peace pipeline or Trojan Horse?', *Mail Online* (28/02/2014); M. Mehyar et al., *A Water and Energy Nexus as a Catalyst for Middle East Peace* (EcoPeace, 2014), 9; Wolf, *Hydropolitics*, 3.

formally non-governmental arrangements designed to bypass the two sides' mutual non-recognition. The treatment plant is governed by retired officials from Nicosia's Greek Cypriot and Turkish Cypriot sewerage boards, operating in a private capacity, and is managed on a day-to-day basis by 'off duty' officers. Meanwhile, the recent upgrade was funded by the RoC and EU and the project implemented by UNDP, to circumvent the need for direct engagement with Turkish Cypriot authorities.[13] Such are the limits of water cooperation in circumstances of frozen conflict.

Or consider, similarly, the so-called Israeli–Jordanian 'picnic table talks' at which, for fifteen years from 1979, water officials from the two countries would meet regularly, if secretly, at the confluence of the Yarmouk and Jordan Rivers to coordinate on technical matters relating to water resource management and supplies. Providing yet another classic illustration of how technical cooperation can develop under the shadow of war, by some accounts these talks also paved the way for the water terms of the 1994 Israeli–Jordanian Peace Treaty, and perhaps even the Treaty as a whole: no case is more regularly cited as proof of how technical cooperation over water can help to build peace. Indeed, there is much to commend the view that the professional relationships and understandings which were established between Israeli and Jordanian water policymakers and technocrats prior to 1994 provided a crucial foundation for the Treaty's water terms and their implementation. However, what is typically passed over in such discussions is that these 'picnic table talks' were but one element of a wide-ranging pattern of unofficial pre–Peace Treaty Israeli–Jordanian coordination, which encompassed issues of agriculture, industry, transport, tourism and health as well as water, and even on occasion extended to discussions about pigeons and mosquitoes. This coordination, moreover, had its roots in a history of accommodation that stretched right back to the 'collusion across the Jordan' before and during the war of 1948–9, when Zionist leaders and Jordan's Hashemite monarchy saw common interests in limiting Egyptian power and preventing the emergence of a Palestinian Arab state; and which, during the Cold War, was keenly promoted and mediated by the United States, for reasons discussed below. In short, the 'picnic table talks' were one element in, and one product of, a long history of Israeli–Jordanian–American under-the-radar collaboration, and do

---

[13] UNDP, 'New Nicosia wastewater treatment plant', www.cy.undp.org/content/cyprus/en/home/projects/NWWTP.html; A. D. Brouma and C. Ezel, 'Water policy networks: a new form of governance for Cyprus? The case of Nicosia', in P. Koundouri (ed.), *Water Resources Allocation* (Springer, 2011), 76–9; Hoffmann interview with Nicosia municipal official (11/11/2010); European Commission, *Council Regulation (EC) No 389/2006* (27/02/2006).

nothing to substantiate the neo-functionalist hope of peacemaking being 'led' by technical cooperation on water specifically.[14]

Staying with the Jordan basin, neo-functionalist reasoning is also often applied to the 1953–6 'Johnston mission' to the Middle East, in which US President Eisenhower's personal representative, Eric Johnston, sought to secure agreement from Israel, Jordan, Lebanon and Syria on the allocation and development of the Jordan River. By some accounts, this mission explicitly followed a neo-functionalist template, with the United States hoping that a regional water plan and ensuing projects 'would be catalysts to peace in the region ... in the fashion suggested by the functionalists'.[15] Johnston and his team conducted four rounds of negotiations with the parties over a two-year period until, ultimately, their Revised Unified Plan was approved by an Arab League Technical Committee. Ultimately, however, the Plan was dashed by Cold War and regional geopolitics, in particular Egypt's 1955 arms deal with the Soviet Union and the growing Arab–Israeli border tensions that culminated in the Suez War of 1956. Instead of helping to bridge geopolitical divisions, the Johnston mission was very much hostage to them.

Other cases suggest likewise. The most famous instance of successful water cooperation under conditions of unresolved inter-state conflict – the regime established through the Indus Waters Treaty of 1960 – is not only a stand-alone regime that has not had any broader positive impact on Indo–Pak relations, but is arguably successful precisely because it divides the Indus River and its tributaries while requiring only limited cooperation; it functions as the hydrological corollary of the territorial partition of India. Moreover, none of the other divided environments examined in this book provide evidence of a water-led spill-over effect. While clearly demarcated 'islands of water cooperation' are sometimes established and maintained under conditions of protracted or frozen conflict, as the Indus and Nicosia cases above illustrate, there is – and with apologies for the mixed spatial metaphors – scant evidence of these 'islands' of water cooperation 'spilling over'. Lothar Brock concluded in a classic 1991 study of environmental peacemaking that the project of '[p]eace through functional cooperation has been just as much a disappointment as the

---

[14] A. Jägerskog, *Why States Cooperate over Shared Water: The Water Negotiations in the Jordan River Basin* (Linköping University, 2003); J. K. Sosland, *Cooperating Rivals: The Riparian Politics of the Jordan River Basin* (SUNY Press, 2007), 1, 108–12, 140, 183, 203–4; M. Zak, 'Israeli–Jordanian negotiations', *Washington Quarterly*, 8:1 (1985), 167–76; A. Garfinkle, *Israel and Jordan in the Shadow of War: Functional Ties and Futile Diplomacy in a Small Place* (Macmillan, 1992); A. Shlaim, *Collusion across the Jordan: King Abdullah. The Zionist Movement and the Partition of Palestine* (Clarendon, 1988); Selby, *Water, Power and Politics*, 58–60.

[15] Lowi, *Water and Power*, 194.

older ideas of the free trade pacifists'.[16] Thirty years of post–Cold War peacebuilding later, we see little reason to disagree.

There are also good theoretical reasons to doubt whether functional cooperation on water could ever 'lead' peacemaking. Ernst Haas' spill-over model was only ever envisaged as applying to the development of regional ties and institutions between democracies already at peace, not to situations of mutual non-recognition or ongoing war. Neo-functionalist thought also contains a gaping contradiction: it presumes that cooperation first develops around relatively uncontroversial and discrete issues where potential gains from cooperation are sufficient to warrant a sacrifice of state autonomy, but then also has to imagine that the issue in question might be so important, and so inextricably linked to other issues, as to propel further integration. Moreover, contrary to Haas and others, the formation of the ECSC was not 'functional politics'; instead, very much like the Johnston mission, it was from the outset a geopolitically mediated project, led by the French, German and US governments with the aim of controlling the two key industries, coal and steel, which had most underpinned Germany's descent into fascist militarism.[17]

By the same token, however, realist IR responses to functionalism are misleading too. Realists typically suppose, as for instance Miriam Lowi does on the failure of the Johnston mission, that 'states that are adversaries in the "high politics" of war and diplomacy do not allow extensive collaboration in the sphere of "low politics"' and that 'the realist critics of functionalism' are therefore 'correct'. But this is not so. Quite apart from the fact that Israel and Jordan *did* engage in 'extensive collaboration' on technical, economic and welfare issues prior to 1994 despite formally being at war, Lowi's characterisation of the Johnston mission as 'low politics' is not right either. For, as already discussed, US policy on the Jordan basin during the 1950s was driven by two quintessentially geopolitical concerns: the resolution of the Arab–Israeli conflict on Israeli terms through permanent resettlement of Palestinian refugees in the Jordan Valley; and Cold War–informed fears about the stability of rural Asia. In Johnston's own words, his mission and the basin-wide development which it was designed to enable were part of 'a global effort to help

---

[16] N. A. Zawahri, 'India, Pakistan and cooperation along the Indus river system', *Water Policy*, 11:1 (2009), 1–20; L. Brock, 'Peace through parks: the environment on the peace research agenda', *Journal of Peace Research*, 28:4 (1991), 414.

[17] E. B. Haas, 'International integration: the European and the universal process', *International Organization*, 15:3 (1961), 366–92; P.C. Schmitter, 'Ernst B. Haas and the legacy of neofunctionalism', *Journal of European Public Policy*, 12:2 (2005), 262–3; T. Judt, *Postwar: A History of Europe Since 1945* (Penguin, 2005), 156–60.

less-advanced peoples ... premised on the assumption that healthy social progress is the most effective antidote to the Communist virus, which, in common with its bacteriological cousins, strikes hardest at run-down, poorly nourished systems'. Like both the 'picnic table talks' and the Green and Concrete Revolutions discussed in the previous chapter, the Johnston mission was not conceived by the United States as a form of technical 'low politics' that at some point would spill over into politics and help catalyse 'peace' – but as an important element of US Cold War containment strategy. More broadly, the very distinction between the 'high politics' of war and diplomacy and the 'low politics' of economics and welfare is anachronistic, deriving, revealingly, from that most anachronistic of intellectual traditions, 'realist' IR.[18] Contrary to what is implied by the low politics–high politics distinction, within our contemporary global capitalist order all issues and practices – from war to wastewater treatment – are simultaneously thoroughly technical and saturated with politics: techno-politics rules right across this supposed divide.

As a final point, the above also suggests that the very distinction between 'conflict' and 'cooperation' should be approached with caution. Most liberal trans-boundary water politics discourse assumes a sharp divide between the two, or at most imagines a continuum stretching from physical violence to integrated management, with multiple steps in between. In parallel, it is also generally assumed that conflict is essentially 'bad', and cooperation 'good', and that it thus 'makes sense to promote and support cooperation of any sort, no matter how slight'. Underpinning these assumptions are rational actor premises: cooperation, it is assumed, results from actors seeking 'to pursue rational and legitimate self-interest' and occurs when it provides 'benefits that add to the aggregate welfare of both sides' – it being for this reason that cooperation is, by definition, imagined as a good. Yet in reality matters are not so simple. 'Cooperation' may be pursued for various reasons beyond rational assessments of absolute gains: because regimes and elites hope to extract resources through it, even if it might bring few benefits to their populations as a whole; because of promised future benefits, which may however never materialise; because of extreme physical, political or economic dependencies, which may make the costs of not engaging in or defecting from cooperation

---

[18] Lowi, *Water and Power*, 196; Lecture by Johnston, Cornell University (06/05/1954), in Lowi, *Water and Power*, 205; S. Hoffmann, 'The European process at Atlantic crosspurposes', *Journal of Common Market Studies*, 3:2 (1965), 85–101; S. Hoffmann, 'Obstinate or obsolete? The fate of the nation-state and the case of Western Europe', *Daedalus*, 95:3 (1966), 862–915; M. Griffiths, *Realism, Idealism and International Politics: A Reinterpretation* (Routledge, 1992); Mitchell, *Rule of Experts*.

inordinately high; or because of anticipated benefits from third parties, especially international mediators and development actors. In turn, both 'cooperation' generally and 'water cooperation' specifically can take any number of forms – some relatively symmetrical, some radically asymmetrical and some little more than a 'dressing up' of conflict, appropriation, hegemony or domination in different guise. 'Peace', equally, is not necessarily any less scarred by violence than war, as studies of the war–peace distinction and war-to-peace transitions have repeatedly emphasised. And 'peace processes' specifically can, especially because they are so protracted, function as means of deferral and delay and through that co-optation, pacification and re-legitimation without conflict resolution.[19] In their different ways, the Israeli–Palestinian, Cypriot and Sudanese peace processes are all cases in point.

The Oslo Process: Dissimulating Decolonisation

Launched in 1993 to worldwide fanfare, the Oslo peace process between Israel and the PLO was widely heralded as 'opening a new era' in which 'peoples who have fought each other for almost a century have agreed to move decisively on the path of dialogue, understanding and cooperation'. Even critical scholars who viewed Zionism as a settler-colonial project judged that remarkable changes were afoot: '[p]eacemaking between Israel and the PLO signals a new, late wave in the decolonization in overseas European societies', argued Gershon Shafir, the foremost exponent of this thesis, a local 'process of decolonization' that 'seems to have passed the point of no return'. Yet, in actuality, Oslo brought neither decolonisation nor political freedom nor even peace, and not only because of its failure. Oslo and its aftermath – that is, the entire period since 1993 – brought, instead, a consolidation of Israeli control and deepening Israeli–Palestinian asymmetries, including in relation to water.[20]

The Oslo process was rooted in three main concerns. Foremost were the demands of Israeli capital: by the early 1990s, economic liberalisation

---

[19] C. W. Sadoff and D. Grey, 'Cooperation on international rivers: a continuum for securing and sharing benefits', *Water International*, 30:4 (2005), 420–7; UNDP, *Beyond Scarcity*, 224, 228; Zeitoun and Warner, 'Hydro-hegemony'; J. Selby, 'Dressing up domination as "cooperation": the case of Israeli–Palestinian water relations', *Review of International Studies*, 29:1 (2003), 121–38; J. Selby, 'Cooperation, domination and colonisation: the Israeli–Palestinian Joint Water Committee', *Water Alternatives*, 6:1 (2013), 2–5; D. Keen, 'War and peace: what's the difference?', *International Peacekeeping*, 7:4 (2000), 1–22; Selby, 'The political economy'.

[20] US President Bill Clinton and Israeli Foreign Minister Shimon Peres, 'Remarks at the Signing of the Israeli–PLO Declaration of Principles' (13/10/1993); G. Shafir, 'Israeli decolonization and critical sociology', *Journal of Palestine Studies*, 25:3 (1996), 30, 33.

was well under way within Israel, the business sector was becoming increasingly influential and Israeli companies were looking to exploit new opportunities in the United States, Europe and Asia; Israel's relative diplomatic isolation was, however, a major obstacle. Addressing this issue thus became a priority for Israeli business elites – as Eli Hurvitz, CEO of Teva Pharmaceuticals expressed it, 'the future is problematic without peace' – and they thus lobbied heavily for an accommodation with the Palestinians. Second, with the Palestinian intifada ongoing, Israel's military establishment desired some sort of sub-contracting arrangement, whereby basic security functions would be performed by Palestinian police rather than directly by the Israel Defense Forces (IDF): Israeli Prime Minister Yitzhak Rabin was candid about this, noting that Palestinians 'will rule by their own methods, freeing, and this is most important, the Israeli army soldiers from having to do what they will do'. Meanwhile, on the Palestinian side, the PLO's Fatah leadership – based 1,000 miles away in Tunis – wanted an agreement that would restore its international diplomatic and financial fortunes and re-establish its leadership of the Palestinian national movement.[21] The Oslo 'peace process' was born, in short, not from the sudden emergence of accommodationist or pacific attitudes, but from a temporary convergence of interests centring on issues of international legitimacy and internal political hegemony.

The structure of the peace process reflected this. Its central conceits were not new, instead reproducing the terms of the 1978 Israeli–Egyptian Camp David Accords, in almost identical language: a five year 'transitional period' during which Palestinians would be granted 'autonomy' in the West Bank and Gaza under an elected 'self-government authority'; a part 'withdrawal' and part 'redeployment' of the IDF; the formation of a 'strong' Palestinian police force to secure areas vacated by these Israeli forces; and 'permanent status' negotiations on all outstanding issues which, upon completion, would bring the transitional period to a close. Indeed, from the perspective of Israeli negotiators, Oslo was 'a major improvement over the Camp David Accords'.[22] For many Palestinians, however, Oslo amounted to capitulation. While fifteen years earlier the PLO had vehemently rejected the Camp David US–Israeli formula for

---

[21] Hurvitz in J. Rossant, 'Israel has everything it needs – except peace', *Business Week* (09/12/1989), 54; M. Bouillon, *The Peace Business: Money and Power in the Palestine–Israel Conflict* (I.B. Tauris, 2004); Rabin in *Ha'aretz* (07/09/1993), in G. Usher, *Dispatches from Palestine: The Rise and Fall of the Oslo Peace Process* (Pluto, 1999), 74.

[22] Egypt and Israel, *Camp David Frameworks for Peace* (1978); Israel and PLO, *Declaration of Principles*; U. Savir, *The Process: 1,100 Days That Changed the Middle East* (Random House, 1998), 67.

peace, now they were willing to accede to it without any promise that their core final status demands would ever be satisfied.

The detailed implementation agreements which followed Oslo, especially the 1994 Gaza–Jericho Agreement and the 1995 Oslo II Agreement, confirmed and concretised this capitulation. Negotiated essentially between Israel's military establishment and Tunis associates of Yasser Arafat, these accords inaugurated one of the most circumscribed versions of 'political autonomy' imaginable. The Palestinians were permitted their own postage stamps, identity cards and other symbolic trappings of statehood. Various responsibilities for public service provision and internal policing were transferred to the newly established PA. But all of these powers were heavily qualified. In 'Area C', covering 60 per cent of the West Bank, Israel retained full military and administrative control, while in the remaining 40 per cent PA powers were limited both through numerous Joint Committees – which effectively functioned as instruments of Israeli veto – and by the aforementioned military orders which, under the terms of the Oslo II Agreement, all remained in force. The PA was granted responsibilities where these did not conflict with Israeli interests, but negligible powers where they did. Moreover, Israel's colonisation of the West Bank and matrix of internal control actually deepened under Oslo. The settlement population of the West Bank grew continuously, from an estimated 264,000 in 1993 to 365,000 in 2000, to 547,000 in 2013 and 675,000 in 2020.[23] A parallel 'bypass road' network was created. And comprehensive permit and external and internal closure systems were established. All of these developments occurred simultaneously to, and consistent with the terms of, the Oslo agreements. Moreover, initially intended as 'interim' arrangements only, they have now been in place for almost thirty years.

Water 'cooperation' after Oslo reflected and was structured by these broader contexts. The PA was granted unilateral control over – or better, unilateral responsibility for – Gaza's run-down water sector and its meagre and over-exploited water resources, while Israel washed its hands of these areas, confident that full local Palestinian responsibility over them posed little threat to its own water security. By contrast, in the West Bank, where Israeli interests were of a different order entirely, occupation-era water control structures and asymmetries were reproduced, courtesy of Article 40 of the Oslo II Agreement. This article – which was drawn up by an Israeli Ministry of Defence official and simply accepted by Palestinian

---

[23] Israel and PLO, *Interim Agreement*, art. 18 (4a); annex III, app. I, art. 28 (4), 29 (2); Foundation for Middle East Peace, 'Comprehensive settlement population 1972–2011'; B'Tselem, 'Statistics on settlements and settler population' (11/05/2015); J. Krauss, 'Report: Israeli settler population surged during Trump era', *AP News* (27/01/2021).

negotiators – essentially codified Israel's prior resource capture. It allocated 87 per cent of the West Bank's trans-boundary groundwater yield to Israel, and just 13 per cent to the Palestinians. It failed to restore Palestinian access to the Jordan River. And it stipulated that Israeli approval would be required for any new well, any increase in abstraction from existing wells, all well rehabilitation, any new supply lines of two-inch diameter or more, and all associated infrastructure. The only major innovations in Article 40 were that the new Palestinian Water Authority (PWA) was granted in principle permission to drill wells into the hitherto barely exploited, but also comparatively slight, Eastern Basin of the Mountain Aquifer; and that Israeli control was now dressed up and institutionalised as an equitable sounding 'coordinated water management' system.[24]

To those unfamiliar with the case, this dismissal of Israeli–Palestinian water cooperation may sound overblown. Indeed, both Article 40 and the Joint Water Committee (JWC) established under it were at least initially lauded as among the most important achievements of the Oslo process. Yet the system of water 'cooperation' that resulted was and remains barely deserving of the name. This system was and remains not just asymmetrical but upside down, applied not to that territory, Gaza, which cannot be self-sufficient in its water resources, and which therefore most requires cooperation; and neither to Israeli territory, since the PA has no say on water management there; but instead to the one space from which Israel wanted to redeploy but maintain ultimate hydro-political control. This space, moreover, became subject to what is surely the most intrusive system of trans-boundary water regulation anywhere in the world: it is hard to imagine that there is anywhere else where the laying of pipelines with a diameter of drainpipes can be vetoed by an 'external' state party. Moreover, under Oslo II Israel came to possess not just single but double veto powers over most West Bank water development, first within the JWC, but then secondarily through its military government, the so-called 'Civil Administration', which maintains full planning powers across Area C. To rephrase Foucault on Clausewitz, this was – and remains – 'cooperation' as a continuation of coloniser's rule by other means.[25]

The consequences for patterns of water development have been profound. As between 1967 and 1995, since then every single Palestinian

---

[24] Israel and PLO, *Agreement on the Gaza Strip and Jericho Area* (1994), annex II, art. 2 (B31a); Israel and PLO, *Interim Agreement*, annex III, app. 1, schedule 10; schedule 8 (1a, 1b, 2d); art. 40 (7, 12).

[25] E. Fletcher, 'Israel, PLO make deal on West Bank water', *San Francisco Examiner* (21/09/1995); M. Foucault, *Power/Knowledge: Selected Interviews and Other Writings 1972–1977* (Harvester Press, 1980), 90; von Clausewitz, *On War*, 119.

JWC application for new production wells in the Mountain Aquifer's Western Basin has been rejected. Many of the new wells that have been approved have been subject to tortuous delays, by as much as eight years in some instances. And while new Palestinian water supply lines have been laid across the West Bank, most are extremely small – only 14 per cent of supply projects submitted to the JWC between 1995 and 2008 included pipelines of 8-inch diameter or more – and many have been approved only on condition that 'no additional water' would be made available. The situation is especially acute in Area C, where Israeli planning policy is, in the words of the Israeli human rights organisation Bimkom, 'systematically designed to restrict the development of Palestinian communities', and where the IDF routinely demolishes Palestinian water infrastructures. As a result, both Palestinian groundwater abstraction and total Palestinian per capita water availability in the West Bank have actually decreased since 1995. And in large part because of this, both the Palestinian agricultural sector and the Palestinian population of Area C have been in sharp relative decline.[26]

It is true, as Israeli officials often point out, that Israel now provides West Bank Palestinians with water well in excess of that promised in 1995. Indeed, water purchases from Israel now account for well over a third of Palestinian West Bank supplies. Yet the sad facts are that these imports are only necessary because of Israeli restrictions on Palestinian well drilling and that 'autonomy' in the West Bank has been correlated with growing external hydro-dependency – in a territory where water is naturally quite abundant. Ramallah, which receives average annual rainfall of over 600 mm, more than London, has effectively been turned into a 'downstream' location with 100 per cent of its supplies being imported from Israel. This is not evidence of generosity or meaningful cooperation, but enforced dependency and control. As Israeli Water Commissioner Uri Shani put it in 2008, capturing the situation perfectly, 'I am the de facto Palestinian water regulator.'[27]

---

[26] Selby interview with Ayman Jarrar, PWA (30/09/2009); Amnesty International, *Troubled Waters*, 35–8; Selby, 'Cooperation, domination and colonisation', 12–14, 19; Bimkom, *Prohibited Zone*, 5, 160–4; Emergency Water, Sanitation and Hygiene Group, *Down the Drain: Israeli Restrictions on the WASH Sector in the Occupied Palestinian Territory and Their Impact on Vulnerable Communities* (2012); Weinthal and Sowers, 'Targeting infrastructure'; World Bank, *Area C and the Future of the Palestinian Economy* (2014); Oxfam, *Addressing the Needs of Palestinian Households in Area C of the West Bank* (2019).

[27] IWA, *The Issue of Water between Israel and the Palestinians* (2009), 34; R. El-Sheikh and F. Bateh, 'Saving Gaza begins with its water', The Century Foundation (14/12/2020), 10; Selby email communication with C. Messerschmid (06/08/21); Messerschmid, 'Nothing new', 437.

Making matters worse still, the Oslo II water regime has also enabled and facilitated continued Israeli settlement-building. For, from 1998 Israel started making its approval of Palestinian JWC projects conditional on simultaneous PA approval of settlement water supply lines, despite these settlements and associated infrastructures being illegal under international law and one of the major impediments to Palestinian statehood. For a dozen years, the PA's pragmatic response – pursued with the full knowledge of Presidents Arafat and then Abbas – was to accept this conditionality and approve every single water supply facility proposed by Israel for its settlements. The consequence was that, during this period, Israel's West Bank settlements were fully integrated into its national water network and were able to expand without experiencing severe water shortages – all with written PA consent. Moreover, this was no equal trade-off. Israeli settlement projects were on average far larger, more likely to be approved and more speedily approved than Palestinian ones. In a single short protocol in 1999, for example, the PA approved five major new transmission lines to settlements with a combined capacity of 25 mcm/y – near double the total volume of water produced by new PA wells during the entire Oslo period. In 2010, the PWA changed policy and decided that it would no longer be willing to approve settlement water facilities. In turn, however, Israel began vetoing all new wells and pipelines for Palestinian communities, leading to complete deadlock within the JWC – until in 2017 the JWC's working procedures were renegotiated to grant Israel carte blanche in the building of settlement water infrastructures. Thus Palestinian water development in the West Bank is now subject to Israeli vetoes while Israel can increase its settlement water networks and supplies without constraint.[28]

Overall, as a result, the period since the signing of the Oslo Agreement has seen a radicalisation of those politically induced patterns of water security and insecurity established post-1967. Gaza remains to all intents and purposes bypassed by Israel's national water supply system – an additional 5 mcm/y promised to Gaza within the Oslo II Agreement was only delivered in 2015, a full twenty years late – leaving it in a situation of enforced and unviable hydro-'self-sufficiency', with the consequence, inevitably, that the enclave's water table has continued to drop and

---

[28] Selby interview with Fadel Kawash, PWA (15/12/2012); 'Protocol of the Israeli–Palestinian Civil Affairs Committee' (02/02/1999); Selby, 'Cooperation, domination and colonisation'; 'Renewal of JWC's activity', agreement between Israeli and Palestinian heads of JWC and CAC (15/01/2017); J. Selby, 'How did Rawabi get its water?', *Middle East Monitor* (16/09/2015); J. Selby, 'Renewing cooperation on water: what hope for a two-state solution?', *Open Democracy* (26/01/2017).

salinity levels to further rise.[29] The West Bank, by contrast, has witnessed a striking combination of uneven infrastructural incorporation into the Israeli water system, settlement colonisation and ongoing repression of Palestinian water consumption – all within the context of, and to some extent facilitated by, a regime of water 'cooperation'. And meanwhile, Israel has attained newfound levels of water abundance and security, as already discussed.

International donors – 'post-conflict peacebuilders' – have been deeply complicit in these developments. Donors have, to be sure, invested heavily in the Palestinian water sector. However, this has always been in line with the manta of 'supporting the peace process' irrespective of its substantive content, and hence has essentially involved acceding to Israeli occupation practices and priorities. Many donors have been well aware that their water projects were only being approved through the JWC because the PWA was simultaneously approving illegal settlement facilities – but preferred to remain silent on the issue. Equally, donors have responded to Israel's veto of Palestinian well projects in the Western Basin by deprioritising well drilling in the West Bank altogether. Indeed, instead of addressing the central cause of Palestinian water shortages and insecurity – Israel's domination of shared water resources – donor efforts have focused on four things, all in accordance with Israeli priorities: wastewater treatment plants, local desalination (in Gaza), supply network rehabilitation and internal Palestinian reform. What the Palestinian water sector really needs, Western donors have effectively said, are high-tech solutions, improved local management, plus the privatisation of local Palestinian supply utilities and installation of pre-paid metres to improve cost-recovery. Like the drought narratives examined in Chapter 3, this is 'anti-politics' par excellence – in which donors' ostensible neutrality means that they not only fail to challenge prevailing power relations, but actually end up buttressing them.[30] Whatever its short-term if limited benefits, international donor support to the Palestinian water sector has structurally been mainly about adapting to Israel's occupation.

Naivasha and After: Self-Determination and Militarisation

Unlike the Oslo Agreement, which reproduced a long-established US–Israeli formula for pacification of the Palestinians through limited self-rule, Sudan's 2005 Comprehensive Peace (or Naivasha) Agreement both

---

[29] El-Sheikh and Bateh, 'Saving Gaza', 10.
[30] Selby, 'Cooperation, domination and colonisation', 21; Ferguson, *Anti-Politics Machine*.

constituted a radical departure from all that had gone before and wholly reconfigured the country's formal political structures – albeit in ways that were far from progressive. Essentially a power-sharing deal, the agreement simultaneously allocated powers between the ruling regime in Khartoum and the opposition SPLM/A and, at the same time – as is often the case within power sharing accords – concentrated power in their two hands. It stipulated that together the two signatories would hold at least 80 per cent of the seats in the executives and legislatures of every level of government across the country, far exceeding their level of popular support; and that together they would monopolise oil revenues, with a mere 2 per cent of earnings to be granted to oil producer regions. It stated that the two signatories' armed forces would continue to exist separately and without any particular downsizing or demobilisation, while implying that all other armed groups in the country would be disbanded. It established an autonomous Government of Southern Sudan with wide-ranging powers, far more extensive than those of the PA. And it stipulated that these arrangements would hold for a six-year interim period, whereupon, in another contrast to the Oslo Agreement, the interim period would be brought to a close through a referendum on South Sudanese independence.[31]

These terms reflected a series of factors. They reflected, first, a convergence between the George W. Bush administration's Orientalist reading of Sudan's civil war, as espoused for instance by its peace envoy, Episcopal minister John Danforth – in which the conflict was read as between southern Christians and northern Muslims, and in which the Islamist military regime of Omar al-Bashir and the equally undemocratic SPLM/A were deemed legitimate representatives of these two groups – and on the other hand the interests of these two Sudanese parties in re-establishing their internal power positions within northern and southern Sudan respectively. It was essentially thanks to this convergence of Western myth and local interests that all other Sudanese actors were excluded from negotiations and that the resulting agreement was framed around a North–South axis. At the same time, second, the agreement was a function of the depth of the political crisis in Khartoum – above all the multiple internal wars, most prominently in Darfur, and international pariah status further exacerbated by the events of 9/11 – which suddenly made the regime amenable to southern autonomy and eventually secession. The SPLM/A, in parallel, found itself pushed towards the option of

---

[31] *Comprehensive Peace Agreement*, ch. 1 (2.5); ch. 2 (2.2.5, 2.5.5, 3.5.1, 3.6.4, 4.4.2, 4.5.1, schedules B, D); ch. 3 (5.5); ch. 6 (1b, 1c, 7a); J. Young, 'Sudan: a flawed peace process leading to a flawed peace', *Review of African Political Economy*, 103 (2005), 99–113; Young, *Fate of Sudan*.

secession by regional actors, especially Uganda, that were interested in weakening and limiting the damage being done by Khartoum. And lastly, the advent of large-scale oil production from 1999 and the resulting boom in government revenues were creating new opportunities for patronage, co-optation and power-sharing. Peacemaking in modern Sudan has always coincided with an upsurge in oil revenues, and Naivasha was no exception.[32]

It is worth emphasising just how distant these terms and motivations were from what the SPLM/A under John Garang had previously espoused. Historically under Garang, the SPLM/A had advocated the creation of a 'New Sudan' through the removal of the Bashir regime and its conservative Islamist agenda, combined with democratic transformation – an agenda which had brought it support nationwide. Speaking of oil and water, for example, Garang had promised that when 'the SPLA liberates our country under SPLM leadership, these two precious liquids shall be developed and used for the benefit of the whole Sudanese people'.[33] However, in the event, national liberation was abandoned in favour of a very different peacemaking model – one which would bring autonomy to South Sudan, and independence from 2011, but in other respects would be far from progressive.

Developments since 2005 within South Sudan bear this out. Within the context established by Naivasha's power- and wealth-sharing protocols, South Sudan became staggeringly dependent on oil exports to the extent that they accounted, in 2011, for 98 per cent of public revenues. Moreover, as so often within poor oil-export economies – or what are somewhat misleadingly lumped together as 'rentier states' – this dependency fed multiple dysfunctions. It led to security sector expansion and militarisation, with the SPLA's payroll expanding from about 40,000 in 2004 to as many as 240,000 in 2011, by which point the security sector accounted for 40 per cent of government spending. It generated extreme forms and levels of corruption: according even to President Salva Kiir, at least US$4 billion, or around one-third of total oil revenues, were diverted by government officials between 2005 and 2011; and from 2011 a system of parallel foreign exchange markets emerged in which only political elites were granted access to 'official' – that is preferential – rates, worsening the

---

[32] J. Danforth, *Report to the President of the United States on the Outlook for Peace in Sudan* (2002); J. Young, 'John Garang's legacy to the peace process, the SPLM/A and the South', *Review of African Political Economy*, 106 (2005), 535–48; J. Young, 'South Sudan: the fractured state', in J. Markakis et al. (eds.), *The Nation State: The Wrong Model for the Horn of Africa* (Max Planck Institute for the History of Science, 2021), 145–65; A. de Waal, *The Real Politics of the Horn of Africa: Money, War and the Business of Power* (Polity, 2015), 69–90.

[33] Garang, *Speeches on the War*, 39.

situation still further. Oil rents were used for building patronage across different regions and ethnic and armed groups. Dependency on oil exports also disincentivised productive investment in industry and agriculture. And it sparked repeated cycles of political and violent conflict. Internally, there have been multiple rebellions by army factions and militia groups seeking increased shares of oil wealth. And in 2012, months of clashes between Sudan and the newly independent South Sudan came to a head when Southern forces seized control of oilfields within the borderland region of Heglig. When Sudan responded to this by shutting down South Sudanese oil exports, as well as bombing South Sudanese oilfields, Juba's revenues and patronage system collapsed, in turn contributing – amid the intensification of long-standing divisions and competition within the SPLM and SPLA in the lead-up to planned national elections – to the country's descent into civil war. A 'corrupt peace' greased by oil revenues broke down, with fighting that began in Juba spreading rapidly to Jonglei and then the country's oil producing states, Unity and Upper Nile. These dynamics have remained largely unchanged since then. For while oil production and exports have since resumed, enabling internal peace agreements in 2015 and 2018, these agreements essentially reproduced the country's highly militarised structures and the concentration of power in the hands of the Salva Kiir regime.[34]

Reflecting all this, neither the water provisions of the Naivasha Agreement nor subsequent agreements or peacebuilding initiatives have done much to enhance the people of South Sudan's water security. Naivasha placed the management of Nile and other trans-boundary waters under Khartoum's exclusive jurisdiction, with only 'water resources other than interstate waters' included in the list of concurrent national powers – which, given that South Sudan falls entirely within the Nile basin, was not exactly generous. Moreover, there have been no post-referendum agreements on water

---

[34] Ø. H. Rolandsen and N. Kindersley, *South Sudan: A Political Economy Analysis* (Norwegian Institute of International Affairs, 2017), 6, 15–18; A. de Waal, 'When kleptocracy becomes insolvent: brute causes of the civil war in South Sudan', *African Affairs*, 113:452 (2014), 355–7, 361–2; K. Adieb, *Public Expenditures in South Sudan: Are They Delivering?* (World Bank, 2013), 7; IMF, *Republic of South Sudan: 2014 Article IV Consultation – Staff Report; Staff Statement; and Press Release* (2014), 11; G. Larson et al., *South Sudan's Capability Trap: Building a State with Disruptive Innovation* (UNU-WIDER, 2013), 22; L. A. Patey, 'Crude days ahead? Oil and the resource curse in Sudan', *African Affairs*, 109:437 (2010), 628–33; S. Spittaels and Y. Weyns, *Mapping Conflict Motives: The Sudan–South Sudan Border* (International Peace Information Service, 2014), 22; P. Le Billon and E. Savage, 'Binding pipelines? Oil, armed conflicts, and economic rationales for peace in the two Sudans', *African Geographical Review*, 35:2 (2016), 146; *Revitalised Agreement on the Resolution of the Conflict*, ch. IV (4.8.1); L. B. D. Kuol, 'The 2018 South Sudanese Peace Agreement: a litmus test of coercive mediation', in C. Zambakari et al. (eds.), *Peacemaking and Peace Agreements in South Sudan* (Zambakari Advisory, 2019), 69–70.

resources, effectively leaving these Naivasha terms in place. The reasons can be easily summarised. On the one hand, Khartoum's renewed hydraulic development ambitions, discussed in Chapter 5, have meant that it has sought to maintain control over Nile flows and limit South Sudanese development to as great an extent as possible. On the other, South Sudan's elites have little interest in developing their abundant water resources, instead being overwhelmingly wedded to the model of oil rent–dependent political control without long-term development as outlined above. And on top of this, international and especially Egyptian demands have clearly been a factor: Egypt, for instance, only supported South Sudan's independence after assurances from Salva Kiir that Nile flow would not be affected.[35]

As a result, there has been only the most limited water-related peace-building and development in South Sudan since 2005. Planned dam-building projects, mostly developed with Egyptian support, have been designed with electricity generation rather than irrigation in mind, thus posing little threat to Egyptian supplies. In any event, no such plans have been realised: plans to construct a hydropower facility at Fula to provide electricity supply to Juba, for instance, were interrupted by the outbreak of civil war in 2013. Moreover, while the Naivasha Agreement was quickly followed by a series of land deals that saw private investors acquiring around 5 million hectares of the country's most water-rich land for agriculture and biofuel projects, few of these projects have been implemented. International support, for its part, has focused mainly on the development of urban water infrastructure, mainly in Juba, as well as small-scale water facilities for livestock and the provision of water and sanitation facilities in response to the war-driven displacement. The recent announcement of a new Nile dam in South Sudan was met with surprise by both the Egyptian and Sudanese governments, with Egypt's head of Nile water affairs cautioning that it would support the plan 'as long as it does not affect Egypt's share of the Nile water' and Sudan's irrigation minister stating that 'South Sudan will not implement plans to build dams on the White Nile without first notifying us and the Egyptians'.[36] Constrained by a combination of

---

[35] *Comprehensive Peace Agreement*, ch. 2, part 5 (schedules A, B, D), ch. 3, part 1; *Mekelle Memorandum of Understanding between the NCP and SLPM on Post-Referendum Issues and Arrangements* (2010); S. M. A. Salman, 'Water resources in the Sudan North–South peace process and the ramifications of the secession of South Sudan', in E. Weinthal et al. (eds.), *Water and Post-Conflict Peacebuilding* (Routledge, 2014) 344–7; H. Verhoeven, 'The rise and fall of Sudan's Al-Ingaz Revolution: the transition from militarised Islamism to economic salvation and the Comprehensive Peace Agreement', *Civil Wars*, 15:2 (2013), 132–3; Selby and Hoffmann, 'Beyond scarcity', 364.
[36] N. Juma, 'Can South Sudan relaunch its Fula Dam project?', *The Niles* (18/03/2021); 'Egypt pushes for further water cooperation with South Sudan', *Daily News Egypt* (14/06/

regional hydro-politics and internal militarisation, conflict and oil-dependency, South Sudan's water sector has not witnessed any 'peace dividend' at all.

The Annan Plan: Dissimulating Unification

As a final example, consider the 2004 Annan Plan, the highpoint – or perhaps nadir – of Cyprus' recent on–off peacemaking history. Unlike the Oslo and Naivasha agreements, this plan was never agreed let alone implemented, being overwhelmingly rejected during the Greek Cypriot referendum on it (though overwhelmingly endorsed by Turkish Cypriots). Yet it is instructive concerning the nature of contemporary peacemaking, nonetheless. The Annan negotiation process, from 1999 to 2004, was above all else launched as a by-product of the EU's Enlargement agenda: its 1999 Helsinki summit had simultaneously guaranteed Cyprus' accession and paved the way for accession talks with Turkey, and this in turn had provoked widespread concern – within the EU, United States and not least the new Justice and Development Party government in Turkey – that an unresolved Cyprus conflict might interfere with these processes. Substantively, however, the eventual Annan Plan reflected a very different set of priorities, of which five are especially noteworthy: widespread Greek Cypriot opposition to any weakening or dilution of what, post-1974, had become an ethnically exclusive Greek Cypriot state; a preference by both Turkey and leading Turkish Cypriot parties for continuation and legitimation of the basic territorial realities established in 1974; British, Turkish and Greek insistences on retaining their 'Guarantor' statuses and military presences on the island, in line with the 1960 Treaty of Guarantee; the United States' long-established geo-strategic practice of accommodating Turkey, made more important by 9/11 and its aftermath; and Greek Cypriot interests in the return of key areas of land and property. Crucially, all but the last of these were 'status quo preferences' that called for the continuation, not transformation, of the status quo ante – in line with the fact that the main local architects of

---

2021); D. Ezzat, 'Egypt furthering relations with South Sudan', *Ahram Online* (07/07/2021); D. K. Deng, *The New Frontier: A Baseline Survey of Large-Scale Land-Based Investment in Southern Sudan* (Norwegian People's Aid, 2011), 8, 22; P. Wight, 'Epilogue of a short-lived land rush: private, rural, and urban land tenure in South Sudan', in L. Cochrane and A. Nathan (eds.), *The Transnational Land Rush in Africa: A Decade After the Spike* (Palgrave Macmillan, 2021), 93–6; Mosello et al., *Improving WASH*; FAO et al., *Water Harvesting for Peacebuilding in South Sudan* (2015); G. Mikhail, 'Egypt, Sudan react to South Sudan's plan for new dam on Nile', *Al-Monitor* (02/07/2021).

the Annan Plan, Rauf Denktaş and Glafcos Clerides, had previously been among the main local architects of Cyprus' division.[37]

The eventual terms of the Annan Plan reflected this, on the one hand offering a 'territorial adjustment for recognition' deal, but in other respects, and overall, promising to reproduce Cyprus' divide. For, even though Annan formally promised to create a reunified independent Cypriot state, in practice this federal state would have possessed exceedingly limited powers while its 'two equal constituent states' would have possessed the exact opposite. Whereas the proposed federal government would have exercised authority in just twelve specified areas – most prominently, external relations and central banking – and possessed a flag, an anthem and other simulations of sovereignty, the two constituent states would have been responsible for everything else (and would have had their own flags, anthems and so on too). The federal government would have operated on a strict power-sharing basis, with a 1960-style consociationalist constitution with disproportionate Turkish Cypriot representation and both sides possessing effective decision-making vetoes. In key areas, including the federal Supreme Court and Central Bank, non-Cypriots would effectively have held casting votes (which suggests that, although Annan presents itself as modelled on the Swiss constitution, the US-brokered Dayton Accords for Bosnia provides the closer parallel). Crucially, moreover, Turkish along with Greek forces would have remained on the island, Britain's sovereign bases would have remained in perpetuity and the Treaty of Guarantee would have remained in effect, while Cypriots themselves would ostensibly have been 'demilitarised'. The main resulting changes would have been some territorial adjustments (with the area of the Turkish Cypriot state dropping to 29 per cent) and recognition of the Turkish Cypriot 'constituent state'.[38]

The water management arrangements proposed under Annan reflected, and illustrate, this overall constitutional architecture. For, even though 'natural resources, including water resources' were defined under the Annan Plan as a federal competence, this headline point is misleading. Under the Annan Plan's proposed (though unpublished)

---

[37] C. Carras, 'The International Relations aspect of the Annan Plan', in: A. Varnava and H. Faustmann (eds.), *Reunifying Cyprus: The Annan Plan and Beyond* (I.B. Tauris, 2009), 53–65; M. Müftüler-Bac and A. Güney, 'The European Union and the Cyprus problem 1961–2003', *Middle Eastern Studies*, 41:2 (2005), 289; C. Palley, *An International Relations Debacle: The UN Secretary-General's Mission of Good Offices in Cyprus 1999–2004* (Hart, 2005).

[38] *Comprehensive Settlement of the Cyprus Problem*, Main arts. 1 (3, 5), 2 (1a), 5, 6 (2), 8; Basic arts. 8, 14 (1, 2), 32 (4); D. Chandler, *Bosnia: Faking Democracy After Dayton* (Pluto, 2000).

federal water law, the responsibilities of the federal water department – to be staffed at least one-third by Turkish Cypriots – would have been limited to ensuring equitable water sharing, monitoring and advising the two constituent states and representing Cyprus internationally. By contrast, the management, conservation, development and distribution of surface and groundwater resources would all have been undertaken separately within the two constituent states. All publicly owned water resources and infrastructures would have become, or in practice remained, the property of one of the two constituent states, thus ensuring that the RoC's post-1974 investments would have been protected for Greek Cypriot use. There would not even have been any joint management of trans-boundary resources. There would have been no federal role in agriculture (a constituent state competence) or in the development of non-conventional water resources. Indeed, symbolising this, Cyprus is bizarrely defined in the Annan Plan water law as comprising 'two river basin districts', corresponding to the areas of the Greek Cypriot and Turkish Cypriot constituent states. Like the rest of the Annan Plan, in sum, what this federal water law offered was a dissimulation of reunification. In practice, the pattern of unilateral ethno-nationalist water development outlined in Chapter 5 would have continued and been consolidated, newly legitimated, under Annan; the Annan water regime would have broadly reproduced, under the mantle of peace, Cyprus' existing hydro-political divide.[39] The one important qualifier here is that the proposed territorial adjustments, especially the transfer of large areas of the Morphou plain, would have deprived the new Turkish Cypriot state of its single most important water source – likely increasing its dependency on Turkey still further.

Within the context of a rejected UN peace plan and multiple other failed peace initiatives, recent 'post-conflict' peacebuilding in Cyprus has had a distinctive cast, oriented not to aiding implementation of any specific agreement but to the promotion of inter-communal dialogue and, above all, technical support for Turkish Cypriot development. The European Commission, for one, has allocated over €500 million to the Turkish Cypriot community since 2006 to facilitate economic development and pave the way for eventual 'reunification' – and implicitly also to reward support for the Annan Plan. This funding has included, in the water sector, support for wastewater treatment plants in Nicosia, Morphou and Famagusta (Gazimağuza, in Turkish), for a desalination

---

[39] *Comprehensive Settlement of the Cyprus Problem*, basic arts. 14 (1e), 30 (2); annex 3, attachment 12: *Federal Law on the Natural Water Resources of the United Cyprus Republic, for Equitable Sharing of These Resources between the Constituent States and for Purposes Connected Therewith* (unpublished document), arts. 4 (2), 6, 8–12, 15.

plant near Morphou, for urban supply and sewerage networks upgrades, for improved irrigation technologies and more. In the absence of a framework peace agreement, however, such technical support has routinely become mired in questions of politics. The European Commission cannot deal directly with the Turkish Cypriot authorities as this would imply recognition. It has not been able to fund the installation of water meters as this would have involved cooperation with them. It cannot support projects involving construction on, or that cross, confiscated Greek Cypriot property. And projects have been stymied by objections from the Turkish military.[40] As a result, the main water 'peacebuilding' project of recent years has, in effect, been one already discussed: the floating water pipeline connecting northern Cyprus to mainland Anatolia which, for all Turkey's 'water for peace' rhetoric, is an instrument of hydro-political division. As in our other cases, so in Cyprus water-related peacebuilding has basically been structured by, and used to reinforce, dominant political imperatives. And in Cyprus specifically it is clear that substantive and meaningful reunification has not occurred at all.

Peace and Climate Change Vulnerability

This chapter has sought to examine both the place of water within contemporary peace processes and the claimed contributions of water peacemaking and water cooperation to peacemaking and cooperation more broadly. Its arguments have essentially been fourfold. First, that contrary to what is often claimed, water does not possess inherently peace-inducing characteristics, since neither its importance to life, nor its disrespect for borders, nor its multi-functionality, nor even its supposedly technical and apolitical qualities are obstacles to conflict and violence over it. Second, that water is generally not high on the agendas of contemporary peace processes, reflecting its limited economic value and political importance under conditions of contemporary high-tech and globally integrated capitalism. Third, that contemporary peace processes are in any case often flawed, barely differing from the periods of 'non-peace' that preceded them and sometimes being even worse, including in relation to water. And fourth, that water is typically not just a marginal but a dependent variable within peacemaking, with patterns of water access, control and (in)security reflecting, and being configured by, broader

---

[40] EU, *Environmental Infrastructure: Investing in a Sustainable, United Cyprus* (2020); European Court of Auditors, *European Union Assistance to the Turkish Cypriot Community*, Special Report No. 6 (2012), 18; Hoffmann interview with rural development consultant (12/03/2011).

peacemaking dynamics – including their self-evident dysfunctions. Thus the Oslo process not only reproduced but exacerbated prior Israeli–Palestinian inequalities, including those relating to water. Naivasha left South Sudan in a state of oil-dependent insecurity and without any sort of peace dividend, including in the water sector. And the failed Annan Plan for Cyprus promised to consolidate division under the banner of unification, with the exact same applying to its waters. Across these cases, the same basic pattern – of hydro-politics as a dependent variable within defective processes of peacemaking – is all too sadly recurrent.

Now, some may object that there does exist evidence that environmental cooperation can lead peacemaking: one recent multi-case analysis, for example, finds that the conclusion of cooperative environmental agreements 'facilitates reconciliation in international rivalries'. Yet in truth this and other such conclusions are confections of neo-positivist method, in which a statistical relationship between environmental cooperation and reduced conflict is identified, before causality is then imputed to the former – irrespective of whether the instances of environmental cooperation in question may themselves have resulted from broader patterns of low-level cooperation, as in the example of the Israeli–Jordanian 'picnic table talks' already discussed. Indeed, in the study just quoted this is, to its credit, acknowledged, the study finding that environmental agreements 'only have such a positive effect when they are embedded into ... already ongoing reconciliation processes'.[41] Just as 'environmental scarcity disputes' rarely spill over into violence, so, by the same token, is there little evidence of environmental issues catalysing or facilitating broader cooperation. Or stated differently, the promises of environmental peacebuilding are not merely hollow; they also mirror and reproduce the characteristic eco-determinism of environmental conflict discourse, just in the opposite direction.

There is much more that could be said here about both environmental peacebuilding and peace processes more broadly, but a few brief remarks will suffice. Because peace processes, on the one hand, are highly politicised, contested and uncertain phenomena, occupying that liminal space between peace and war, it follows that they are often characterised by extreme socio-ecological contradictions: by sudden inflows of capital, whether from international 'peacebuilders', power-sharing formulae or diaspora returnees; by attendant property bubbles and speculative investments; by continuing political and violent conflict, directed now to

---

[41] T. Ide, 'Does environmental peacemaking between states work? Insights on cooperative environmental agreements and reconciliation in international rivalries', *Journal of Peace Research*, 55:3 (2018), 361.

'winning the peace'; by regulatory uncertainties, even lawlessness; and, as an inevitable consequence, by unprecedented environmental degradation and landscape change. Both northern Cyprus during the Annan Plan period and the West Bank under the Oslo Agreement provide plentiful evidence to this effect. And environmental peacebuilding projects specifically can have any number of adverse consequences, including contributing to environmental degradation, displacement and conflict, and simultaneously obscuring, through de-politicisation, the inherently political causes of environmental crises and injustices. The proposed Lake Chad water transfer, for instance, has been presented as an initiative that would simultaneously reverse the lake's 'decline' and 'restore peace' – even though it is not evident how it might do this and the transfer could itself cause large-scale displacement, dispossession and ecological destruction. The Red Sea–Dead Sea Canal project between Israel and Jordan, which we unfortunately cannot discuss in further detail, has been marketed in similar terms – despite being structured to marginalise and exclude the Palestinians. And Turkey has also deployed 'water for peace' rhetoric in relation to its northern Cyprus water pipeline, as already discussed.[42]

To understand the ramifications of all this for a world of rising climate vulnerabilities, we simply recommend that these conclusions be read alongside those of the previous chapter. For if, as argued there, war is such an important cause of environmental vulnerabilities that future vulnerabilities to climate change will in large part be shaped by the future of war, and if, as argued here, contemporary peace processes often reproduce, and sometimes even intensify, wartime patterns of inequality and insecurity, then the implications are easily inferred: that future climate change vulnerabilities will be significantly affected by the future politics of peace-making. Quite what this future consists of, is of course beyond the scope of this book. Yet it is hard to feel optimistic about it. The peace processes analysed above were, for all their faults, products of a post–Cold War era of

---

[42] V. Yorucu and R. Keles, 'The construction boom and environmental protection in Northern Cyprus as a consequence of the Annan Plan', *Construction Management and Economics*, 25:1 (2007), 77–86; R. Shehadeh, *Palestinian Walks: Notes on a Vanishing Landscape* (Profile, 2007); T. Ide, 'The dark side of environmental peacebuilding', *World Development*, 127 (2020), 104777; Government of Nigeria and LCBC, *Roadmap on Saving the Lake Chad to Revitalise the Basin Ecosystem for Sustainable Livelihood, Security and Development* (2018), 1; CIMA International, *Feasibility of the Water Transfer Project from the Ubangi to Lake Chad* (2011), 18–20; E. Abitbol, 'Developing water and marginalising Israeli/Palestinian peace: a critical examination of the Red Sea–Dead Sea Canal feasibility study process', *Journal of Peacebuilding and Development*, 5:1 (2009), 35–49; K. Aggestam and A. Sundell, 'Depoliticising water conflict: functional peacebuilding in the Red Sea–Dead Sea Water Conveyance project', *Hydrological Sciences Journal*, 61:7 (2016), 1302–12.

relative faith in US-led and UN-legitimised peacemaking. By contrast, writing after ten years of war in Syria, twenty years of on–off war in Israel–Palestine and just following the United States' rapid retreat from and evacuation of Afghanistan – the product of sordid United States–Taliban peace terms – the landscape of international peacemaking looks very different. Both for the sake of climate vulnerability and much else besides, we can only hope that future peacemaking does not follow in the footsteps of the fall of Kabul.

9 Transformations and Circulations

Nexus, Nexus, Nexus

Over the last decade or so, the claim that there exists a complex 'nexus' – or indeed a series of interlocking 'nexuses' – connecting water and other global challenges has become a commonplace of international expert and policy discourse on resource governance. Thus water management challenges are said to be entwined with questions of food supply. The water and energy sectors are said to be mutually dependent. Food and energy are said to be interlinked too. All are held to be imperilled by conflict, by mounting resources pressures and by climate change. We live, it is claimed, in a 'hyperconnected world' which is 'more complex than ever before' and where the 'challenges share one critical feature: they're all inter-connected'. What is needed to meet these challenges, it is said, are new modes of both analysis and governance – ones that attend to the 'feedbacks', 'interplays', 'interactions', 'trade-offs' and 'synergies' between issues and sectors, and which achieve this through improved policy 'integration', new 'partnerships', cross-sectoral 'collaboration' and 'coordination' and a general disregard of prevailing disciplinary and organisational 'siloes'. What is needed, it is frequently suggested, in sum, is a 'nexus approach' or 'perspective' on the challenges of water security.[1]

And at first glance it is hard to disagree. The fundamental problem with eco-determinist water security discourse, for instance, is its general failure to attribute any great significance to factors beyond local water resource–demand imbalances – that is, to consider how modern patterns of water (in)security are shaped by technologies, by energy supplies, by agricultural

---

[1] WEF, *The Network of Global Agenda Councils*, promotional film (2014); H. Hoff, 'Understanding the nexus – background paper for the Bonn 2011 Nexus Conference: the water, energy and food security nexus' (Stockholm Environment Institute, 2011); Bonn 2011 Conference, *The Water Energy and Food Security Nexus: Solutions for a Green Economy* (2012); R. Cairns and A. Krzywoszynska, 'Anatomy of a buzzword: the emergence of the "water-food-energy security nexus" in UK natural resource debates', *Environmental Science and Policy*, 64 (2016), 164–70.

policies, by trade, by violence and, more broadly, by all those processes of development and state-building discussed at length in the preceding chapters. Conversely, it has long been recognised within the water sector that 'integrated' thinking and governance are required if the diverse needs and interests associated with water are to be accommodated, as captured by the notion of 'Integrated Water Resource Management'. Moreover, as an approach, political ecology is premised on the view that long-entrenched disciplinary boundaries – and even the 'two cultures' divide between the natural and the social sciences – are obstacles to understanding contemporary eco-social crises. Marxist thought, in particular, has long been anti-disciplinary – or 'pre-disciplinary' – in both instinct and method, committed to understanding the metabolic relations between ostensibly distinct spheres.[2] Who, from such perspectives, could seriously disagree with the demand for more 'joined-up' and cross-disciplinary engagement?

Yet at the same time, it is this very unarguability of nexus talk – its 'luminous obviousness' – which should give pause for thought. For, however compelling the rhetoric of joined-up thinking and boundary transgression may be, beneath it lie a series of much more problematic commitments. Water nexus discourse was born, first of all, within the World Economic Forum (WEF), an institutional space which, as Stephen Gill, Kees van der Pijl and fellow neo-Gramscians have long argued, is central to the contemporary articulation and circulation of neo-liberal elites, interests and ideas. The WEF's Water Initiative specifically, established in 2006 as one of several dozen such issue networks, has a spread of questionable corporate partners: Coca-Cola, repeatedly implicated in groundwater pollution and depletion, especially in India; Nestlé, the number one player in the global bottled water industry and widely boycotted since the 1970s for its aggressive promotion of powdered baby milks in the global South (use of which has been linked, among other things, to widespread use of non-potable water and consequent infant malnutrition and mortality); and Dow Chemicals, owner of the company responsible for the 1984 Bhopal gas leak, the world's single most deadly industrial disaster, which left a legacy of polluted groundwater that endures to this day. In turn, the key output of the WEF's Water Initiative, *Water Security: The Water–Food–Energy–Climate Nexus* – which launched the idea of water nexuses onto the international stage – strongly reflects corporate interests and neo-liberal premises. Thus the report discusses water-related risks to corporate operations and how businesses can be partners to

---

[2] D. Benson et al., 'Water governance in comparative perspective: from IWRM to a "nexus" approach?', *Water Alternatives*, 8:1 (2015), 756–73; C. P. Snow, *The Two Cultures and the Scientific Revolution* (Cambridge University Press, 1959); B. Jessop and N.-L. Sum, 'Pre-disciplinary and post-disciplinary perspectives', *New Political Economy*, 6:1 (2001), 89–101.

government, but barely hints at how they are frequently also part of the problem. The report advocates market-based solutions to water problems, but does not even consider water as a human right (and perhaps unsurprising so, given that Peter Brabeck-Letmathe, former chairman and CEO of Nestlé and a key figure within the WEF's Water Initiative, has labelled as 'extreme' the view that 'as a human being you should have a right to water'). It emphasises, more specifically, the importance of innovation, water pricing, improved resource-use efficiency and trade liberalisation, going so far as to suggest that successful completion of the Doha round of world trade negotiations is important to resolving the world's water problems.[3] A fundamentally problematic set of commitments is smuggled in under the benign-sounding objective of promoting joined-up policy and analysis.

Moreover, reflecting its roots in the world of policy, nexus discourse is much more normative than analytical in content, jumping quickly to offering neo-liberal answers but providing only the most superficial – and in certain respects misleading – explanations. There are several issues here. For one, throughout nexus discourse there is an assumed but uninterrogated water-centrism, according to which it is imagined, for instance, that nations with 'a legacy of difficult hydrology have remained poor', that water is 'the only natural limit to economic growth' and that water security is 'the gossamer that links together the web of food, energy, climate, economic growth and human security'. Linked to this, contemporary water challenges are assumed to arise, at root, from scarcity, for instance from the fact that '[t]hirst is now global' and from the 'world's vulnerability to the shock of diminishing resources'. Eco-determinist crisis rhetoric and neo-liberal policy recommendations are here uneasily, if not atypically, fused. In addition, it is not clear why certain factors, most notably water, food and energy, are identified as part of the 'nexus', while others – land, labour, soil, politics or gender, to name but a few – are barely mentioned at all.[4] And last, despite or perhaps because of its emphasis on 'multi-directionality', nexus discourse

---

[3] J. Toye, 'Poverty reduction', *Development in Practice*, 17: 4–5 (2007), 505; Cairns and Krzywoszynska, 'Anatomy of a buzzword', 166; S. Gill (ed.), *Gramsci, Historical Materialism and International Relations* (Cambridge University Press, 1993), van der Pijl, *Transnational Classes*, 133–5; K. R. Raman, 'Transverse solidarity: water, power, and resistance', *Review of Radical Political Economics*, 42:2 (2010), 251–68; M. Muller, *The Baby Killer* (War On Want, 1974); U. Schnell (dir.), *Bottled Life: Nestlé's Business with Water* (2012); Amnesty International, *Clouds of Injustice: Bhopal Disaster 20 Years On* (2004); WEF, *Water Security*, 75, 79, ch. 7; P. Brabeck-Letmathe in E. Wagenhofer (dir.), *We Feed the World* (2005).

[4] WEF, *Water Security*, xxi, xix, 1, 88; Hoff, 'Understanding the nexus', fig. 2, 16; J. Allouche et al., 'Technical veil, hidden politics: interrogating the power linkages behind the nexus', *Water Alternatives*, 8:1 (2015), 610–26; D. Wichelns, 'The water–energy–food nexus: is the increasing attention warranted, from either a research or policy perspective?', *Environmental Science and Policy*, 69 (2017), 117–18.

ends up saying very little about which of the links between water, food, energy and so on are the most causally weighty or significant. As a contribution to explaining the nature and causes of water (in)security, nexus discourse is thus thin gruel indeed.

But how, then, should we understand the relations between water and cognate areas? This final substantive chapter of the book seeks to provide an alternative answer to this question, by examining what we take to be the four key sets of relations underpinning water security and insecurity today: the relationships between water and trade; between water and agricultural production; between water and energy; and between water and capital. The chapter considers these four sets of relations in turn, in each case providing an overall mapping of the transformations and circulations that define them and an assessment of how these relations structure patterns of water-related security and insecurity, especially in our five divided environments. And it ends, in line with previous chapters, by turning to the future and climate change in particular.

Through all of this our core arguments are fourfold. Overall we show, in basic agreement with nexus discourse, that acts of transformation or transmutation – the turning of one substance into another – are key underpinnings of water security and insecurity. And yet simultaneously we contend, against the prevailing water-centrism of nexus discourse, that water is much more a dependent than an independent variable within nexus relations, such that contemporary patterns of water resource degradation and vulnerability are much more a function of trade relations, fuel prices and so on, than the other way around. More specifically, we demonstrate, in line with world systems–informed theorisations of political economy and ecology, that contemporary patterns of water (in)security, shaped as they are by nexus relations, are determined neither by natural resource availability nor relative market efficiencies but instead by countries' particular positionings within a structurally unequal and hierarchical capitalist world order.[5] And this, we argue in conclusion, will in all likelihood remain the case as our planet continues to warm.

The International Political Ecology of Virtual Water

Nothing illustrates nexus discourse better than the thesis that the international trade in 'virtual water' is key to understanding the world's water problems and the prospects for mitigating them. Agricultural and

---

[5] I. Wallerstein, *The Modern World System*, Vols. 1–3 (Academic Press, 1974, 1980, 1989); A. Hornborg, *Global Ecology and Unequal Exchange: Fetishism in a Zero-Sum World* (Routledge, 2011); Moore, *Capitalism*.

industrial production, the argument goes, together account for well over 80 per cent of human water use. Many of the resulting agricultural and industrial commodities are traded across borders. And this means, according to the thesis' proponents, that international trade effectively involves the export of water resources in 'embedded' or 'virtual' form. In the view of these proponents, moreover, international trade enables water-poor regions and countries to import virtual water from water-rich ones, reducing pressures where water is locally scarce and at the same time – through the market's invisible hand – increasing global resource-use efficiencies. Many water-poor countries, it is argued, already do this, especially by importing food staples like wheat and soya and reallocating their limited internal supplies to higher value productive and domestic uses. Indeed virtual water imports, it is often claimed, are perhaps the key reason why water-poor Middle Eastern and North African states have not yet gone to war over water, trade having offered them a far cheaper and more reliable alternative. As the concept's originator, Tony Allan, has put it, 'the concept of virtual water ... provides an analytical perspective on how economies achieve water security'. Furthermore, for many proponents of virtual water ideas, this existing pattern provides a model for the future, suggesting, as the WEF's Water Initiative has it, that the 'water and trade nexus will be especially important for those water-constrained but expanding economies across North Africa, the Middle East, and Asia over the coming decades' and even that there should be 'a major reconfiguration of international trade to enable country-to-country trades in "virtual water" to alleviate domestic water constraints'. There now exists an extensive academic literature on these themes.[6]

Both the virtual water thesis and the related notion of 'water footprints' have, of course, been challenged. Both have been critiqued as water-centric – criticised for approaching trade and consumption patterns through the lens of water, for supposing that states engage in trade to secure it, and for suggesting, surely implausibly, that global production and trade patterns should be reorganised to maximise water-use efficiencies and ameliorate water scarcity. From perspectives close to our own, virtual water doctrines have also been critiqued for their pro-liberalisation biases, including their grave economic implications for agriculture, and especially small farmers, in relatively water-poor states.

---

[6] T. Allan, 'Virtual water: the water, food, and trade nexus. Useful concept or misleading metaphor?', *Water International*, 28:1 (2003), 111; WEF, *Water Security*, 68, 75–6; J. Chenoweth et al., 'Quantifying the human impact on water resources: a critical review of the water footprint concept', *Hydrological and Earth System Sciences*, 18:6 (2014), 2325–42.

And it has been correctly observed that the phrase 'virtual water' is misleading since there is nothing 'virtual' – that is, unreal or simulated – about it. 'Virtual water' is in truth less a scientific concept than a policy buzzword.[7]

The biggest problem with the virtual water thesis, however – which for all its obviousness barely figures in debates on the subject – is that the structure of the global virtual water trade is organised less around a logic of scarcity than around patterns of development, affluence and power. This trade is, first of all, dominated not by water-poor states but by the major trading blocks, specifically the United States, the EU and East Asia. While the volume of this trade has expanded hugely in recent decades, this has essentially been a function of the increasing density of international trade relations, not a response to water scarcities. Globally, moreover, there are sharp differences in per capita national 'water footprints' – the total volume of water plus embedded water consumed by the inhabitants of a country – with the average person in the highest water-consuming country, Luxembourg, consuming by one calculation over 100 times the amount in southern central Africa. As a result, at a global level net virtual water flows are less from water-rich to water-poor economies than between major trading economies, and from poor and low-water footprint countries to affluent high-water footprint ones. With each doubling of income, per capita virtual water imports increase by more than 80 per cent (suggesting that claims that there has been a 'decoupling' of water consumption and economic growth are misplaced). The UK, at the extreme, has a global water footprint six times larger than its internal water withdrawals. Hence as Ranran Wang and colleagues summarise it, virtual water 'flows uphill toward money' – just like the real water piped from the Colorado River across Arizona.[8]

---

[7] Chenoweth et al., 'Quantifying the human impact'; D. Wichelns, 'Virtual water and water footprints do not provide helpful insight regarding international trade or water scarcity', *Ecological Indicators*, 52 (2015), 277–83; E. Gawel and K. Bernsen, 'What is wrong with virtual water trading? On the limitations of the virtual water concept', *Environment and Planning C*, 31:1 (2013), 168–81; Barnes, 'Water, water everywhere'; S. Gialis and S. Mavroudeas, 'Virtual water: more heat than light?', *Capitalism Nature Socialism*, 25:2 (2014), 60–74; S. Merrett, 'Virtual water and Occam's razor', *Water International*, 28:1 (2003), 103–5.

[8] C. Zhan-Ming and G. Q. Chen, 'Virtual water accounting for the globalized world economy: national water footprint and international virtual water trade', *Ecological Indicators*, 28 (2013), 145–8; C. Dalin et al., 'Evolution of the global virtual water trade network', *Proceedings of the National Academy of Sciences*, 109:16 (2012), 5989–94; I. Arto et al., 'Global use of water resources: a multiregional analysis of water use, water footprint and water trade balance', *Water Resources and Economics*, 15 (2016), 1–14; R. Wang et al., '(Virtual) water flows uphill toward money', *Environmental Science and Technology*, 50:22 (2016), 12327.

Certain important qualifiers are admittedly required. While the United States is a net virtual water importer overall, with regard to crop production alone it is the world's biggest exporter, reflecting the scale of its wheat exports in particular. China, conversely, though a net exporter – reflecting its status as the workshop of the world – is a net importer in terms of food. The structure of the global virtual water trade is not static: South America, for instance, is an increasingly important exporter, thanks in particular to its ever-expanding production of soya. Moreover, water resource availability is evidently one factor in virtual water imports: that most Middle Eastern and North African countries, for instance, are net importers of embedded water is clearly in part because of their limited water supplies. Yet for all this, the central point still stands. Globally, there is no correlation between water scarcity and virtual water import dependency, as virtual water analysts are often compelled to admit. To the contrary, the structure of the virtual water trade instead principally reflects differences in wealth and power, and is essentially organised along North–North and South–North axes. That this is so often missed is partly because the 'virtual water' idea was initially coined by Tony Allan to capture the wheat-importing behaviour of water-poor Middle Eastern states and the absence of 'water wars' between them – and also no doubt because of the ideological premises of water nexus discourse, as discussed above.[9]

Research on the virtual trade in 'scarce water' – which here refers to countries in which a high proportion of the renewable water resources available are withdrawn, i.e. where there is a high level of water stress as per SDG indicator 6.4.2 – provides further corroboration of this. The biggest net importers of scarce embedded water are all developed countries: Japan, Germany, the United States, the UK and France. Conversely, the biggest exporters are almost exclusively developing states, with the top three gross exporters being India, Pakistan and China. Noteworthy also is that a far higher proportion of scarce water withdrawals as against 'non-scarce' water withdrawals are traded, reflecting the fact that scarce water is disproportionately exploited for export. Groundwater resources, in particular, are being depleted at an ever-increasing rate for agricultural

---

[9] A. Y. Hoekstra and P.Q . Hung, 'Globalisation of water resources: international virtual water flows in relation to crop trade', *Global Environmental Change*, 15:1 (2005), 50, 54–5; Zhan-Ming and Chen, 'Virtual water accounting', 146–8; J. A. Carr, 'On the temporal variability of the virtual water network', *Geophysical Research Letters*, 39:6 (2012); M. D. Kumar and O. P. Singh, 'Virtual water in global food and water policy making: is there a need for rethinking?', *Water Resources Management*, 19:6 (2005), 765–7; E. Ansink, 'Refuting two claims about the virtual water trade', *Ecological Economics*, 69:10 (2010), 2027–32; T. Allan, *The Middle East Water Question: Hydropolitics and the Global Economy* (I.B. Tauris, 2000); Allan, *Virtual Water*, 48.

exports, especially in key Southern agricultural producer states – most notably India, Iran, Pakistan and China – but also in the United States.[10] Far from being means for water-poor countries to save water and mitigate scarcities, the global trade in virtual water – or more precisely, the overall structure of the global commodity trade and the various national export and import strategies which underpin it – is a major contributor to water degradation, depletion and scarcity, especially in the global South.

And to be brutally honest: why should we expect anything else? The global economic order established by Britain and other European colonial powers during the eighteenth and nineteenth centuries was above all predicated on exploitation and 'ecologically unequal exchange' – that is, on ensuring that environmental resources and agricultural products, plus the vast profits derived from indigenous, indentured and slave labour, would flow overwhelmingly from colony to metropole, from 'South' to 'North'. Ireland provided the model, Marx observing in *Das Kapital* that 'for a century and a half' after 1691, England 'indirectly exported the soil of Ireland without even allowing its cultivators the means for replacing the constituents of the exhausted soil', simultaneously 'robbing the soil' and 'robbing the worker'. The Baltic region was exploited too, albeit without formal colonisation, its abundant forests plundered to provide an 'environmental overdraft' of timber for a largely denuded Britain. And as Europe's economy developed and its own internal resources were degraded so this pattern was globalised, reproduced across the non-European world from the Caribbean and South America to the Indian subcontinent and central Africa. Decolonisation for the most part saw a continuation of these relations of economic subordination and dependency. That present day Europe is a – no, *the* – biggest net virtual water importing region, while most Asian, African and Latin American societies are net virtual water exporters, is but one instance of the more general pattern of primary commodity flows, a pattern established during the era when Europe ruled the world.[11]

Our particular divided environments bear this out. Sudan, to start with, is by most assessments a large net exporter of virtual water. Thus,

---

[10] M. Lenzen et al., 'International trade of scarce water', *Ecological Economics*, 94 (2013), 81–2; C. Dalin et al., 'Groundwater depletion embedded in international food trade', *Nature*, 543 (2017), 700–4.

[11] Hornborg, *Global Ecology*; A. Hornborg and J. Martinez-Alier, 'Ecologically unequal exchange and ecological debt', *Journal of Political Ecology*, 23:1 (2016), 328–33; Marx, *Capital*, 637–8, 860; Pomeranz, *Great Divergence*, 220–5; Clark and Foster, 'Ecological imperialism', 316; Moore, 'Sugar and the expansion of the early modern world-economy'; Arto et al., 'Global use of water'; Zhan-Ming and Chen, 'Virtual water accounting', fig. 5.

according to the formative work of Arjen Hoekstra and colleagues, in the late 1990s Sudan was exporting over 1.7 bcm of virtual water each year in food crops, more than three times the level of its virtual water crop imports. The virtual water embedded in its livestock and meat exports was over twenty times the level of its livestock and meat imports. Sudan was also, of course, a large net exporter of virtual water through cotton. And its external water footprint was, as a proportion of its total water footprint, one of the smallest in the world. Typical of Southern states, the majority of its virtual water exports were and remain exported not to elsewhere within the global South, let alone to other Nile basin riparians, but instead to Europe, North America and the richest states in the Persian Gulf.[12] Sudan's rapidly expanding livestock sector, in particular, now effectively exports huge volumes of virtual water to Saudi Arabia and the other Gulf emirates. Moreover, viewed thus, the recent land investments in Sudan by Saudi Arabia and other Gulf states – discussed in Chapter 6 as instances of 'land grabbing' and 'water grabbing' – are but extensions of a long-established pattern of net resource transfer from poorer to more affluent states.

Syria presents an even more striking instance of this pattern. More so even than Sudan, Syria was – at least until the civil war – a conspicuously large net exporter of virtual water in food crops: during 1995–9, for instance, Syria was exporting over 5 bcm annually in crops, just under six times the level of its imports, and was the fourteenth biggest crop virtual water net exporting country overall (and one of the very biggest when population or territorial area are also factored in). Syria was also exporting virtual water through cotton to the tune of 1.75 bcm/y, making it the world's twelfth biggest virtual water cotton exporter. And it had the seventh highest overall per capita blue water footprint – that is, footprint of surface and groundwater used in irrigation – in the world. Even more tellingly, in 2000 Syria was the fifth biggest gross exporter and the third biggest net exporter of 'scarce' virtual water, behind only Pakistan and China on the latter score, and in per capita terms, was the largest major exporter of scarce virtual water in the world.[13] We will have more to say on

[12] Hoekstra and Hung, 'Globalisation of water', 51; A. Y. Hoekstra and P. Q. Hung, *A Quantification of Virtual Water Flows between Nations in Relation to International Crop Trade* (IHE Delft, 2002), 26, 107–13; A. K. Chapagain and A. Y. Hoekstra, *Virtual Water Flows between Nations in Relation to Trade in Livestock and Livestock Products* (IHE Delft, 2003), 177, 193; A. Y. Hoekstra and M. M. Mekonnen, 'The water footprint of humanity', *Proceedings of the National Academy of Sciences*, 109:9 (2012), 3234; M. Zeitoun et al., 'Virtual water "flows" of the Nile Basin, 1998–2004: a first approximation and implications for water security', *Global Environmental Change*, 20 (2010), 229–42.

[13] Hoekstra and Hung, 'Globalisation of water', 51; Hoekstra and Hung, *Quantification in Relation to International Crop Trade*, 26, 107–13; A. Chapagain et al., 'The water footprint of cotton consumption: an assessment of the impact of worldwide consumption of cotton

these exports later in the chapter. Suffice to say now that there is scant evidence here of virtual water imports being used to mitigate scarcity. Rather, and as evidenced in previous chapters, an agriculturally-centred development strategy dictated that Syria would continue exporting its once-abundant surface and especially groundwater resources to the rest of the world, even as they became more and more severely degraded.

By contrast, the more wealthy among our divided environments are clear net importers, not exporters, of virtual water. The Republic of Cyprus' virtual water crop imports exceed exports by as much as six times. Overall, it has a particularly high level of external water dependency. And the volume of Cyprus' crop virtual water exports has also been declining over time, as an ever-rising proportion of its water resources have been allocated to internal domestic and municipal uses, especially to support burgeoning tourist demand.[14] Despite having a level of water resource availability which is broadly comparable to that of Syria, just across the eastern Mediterranean (Syria has an availability of just under 1,000 cm/cap/y and Cyprus just over 650 cm/cap/y, but with much lower exploitation of available resources), the two countries have diametrically opposed virtual water profiles.

Israel, equally, is a large net importer of virtual water. In 1999, for instance, by Hoekstra and colleagues' calculations, Israel exported around 0.7 bcm of crop virtual water but imported over ten times that amount – 7.4 bcm – making it one of the top twenty-five net virtual water importing countries worldwide. It was also one of the twenty leading net importers of livestock and meat products. And Israel had, and continues to have, an exceptionally high level of external water dependency, higher even than that of Cyprus. Pro-Israeli commentators regularly emphasise how the country's supposed 'agricultural miracle' has enabled it to develop a 'multi-billion dollar agricultural export industry' and to become 'an agricultural exporter of significance' – while conveniently failing to mention its dependence upon agricultural (and other) imports. In reality, agricultural exports are both a minor part of Israel's overall export profile – accounting, in 2017, for less than 3 per cent of total exports – and small by comparison with its agricultural imports. Israel is a large net importer of food across all categories except fruit and vegetables: that it is currently

---

products on the water resources in the cotton producing countries', *Ecological Economics*, 60:1 (2006), 194; Hoekstra and Mekonnen, 'The water footprint', supporting information, fig. S5; Lenzen et al., 'International trade', 82.

[14] Hoekstra and Hung, *Quantification in Relation to International Crop Trade*, 105, 108; Hoekstra and Mekonnen, 'The water footprint', 3235; C. Zoumides et al., 'Policy-relevant indicators for semi-arid nations: the water footprint of crop production and supply utilization in Cyprus', *Ecological Indicators*, 43 (2014), 205–14.

a land of water plenty is much more a function of these imports than of the country's impressive wastewater reuse systems or the wonders of drip irrigation. Viewed in this light, the idea that Israel provides a model 'solution for a water-starved world' is even more fatally flawed than already discussed.[15]

More than this, Israel is also instructive regarding the basic structure of the political economy of virtual water – as well as the shortcomings of virtual water discourse. Israel is often held up as a paradigmatic example of virtual water importing in action: Tony Allan has claimed, for instance, that he derived the term from the work of Israeli economists who 'had by the mid-1980s spotted that it was less than sensible from an economic perspective to export scarce Israeli water', while others, building upon this observation, have claimed that 'the real origin' of the virtual water concept can be traced to these economists' proposals that 'Israel should restructure and reorient its primary production toward less water-intensive products and foods' and import these products instead – proposals which were apparently 'well implemented'. Such is the foundational narrative of virtual water discourse. However, this narrative is largely built on myth. The Israeli economists who ostensibly inspired virtual water theory did not advocate a reorientation of the country's agricultural system to 'less water-intensive products' or away from exporting 'scarce Israeli water', quite the contrary. There is also no evidence, so far as we are aware, that any such reorientation was ever considered, let alone 'implemented', by Israeli policymakers. And, most importantly, neither the Israeli agricultural sector nor the country's import and export profiles have shifted towards 'water saving'; the opposite is in fact the case. While the volume of agricultural water consumption did decline during the late 1980s specifically, outside of this period Israel's agricultural water use has been either stable or rising and is currently at record levels (thanks to the wastewater and desalination investments discussed in Chapter 5). Israel's agricultural production has increased by both value and volume, with the main shift since the 1970s being a progressive decline in cereal crop production and corresponding increases in vegetables and livestock – an overall shift, that is, towards more, not less, water-intensive production. This shift, we may add, was evidently a result of economic calculations within the context of shifting international terms of

[15] Hoekstra and Hung, 'Globalisation of water', 52; Hoekstra and Hung, *Quantification in Relation to International Crop Trade*, 26, 29, 106, 109; Chapagain and Hoekstra, *Virtual Water*, 30; Hoekstra and Mekonnen, 'The water footprint', 3235; Siegel, *Let There Be Water*, 70, 82; Israel CBS, 'Imports and exports, by the Standard International Trade Classification – SITC (revision 4)', *Statistical Abstract of Israel 2018* (2018).

trade, and was not a function of any political decisions (as Tony Allan rightly emphasises, the virtual water trade is 'politically silent'). Agricultural exports have also continued to increase, albeit slowly. Plus it is worth stressing that Israel has always been reliant on virtual water embedded in food and other imports.[16] Overall, Israel has remained committed, for the political reasons examined in previous chapters, to its hydraulic and agricultural missions. Its agricultural sector has sought, within this, to shift towards more high-value production, taking advantage of the country's numerous technological advantages. In aggregate, this has involved using ever-more water, not less. And this has been accompanied by a deepening but hardly new-born import dependence. In sum, there is no evidence that Israel at any point adopted a strategy of saving water through imports – and still less that virtual water importation is essentially about 'saving water'. If the Israeli case illustrates anything it is rather the mythical character of the virtual water narrative.

Now, the four cases discussed above could be interpreted as collectively conforming to a virtual water logic, with Sudan and Syria, the two countries with the highest per capita water resource bases, both exporting virtual water, and Cyprus and Israel, which both have much smaller resource bases, both importing it. And to be sure, it clearly is the case that states with very limited water availability are unlikely to be major exporters of virtual water. That said, across these cases, as globally, relative affluence and power are much the more important determinants of virtual water flows. Israel and Cyprus are large net importers because of their high per capita incomes and consequently large water footprints. Sudan and Syria, on the other hand, are net exporters because of their much smaller water footprints and typical poor-state dependence on agricultural export markets. Cyprus has a completely converse virtual water profile to Syria despite a not dissimilar per capita resource base. Syria continued exporting virtual water in the run-up to its civil war despite suffering extreme groundwater depletion. Sudan has continued increasing its virtual water exports, despite the trail of frontier devastation

---

[16] Allan, 'Virtual water', 106; Gialis and Mavroudeas, 'Virtual water', 62; Barnes, 'Water, water everywhere', 375; G. Fishelson, 'The allocation and marginal value of water in Israeli agriculture', in J. Isaac and H. Shuval (eds.), *Water and Peace in the Middle East* (Elsevier, 1994), 427–40; M. Zeitoun, *Power and Water in the Middle East: The Hidden Politics of the Palestinian–Israeli Water Conflict* (I.B. Tauris, 2008), 133–4; Israel CBS, 'Livestock', 'Input, output and volume of price indices', 'Agricultural output, by industry and product', 'Agricultural crop area', 'Exports, by industry', in *Statistical Abstract of Israel 2018* (2018); J. A. Allan, 'Virtual water – economically invisible and politically silent: a way to solve strategic water problems', *International Water and Irrigation Journal*, 21:4 (2001), 39–41.

left in these exports' wake. And the West Bank and Gaza, we may add, are much smaller net importers of virtual water than Israel – this being a function of their low per capita incomes and despite the many restrictions on both Palestinian access to local water resources and Palestinian agricultural production and exports.[17] During 1971–4, as large swathes of the global South reeled from global grain price increases and many areas stood on the cusp of famine, the UK, Germany and Italy each imported more grain than the whole of Africa put together.[18] Both in this extreme example and globally, virtual water predominantly flows not to where it is most needed, but to money and power. Or put differently: virtual water and the water–trade nexus are radically differentiating – simultaneously bringing relative water security to some, but widespread conflict, vulnerability and environmental degradation to others.

Rebound in the Negev

As an afterword to this discussion it is also worth considering a second nexus, the relationship between water and agricultural production. The standard proposition here is straightforward: that in view of global and local water resource scarcities and given that the large majority of water withdrawals are for agricultural production, it follows that improved resource-use efficiency, made possible by technological innovation, is crucial. The development of drought-resistant crops, the much more widespread use of sprinkler and drip irrigation, the improved use of pesticides to ensure that water is delivered to crops instead of to weeds and pests, optimal use of fertilisers to reduce other productivity constraints, and a panoply of water price, extension service and regulatory reforms: all are seen as vital. 'More crop per drop' runs the liberal-technical mantra – key both to saving water and to ensuring current and future water and food security.[19]

The problem, though – which is well established in environmental and ecological economics – is that improved resource-use efficiency does not typically lead to reduced resource use overall, owing to what is commonly referred to as the 'rebound effect' or 'Jevons paradox'. In his 1865 book *The Coal Question*, William Stanley Jevons famously argued that it '*is*

---

[17] D. W. Nazer et al., 'Water footprint of the Palestinians in the West Bank', *Journal of the American Water Resources Association*, 44:2 (2008), 449–58; M. J. Beltrán and G. Kallis, 'How does virtual water flow in Palestine? A political ecology analysis', *Ecological Economics*, 143 (2018), 17–26.

[18] Garcia, *Drought and Man*, Vol. 1, 23.

[19] D. Seckler, *The New Era of Water Resources Management: From 'Dry' to 'Wet' Water Savings* (International Irrigation Management Institute, 1996); WEF, *Water Security*, ch. 1.

wholly a confusion of ideas to suppose that the economical use of fuel is equivalent to a diminished consumption. The very contrary is the truth.' 'Nor is it difficult to see how this paradox arises', Jevons continued:

[I]f the quantity of coal used in a blast furnace, for instance, be diminished in comparison with the yield, the profits of the trade will increase, new capital will be attracted, the price of pig iron will fall, but the demand for it increase; and, eventually, the greater number of furnaces will more than make up for the diminished consumption of each . . . . It needs but little reflection, indeed, to see that the whole of our present vast industrial system, and its consequent consumption of coal, has chiefly arisen from successive measures of economy.[20]

Technical efficiency improvements, Jevons concluded, while leading to input savings at the level of individual production processes, nonetheless have the very opposite effect at the macro-economic level; indeed, in Jevons' formulation, such efficiency improvements are *the* central motor of capitalist development and its rapacious thirst for resources. Or, as the more recent and diluted version of this thesis has it, there is typically a 'rebound' from efficiency measures, with a proportion and perhaps most of the direct savings from them being lost through increased consumption elsewhere.[21]

Now, there are legitimate questions to ask about whether, and to what extent, insights developed regarding the role of King Coal in British industrialisation are applicable to the contemporary political ecology of water. Yet a growing body of work concludes that, broadly speaking, they are. The adoption of sprinkler and especially drip irrigation, this work suggests, may facilitate shifts towards more high-value and water-intensive crops, in turn increasing water demand. It may lead to multi-year cropping, increasing demand still further. And the increased economic efficiency of irrigation may prompt farmers to expand the area under irrigation, either through the cultivation of hitherto open land or by switching from rain-fed to irrigated production. Far from being a marginal issue, institutions ranging from the European Commission to UNEP are aware of these problems. In addition – and this is specific to water – the adoption of more sophisticated irrigation systems with comparatively small 'water losses' has the effect, in practice, of reducing return flows from irrigation sites to soil, groundwater, streams, rivers and downstream water users (or stated the other way around, water is never actually 'lost'). Especially in the absence of strong regulation,

---

[20] W. S. Jevons, *The Coal Question: An Inquiry Concerning the Progress of the Nation, and the Probable Exhaustion of Our Coal-Mines* (Macmillan and Co, 1865), 103–5 (italics in original).

[21] B. Alcott, 'Jevons' paradox', *Ecological Economics*, 54:1 (2005), 9–21; S. Sorrell et al., 'Empirical estimates of the direct rebound effect: a review', *Energy Policy*, 37:4 (2009), 1356–71; Foster et al., *Ecological Rift*, ch. 7.

irrigation modernisation may thus have significant negative environmental, economic and also distributive consequences including further intensifying agricultural production, facilitating the 'opening up' or colonisation of new land, reducing return flow and water sharing with other users, and increasing overall water demand. This is in addition to problems directly associated with drip irrigation, especially increased soil salinity arising from the absence of flood flows to wash salts out of the soil.[22]

That these paradoxical effects apply within the divided environments considered here does not need belabouring: we limit our remarks to Israel, as the birthplace of drip irrigation technology. Modern drip irrigation using perforated plastic pipes was invented during the late 1950s by leading Israeli water engineer Simcha Blass, and initially developed out of a kibbutz in the northern Negev and the company established by it – Netafim, still the global leader in drip irrigation sales. Drip irrigation was adopted throughout Israel almost immediately and is widely credited as having revolutionised its agriculture for the better; today 75 per cent of Israeli irrigation involves drip systems. And yet, despite the 'crop per drop' savings, Israel's drip irrigation revolution has not, in aggregate terms, self-evidently saved water. The country's total agricultural water consumption increased steadily between the 1960s and the mid-1980s as drip irrigation was adopted, and is currently at record levels, as already noted. Moreover, studies have identified productivity gains and profit maximisation, rather than water saving, as the primary motivations behind farmers' adoption of the technology. While drip irrigation could, in principle, have enabled reduced exploitation of the country's rivers and aquifers – or even redistribution to Palestinians or to neighbouring states – in practice what it prompted was agricultural expansion and intensification. The adoption of drip irrigation coincided with, and facilitated, the shift from cereals to the higher value and more capital-intensive vegetable production discussed in the previous section. It spurred increased cultivation of marginal land, such that Israel's total irrigated area increased

---

[22] L. Pfeiffer and C.-Y.C. Lin, 'Does efficient irrigation technology lead to reduced groundwater extraction? Empirical evidence', *Journal of Environmental Economics and Management*, 67:2 (2014), 189–208; R. Q. Grafton et al., 'The paradox of irrigation efficiency', *Science*, 361:6404 (2018), 748–50; C. Paul et al., 'Rebound effects in agricultural land and soil management: review and analytical framework', *Journal of Cleaner Production*, 227 (2019),1054–67; European Commission, *Communication from the Commission to the European Parliament, the Council, the European Economic and Social Committee and the Committee of the Regions: A Blueprint to Safeguard Europe's Water Resources* (2012), 11–12; UNEP, *Measuring Water Use in a Green Economy* (2012), 27; A. J. Clemmens et al., 'Technical concepts related to conservation of irrigation and rainwater in agricultural systems', *Water Resources Research*, 44:7 (2008); C. Batchelor et al., 'Do water-saving technologies improve environmental flows?', *Journal of Hydrology*, 518 (2014), 140–9; Barnes, 'Water, water everywhere'.

throughout the period when drip irrigation was being adopted. 'Crop per drop' efficiency savings were more than cancelled out by this intensification plus horizontal expansion. As a result, up until hitting a crisis point in 1987 – when irrigation water quotas suddenly had to be cut – groundwater levels in Israel were in continuous decline. Moreover, even though Israel's total water budget has since increased, this has not translated into environmental recovery: Israel's promise of increased flow into the lower Jordan River has not materialised, and neither, to the best of our knowledge, have improved groundwater levels. Indeed, looking to the future, the IWA projects a sustained decade-on-decade rise in agricultural water consumption right through to 2050. In Israel, in short, the supply-side hydraulic mission remains in full swing: drip irrigation and other water-saving technologies have been, and in all likelihood will continue to be, correlated not with decreased but ever-increasing exploitation and use of water resources.[23]

Just as virtual water is no panacea for water scarcity, so the same applies to irrigation efficiency too, our global capitalist order dictating that both are currently structured around logics of capital much more than by concerns about scarcity. More than this, though, water use efficiency is, like virtual water, radically differentiating at the levels of both discourse and practice. Discursively, efficiency talk is premised on a differentiation between the 'efficient' and the 'inefficient', to the extent that it may justify inequalities, or even resource expropriation, on the basis of the former. And in practice, technologies like drip irrigation have parallel effects. The capital required for drip irrigation and other such innovations inevitably bestows the mantle of 'inefficiency' on poor smallholder farmers, legitimating a particular model of agriculture: high-tech export-oriented agribusiness. The scientific targeting of water droplets directly at the roots of designated crops inevitably means that competitors – such as the plant *khubbiza*, which in the West Bank commonly grows as a weed in irrigated fields but has traditionally been an important component of poor Palestinians' diets – must do without.[24] And at a different scale, the widespread adoption of drip irrigation in the upper Negev must have had the effect of reducing irrigation return flow into the Gaza Strip (though by how much we cannot say). Like so many of the other

---

[23] Siegel, *Let There Be Water*, ch. 4; A. Tal, 'Rethinking the sustainability of Israel's irrigation practices in the drylands', *Water Research*, 90 (2016), 389; Zeitoun, *Power and Water*, 133–4; E. Feinerman and D. Yaron, 'Adoption of drip irrigation in cotton: the case of kibbutz cotton-growers in Israel', *Oxford Agrarian Studies*, 18:1 (1990), 51; C. Messerschmid, 'Irrigation efficiency and water resource crisis – lessons from Israel' (unpublished manuscript, 2015); Selby email communication with G. Bromberg (06/08/21); IWA, *Long-Term Master Plan*, 14.

[24] Trottier and Perrier, 'Challenging the coproduction', 88.

technologies encountered in this book, drip irrigation is techno-political – and redistributive – through and through.

Fossil Fuels, Fossil Capital and Water Security

Whereas on both trade and agricultural efficiency, nexus talk is neoliberal in its premises – and is seriously wanting for that – the same does not apply in relation to energy. Instead, the standard central motif of discussions on the 'water–energy nexus' is simply that the water and energy sectors, or water and energy security, are interconnected in multiple different ways, water being vital within energy production just as energy supplies are vital to the provision of water.[25] And, of course, at this most generic level we cannot but agree. However, as touched on at the outset of this chapter such allusions to the multi-directional character of nexus relations are ultimately unsatisfactory, leaving unaddressed the question of the relative causal weight or significance of different linkages and factors. Going beyond such emphases, what we need to consider, to put it baldly, is which of water or energy is the dominant variable within the contemporary political ecology of water – as well as how and why. And to our minds the answer is clear.

Consider, to start with, some of the many passing references to fossil fuels and energy supplies over the course of the preceding chapters. In Darfur, north-east Nigeria and Syria alike, the adoption of diesel-powered tubewells from the 1960s onwards intensified groundwater abstraction and opened up new agricultural frontiers – to the extent that, as long ago as 1988, access to fuel had become the biggest constraint on irrigation expansion in Darfur. In Syria, particularly high fuel subsidies drove equally exceptional levels of groundwater mining, while their sudden withdrawal in 2008 was perhaps the principal immediate trigger of rural out-migration during the run-up to the civil war. In Sudan, similarly, the 1970s Breadbasket Strategy, which was underpinned by subsidised fuel, collapsed as subsidies were cut in the face of high international oil prices and spiralling foreign debt.[26] In the West Bank, household diesel generators are crucial to filling the rooftop water tanks needed for coping with supply cuts. And from Syria to South Sudan and Gaza, the wartime destruction of energy supply infrastructures has repeatedly plunged hundreds of thousands if not millions of people into water insecurity, as wells,

---

[25] WEF, *Water Security*, ch. 2; K. Hussey and J. Pittock, 'The energy-water nexus: managing the links between energy and water for a sustainable future', *Ecology and Society*, 17:1 (2012), 31.

[26] Jebel Marra Rural Development Project, 'Irrigation Survey 1988'; De Châtel, 'The role of drought', 526; D'Silva, *Sudan*, 26; Brown, *Public Debt*, 120–3.

distribution networks and wastewater plants find themselves without the power needed to function properly.

To these examples others may be added. Coal, as put to use in steamboats in particular, provided the crucial energetic underpinning for the imperial conquests and racialised colonial divisions analysed in Chapter 4. Electric power, most usually from coal and gas, is crucial to the functioning of modern hydraulic systems: Israel's water sector, to give but one example, accounts for around 10 per cent of the country's total electricity consumption, mostly for pumping water uphill through the National Water Carrier and for desalination. At the other end of the development spectrum, in South Sudan only 12 per cent of urban areas and less than 4 percent of rural ones had electricity in 2012, mostly from small diesel generators – with all-too-obvious consequences for sanitation and water supplies. Indeed, lest there be any doubt over the extreme dependency of modern water sectors on constant flows of energy, consider the situation following Hamas' takeover of the Gaza Strip, as recounted by the Centre on Housing Rights and Evictions:

In the first week of May 2008, all 135 [domestic] water wells in Gaza were out of fuel. Twenty ceased to operate. The remaining 115 were relying on intermittent electricity supplies and are therefore non-functional for around eight hours each day. By June 2008, 15% of Gaza's population had access to water for 4–6 hours per week, 25% had access to water every four days and 60% had access to water every other day. Only 10 out of every 37 sewage pumping stations were functioning. All three waste-water treatment plants had run out of fuel and were totally dependent on the intermittent electricity supply. Most water and wastewater vehicles had stopped operating. Around 70% of Gaza's agricultural wells were running out of fuel supplies.

Just as fossil fuels came to be the 'prime movers' of nineteenth-century industrial capitalism – valued for their reliability, mobility and fungibility, even when dearer than non-fossil alternatives – so today fossil fuel energy is still the prime mover of our global capitalist order, including its systems of water and sanitation management and provision and attendant patterns of water (in)security. Overall, modern water systems are much more dependent on fossil fuel energy than vice versa.[27]

---

[27] D. R. Headrick, *The Tools of Empire: Technology and European Imperialism in the Nineteenth Century* (Oxford University Press, 1981); E. Spiritos and C. Lipchin, 'Desalination in Israel', in N. Becker (ed.), *Water Policy in Israel: Context, Issues and Options* (Springer, 2013), 106–7; Ministry of Environment and UNEP, *South Sudan: First State of Environment and Outlook Report* (2018), 230–2; Centre on Housing Rights and Evictions, *Hostage to Politics*, 3; Malm, Fossil Capital; A. Siddiqi and L. D. Anadon, 'The water–energy nexus in Middle East and North Africa', *Energy Policy*, 39:8 (2011), 4529–40.

Yet the links between hydrocarbons and water security are not limited to such direct material transformations, for in addition to this is the extensive use of revenues from oil and gas exports to transform, mobilise and in the process often decimate natural waterscapes and natural water resources. Nigeria, Syria and Sudan all provide cases in point. Oil production began in Nigeria in 1957, and by the early 1970s it accounted for over 90 per cent of all export earnings, rising to 96 per cent in 1980 following the price shock caused by the start of the Iran–Iraq War. It was these enormous oil revenues which enabled the major hydraulic and irrigation developments detailed in Chapter 6, including the Baga Polder and South Chad Irrigation projects, the numerous dams along the Komadougou-Yobe River and its tributaries, and the country's expanded agricultural subsidy regime. North-east Nigeria's transformation into a breadbasket region, in sum, can be read as a conscious strategy of 'turning oil rents into cereals'. At the same time, oil revenues have also had broader political–economic consequences, facilitating elite enrichment, militarisation and the country's notorious levels of corruption and, as a result, contributing to violent conflict – whether directly, as in the Niger Delta, or indirectly, as in the north-east. There are, in short, clear links between reduced river flow into Lake Chad (and the regional Lake Chad crisis) and the political economy of oil revenues.[28]

Sudan presents an even more twisted tale, shaped as its recent history has been by a combination of domestic and regional oil revenues. Sudan's Breadbasket Strategy was, to start with, financed in large part thanks to oil, with much of the financing for it coming from Arab Gulf investors, linked to the broader post-1970 reorientation of Sudan's political economy towards the booming oil economies of the Middle East. Sudan then became an oil producer itself in 1999, such that over the decade that followed oil accounted for an average 85 per cent of total exports and nearly 50 per cent of government receipts. Gulf (and Chinese) capital in turn poured in to take advantage of this new oil boom. And together these developments fundamentally reconfigured Sudan's hydro-politics, as witnessed by the establishment of the powerful DIU, the construction of Merowe and other large dams, the new land leases and irrigation schemes

[28] Okolie, 'Oil rents', 201–4; A. T. Salau, 'River basin planning as a strategy for rural development in Nigeria', *Journal of Rural Studies*, 2:4 (1986), 321–35; Adams, 'Large scale irrigation', 297–8; Ngaressem and Magrin, 'Investissements publics et accès aux services: un Lac sous équipé?', in Lemoalle and Magrin, *Développement du Lac Tchad*, 227; Watts, 'Frontiers'; K. Nwajiaku-Dahou, 'The political economy of oil and "rebellion" in Nigeria's Niger Delta', *Review of African Political Economy*, 39:132 (2012), 295–313; C. I. Obi, 'Oil extraction, dispossession, resistance, and conflict in Nigeria's oil-rich Niger Delta', *Canadian Journal of Development Studies*, 30: 1–2 (2010), 219–36; Mustapha, 'Understanding Boko Haram', 171–6; Zhu et al., 'Relative contribution'.

and the associated displacement and dispossession detailed in Chapter 5. China – the key player in Sudan's oil industry and the destination for nearly 65 per cent of its oil exports – became central to this, with the state-owned Sinohydro company leading construction of the Merowe and Roseires projects and other Chinese companies providing support. Companies owned by the ruling National Congress Party and Sudan's army and intelligence service in turn became standard recipients of hydro-construction and agricultural development contracts, both party and military having become key commercial actors within Sudan's economy. As in Nigeria, hydraulic development thus became an arena for petrodollar recycling, for the expansion of crony capitalist networks and for the extraversion – the capture and hollowing out of government by private interests – of the Sudanese state.[29]

Syria's hydraulic and agrarian revolution under Hafez al-Assad was likewise only made possible by international rents. Initially, these were geopolitical in origin as much as oil-related, arising from the Syrian regime's military–political ties with the Soviet bloc and, later, the Gulf; following the 1973 Yom Kippur War and oil shock, this external aid accounted for over 50 per cent of Syrian government income. Thereafter, however, oil took over. For although Syria never came anywhere near the first rank of international producers, oil accounted, at peak, for over 75 per cent of its exports and just under 60 per cent of government revenues. It was these revenues, in combination with the regime's agrarian-socialist ideological commitments, that underpinned the country's rapid hydraulic and agricultural development under al-Assad – funding all the dam building, input subsidies, loans, price controls and debt write-offs – as well as the extreme degradation of its water resources. By no coincidence, the rise in production of wheat, Syria's key strategic crop, closely followed, and mirrored, rising oil exports and revenues.[30]

Conversely, Syria's pre–civil war agrarian crisis was linked to a steep decline, even collapse, in the hydrocarbon foundations of its political economy. From 2002 onwards, domestic oil production dropped rapidly and, with domestic consumption rising, exports fell even more sharply. In turn, oil's contribution to Syrian export earnings and government

---

[29] Brown, *Public Debt*, 110; World Bank, *Sudan Country Economic Memorandum: Realizing the Potential for Diversified Development* (2015), 125–34; Verhoeven, *Water, Civilisation and Power*, 10, 116–50, 168–82, 192, 211–15; Verhoeven, 'African dam building'.

[30] J. Leca, 'Social structure and political stability: comparative evidence from the Algerian, Syrian and Iraqi cases', in G. Luciani (ed.), *The Arab State* (Routledge, 1990), 170; D. Carr, 'Capital flows and development in Syria', *Middle East Journal*, 34:4 (1980), 455–67; IMF, *Syrian Arab Republic: IMF Country Report No. 09/55* (2009); Selby, 'Climate change and the Syrian civil war, part II', 268–9.

revenues dropped by more than two-thirds between 2001 and 2010. In 2007, for the first time since its oil boom, Syria became a net importer of petroleum products by value. This inevitably had wide-reaching consequences, including necessitating the dramatic 2008 fuel subsidy cuts discussed above, which were equivalent to 7 per cent of GDP. As oil rents collapsed and state subsidies dried up, so too did strategic crop production. Syria's brief foray as a net wheat exporter ended in 2008, and while the immediate cause of this was the 2007/8 drought, it is also no coincidence that it occurred just one year after the country stopped being a net oil exporter. Indeed, in long-term prospectus the collapse of Syria's wheat production and exports is best attributed, not to drought, but to the country's declining oil income and the unsustainability of its oil-dependent agricultural development strategy. Syria's pre–civil war agrarian crisis and the mass migration from rural to peri-urban areas occurred when the props which had artificially maintained an over-extended agricultural production system – oil export rents, a pro-agrarian ideology and their associated price controls – were suddenly and decisively removed.[31]

Generalising from these cases, it is clear that there exists a widespread tendency for fossil fuel rents to be mobilised to drive hydraulic development, at least within arid and semi-arid post-colonial states. From Iran and Saudi Arabia to Libya and Algeria, the pattern has been for these rents to be used to finance hydraulic mega-projects, to subsidise groundwater depletion and to enable wholly unsustainable forms of agricultural development. Between 1980 and 2005, Saudi Arabia spent an estimated US$85 billion, 18 per cent of its oil revenues, producing wheat from fossil aquifers. Iran so subsidises water and agriculture that it has become, in gross terms, the second biggest depleter of groundwater resources worldwide, with only India ranking above it (which is astonishing given its size). Fossil fuels thus make for a doubly violent extractivism – first through the landscape destruction, pollution, social dislocation and often racialised labour exploitation directly associated with their production; and second through those ancillary forms of extractivism, including of water resources, that the vast revenues from them enable. In turn, fossil fuels also make for multiple waves of resource export – first directly, of the fossil fuels themselves; and second of other resources, water resources included. Exporting their long-stored reserves of fossilised carbon and using the proceeds therefrom to exploit, commodify and export their often limited reserves of water, the aforementioned hydrocarbon producers are thus 'nature-exporting societies' par excellence. Conversely,

---

[31] IMF, *Syrian Arab Republic: No. 09/55*; IMF, *Syrian Arab Republic: IMF Country Report No. 10/86* (2009), 10, 20; Selby, 'Climate change and the Syrian civil war, part II', 269.

Europe is not only the biggest net virtual water importing region, but with East Asia, the biggest importer of fossil fuels – and arguably the number one 'nature-importing' region in the world.[32]

Implicit within the above is that, contrary to periodic claims that water is 'the new oil' – a crucial source of power and looming cause of conflict – the truth is rather different: that within the context of a highly energy-dependent capitalist global order, oil and other fossil fuels are much more economically and politically important than water, with much the bigger effects on patterns of development, state-building and, in turn, conflict. To say this is not to resort to resource determinism. Contrary to 'rentier state' theory, there is no singular model of how hydrocarbon rents affect state-building and development: Russia, Nigeria and Qatar – let alone the 'carbon democracies' of the global North – all vary hugely from one another in ways that reflect specific social, international and historical trajectories and which cannot be read from their resources alone. Moreover, contrary to widespread claims on the left, Western interventions such as the United States' 2003 invasion of Iraq should not be read as essentially about oil; to argue thus would be to apply a fallacious resource determinism to oil, while simultaneously resisting it in relation to water.[33] That said, the contemporary political ecology of fossil fuels is quite different from that of water, fossil fuels being the 'prime movers' of mass industrial production, mass consumption and exchange, being generators of far higher rents and profits, and in turn having far weightier and more wide-ranging political and economic consequences, including facilitating elite enrichment, militarisation and the hyper-exploitation of water resources. At present, water is very much the secondary and dependent variable within the oil–water relationship.

Capital into Water, Water into Capital

We have already ventured, in the previous section, into discussion of our fourth and final nexus, the relationship between water and capital; a few quick additional notes are required here. As with energy, there are two

---

[32] T. C. Jones, *Desert Kingdom: How Oil and Water Forged Modern Saudi Arabia* (Harvard University Press, 2010), 230; Dalin et al., 'Groundwater depletion', 701; F. Coronil, *The Magical State: Nature, Money, and Modernity in Venezuela* (University of Chicago Press, 1997), 7.

[33] J. Selby, 'Oil and water: the contrasting anatomies of resource conflicts', *Government and Opposition*, 40:2 (2005), 200–24; H. Beblawi and G. Luciani (eds.), *The Rentier State* (Routledge, 1987); C. Hoffmann, 'Beyond the resource curse and pipeline conspiracies: energy as a social relation in the Middle East', *Energy Research and Social Science*, 41 (2018), 39–47; T. Mitchell, *Carbon Democracy: Political Power in the Age of Oil* (Verso, 2011); R. Vitalis, *Oilcraft: The Myths of Scarcity and Security That Haunt US Energy Policy* (Stanford University Press, 2020).

sides to the water–capital nexus. On the one hand, and most obviously, patterns of water development and water (in)security are significantly shaped by access to finance, it being for this reason that 'water flows uphill towards money'. We need not repeat the evidence of oil export-driven hydraulic development or British, Soviet and Chinese-financed dam building already discussed, since there are other examples aplenty. Thus Israel's National Water Carrier, launched in 1952, was in large part financed via the country's 1952 reparations agreement with Germany and other investments by the international Zionist movement (both Mekorot and Israel's water planning company, Tahal, were established and initially part-owned by the Jewish Agency and JNF). Turkey's floating northern Cyprus supply pipeline was constructed by a private Turkish–Malaysian consortium but, as per normal in Turkey, financed through access to cheap loans from the country's many state-owned banks. And the GERD, denied international development bank financing given Egyptian and Arab League opposition, has been funded mostly through specially created GERD bonds, which Ethiopian citizens and the diaspora have been pressured to purchase.[34] Whether the source be petrodollar recycling, domestic taxation, gift aid or financial markets, both major hydraulic developments and water-related subsidies always depend upon the availability of capital, and thus always in one way or another reflect patterns of economic and political power.

Yet the converse is also true, namely that water is routinely turned into money. Again, we need not repeat ourselves: the transformation of abundant waters into agricultural export commodities and profits has been a recurring theme of the foregoing pages. But this is not the only way in which water can be transmuted into gold. In addition, access to political authority and power has – within the context of a capitalist system where the state has powers way beyond those of classical liberal imagination – become key to the generation of water-related profits. This applies, for instance, to access to agricultural land concessions and subsidies within countries like Sudan. It applies to the pervasive corruption that governs local access to water supplies in contexts like pre-war Syria. It applies, worldwide, to major water consultancy and hydro-engineering contracts, where profits are regularly greased by clientelism – as, for instance, in the construction of the Turkey–northern Cyprus pipeline – and legacies of colonial rule. It applies to urban water supply provision, which, since the 1990s, has provided rich pickings for multinational water companies like

---

[34] B. Özgür, 'A water supply project in Northern Cyprus and Turkish clientelism', *DuvarEnglish* (22/06/2020); Abtew and Dessu, *Grand Ethiopian Renaissance Dam*, 161–9; D. Herzog, Contested Waterscapes: Constructing Israel's National Water Carrier (PhD thesis, New York University, 2019), 90, 157, 161–70.

Suez and others such as Bechtel, sometimes directly causing conflict. And it applies similarly at the municipal level, where the contracting of water supply services to local capitalists is perfect territory for monopoly profits. Indeed, even the WEF acknowledges this, and is unabashed about it, using nothing less than its landmark *Water Security* report to highlight the advantages of 'natural monopolies' over water – such as the arrangement made in 1799 by Alexander Hamilton, then US treasury secretary, and Aaron Burr, US vice president, to become monopoly providers of New York City's water supply, which generated so much capital that they were able to launch a new bank on the back of it, what is now JPMorgan Chase.[35]

In an era of 'disaster capitalism' – a time when disasters, conflicts and insecurities are much less threats to prevailing orders than opportunities for capital – supposed water scarcity and supply crises have also become an important dimension of the water–money nexus. In such cases, profit-making opportunities typically rest not only upon monopoly agreements with the state, but also upon 'crisis' discourses – narratives of existing 'droughts' and looming 'scarcities' – to legitimate them. Indeed, this dual tendency is widespread, underpinning many of the forms of water-related profit-making just discussed. Among our cases, however, nothing illustrates it better than the desalination industry. Within Israel and Cyprus alike, misleading policy and media narratives of drought-induced crisis have, since the early 2000s, repeatedly been rolled out to legitimate major investments in seawater desalination. In both cases, in turn, the state has entered into concession agreements for the construction and operation of new plants which, in addition to manufacturing over-capacity, have committed it to purchasing set quotas of desalinated water (or else paying compensation) even when rains are good and conventional water resources are sufficient, thus guaranteeing desalination industry returns. Water price increases have followed, as the state passes on concession costs to consumers; water conservation initiatives have collapsed, as water flows aplenty; and domestic water provision has effectively been privatised and turned over to major corporate interests. Illustratively, the major player within both Israel and Cyprus's desalination sectors is IDE Technologies, which is jointly owned by the Delek Group and Alfa Partners (and previously Israel Chemicals). Delek is a conglomerate owned by leading Israeli magnate Yitzhak Tshuva, which, alongside Texas-based Noble Energy, holds the private monopoly on production from Israel's new natural gas fields – the gas from which is to be used in

---

[35] Özgür, 'A water supply project'; O. Olivera, *Cochabamba! Water War in Bolivia* (South End Press, 2004); WEF, *Water Security*, 172–3.

IDE's desalination plants (plus Delek is also involved in gas production in Cyprus). Israel Chemicals, for its part, is owned by the Ofer family, at one time the second biggest corporate group in Israel and infamous for its use of political connections to obtain state concessions and companies at knock-down prices. Of course, water is not within the grand scheme of things that economically important, as already discussed. However, within the practice of 'accumulation by desalination', constructed 'scarcities' allied with tight state–corporate relations and ensuing monopoly deals can turn water into a highly profitable commodity – at environmental and consumer expense. More broadly, bringing solutions to a 'water-starved world' has become an important international sales pitch, promoted especially by Israeli corporate and political leaders and their media allies.[36] Today, water 'crisis' has become no less than a major business opportunity and element of capitalist reproduction.

Capitalism over Climate

This chapter has sought to explore the four key nexus relations in which water is implicated today – its nexuses with trade, agriculture, fossil energy and capital – and through this to show how patterns of water security and insecurity worldwide are shaped by interlocking 'circulations and transformations' on a global scale. Its central target has been liberal nexus discourse, according to which the virtual water trade is a potential panacea, high-tech irrigation investments are the same, business and capital are unalloyed goods and energy is simply in an important two-way relationship with water. The reality, we have argued, is quite different. Thus the international political ecology of virtual water involves, above all, its transfer not from water-rich to water-poor countries, but from poor to rich ones, including from countries like Syria where water resources have become highly degraded as a result. Technologies like drip

---

[36] N. Klein, *The Shock Doctrine: The Rise of Disaster Capitalism* (Knopf, 2007); S. Kassirer, Depoliticisation of Desalination Discourse: Media Analysis of Hydro-Policies for (Climate) Resilience in Israel, 2001–2018 (PhD thesis, University of Brighton, 2020); A. Bar-Eli, 'With rainfall aplenty, Israel to rely less on expensive desalinated water this year', *Ha'aretz* (03/01/2014); P. Darr, 'Save, don't desalinate', *Ha'aretz* (05/02/2010); Z. Rinat, 'Israelis asked to cut down shower time in bid to combat five-year drought', *Ha'aretz* (22/05/2018); B. Browne, 'Water down the drain for Paphos desalination plant', *Cyprus Mail* (07/11/2013); S. Surkes, 'Public losing money, again, as gas kingpin Yikzhak Tshuva teeters on abyss', *Times of Israel* (06/04/2020); A. Rabinovitch, 'Leviathan gas for new Israeli desalination plant', *Offshore Engineer* (27/05/2020); J. Nitzan and S. Bichler, *The Global Political Economy of Israel* (Pluto, 2002), 86; I. Abudi (dir.), *The Shakshuka System* (2008, in Hebrew); E. Swyngedouw and J. Williams, 'From Spain's hydro-deadlock to the desalination fix', *Water International*, 41:1 (2016), 63–5; Siegel, *Let There Be Water*, chs. 8, 10.

irrigation, far from saving water, instead cut costs and spur ever-increasing resource demand and exploitation. Hydrocarbon energy, far from being locked in a mutual embrace with water, is in fact its dominant partner, with the availability of fossil fuels and the rents therefrom being foundational to modern capitalist hydraulic systems and projects and their attendant water (in)securities. And capital, far from being an unalloyed good – the sine qua non of modern water security – is decidedly double edged, simultaneously a lover of monopoly arrangements, motor of dispossession and profiteer from constructed fears of crisis.

It is also worth stressing that these four sets of nexus relations are in structured, dynamic and dialectical interaction. At the extreme, in Israel, capital availability, technological innovation and close state–finance coordination have enabled near-continuous hydraulic and agricultural development, in which money is recycled into hydraulic engineering projects and these in turn spur ever-increasing water supply, agricultural production and agricultural exports (and require ever-increasing energy inputs too), which at each stage can be recycled into more money: what we may summarise, using a variation of Marx's general formula, as Money–Water–Money, M–W–M. Here what we encounter is a continual ratcheting up of both water supply and demand, in which new desalination plants are constructed, the resulting water is quickly apportioned, the spectre of 'drought' is mobilised once again and more desalination is then required.[37] Meanwhile, globally what we encounter is a system of ecologically hierarchical exchange, in which Northern economic and financial power enables huge virtual water imports, while poor Southern producers like Syria and Sudan are compelled to adopt aggressive agricultural export strategies and super-exploit their precious water resources to this end. Suffice to say that the primary limiting factor across all of this is not water itself – but economic power and the availability of capital.

Implicit within the above is that contemporary international circulations of food, energy, industrial commodities and money are so structurally and hierarchically embedded as to cause water insecurities – that, contrary to neo-liberalism, international circulations are part of the problem. Yet it needs also to be noted that temporal disruptions to these circulations are often problematic too. We have already seen, at national and regional scales, that supply disruptions associated with technical failures, electricity cuts, local environmental shocks and war can have significant water vulnerability consequences in ways which always reflect and are mediated by patterns of infrastructural dependency. However,

---

[37] Marx, *Capital*, ch. 4; Rinat, 'Israelis asked to cut down shower time'; 'IDE secures contract for Israel's Sorek B desal plant', *WaterWorld* (29/05/2020).

the same also applies globally, in relation to patterns of global finance and trade. Today, commodity price shocks, oil and gas shortages, blocked shipping lanes, extreme weather events and, not least, health crises can all have repercussions that reverberate globally, compounding existing vulnerabilities as well as creating new ones, including in relation to water. Covid-19's impacts during 2020–1 provide the obvious case in point. From Nigeria to Syria, collapsing oil revenues during spring 2020 – as worldwide lockdowns resulted in unprecedented drops in global oil demand, leading the price of oil futures to go negative for the first time on record – sparked sudden economic and budgetary downturns, leading to fuel subsidy cuts, increased electricity prices and reduced water access. Within these countries as elsewhere, local restrictions on movement introduced to limit Covid-19 transmission impeded local water collection and forced people to turn to more expensive water tanker and other private vendor options. At the same time, especially in humanitarian settings such as IDP and refugee camps, preventive measures contributed to increased water use, with many health agencies advising that 8–10 l/c/d of clean water should be used for handwashing. Water prices rose in many places: in Nigeria they doubled in early 2020 after the introduction of lockdown measures. And these and related pressures placed particular burdens on women and girls, given their disproportionate roles in household water collection and management.[38] Global disruptions – which are inevitable corollaries of a highly integrated capitalist world order – can have any number of ripple effects on water (in)security, though primarily, as here, by compounding existing patterns of inequality and vulnerability.

While we can only speculate, of course, on how these dynamics may shift as the planet continues to warm, we nevertheless wish to highlight four things. First, it bears emphasising that the transformations and

---

[38] P. Davies, 'Nigeria hikes petrol prices as COVID-19 bites budget' *Africanews* (14/09/2020); FAO, *2020 FAO Crop and Food Supply Assessment Mission (CFSAM) to the Republic of the Sudan* (2021), 7, 16–20; iMMAP, *COVID-19 Situation Analysis: Syria* (2021), 46; S. Agada, 'Nigeria's informal water providers: filling in the public gaps', in D. A. McDonald et al. (eds.), *Public Water and Covid-19: Dark Clouds and Silver Linings* (Municipal Service Project et al., 2021), 281–90; J. Stoler et al., 'Household water insecurity will complicate the ongoing COVID-19 response: evidence from 29 sites in 23 low- and middle-income countries', *International Journal of Hygiene and Environmental Health*, 234 (2021), 113715; K. Harrisberg and L. Ndlovu, 'In parched southern Africa, coronavirus spurs action on water supply', *Reuters* (06/05/2020); O. C. Staddon et al., 'Water insecurity compounds the global coronavirus crisis', *Water International*, 45:5 (2020), 416; A. Hoban, 'The impact of COVID-19 on Nigeria's economy', The Borgen Project (20/08/2020); E. A. Adams et al., 'Water, sanitation, and hygiene (WASH) insecurity will exacerbate the toll of COVID-19 on women and girls in low-income countries', *Sustainability: Science, Practice and Policy*, 17:1 (2021), 86–90.

circulations considered in this chapter, and the patterns of water security, insecurity and ecological degradation resulting from them, are in no way caused by climate change. Their fundamental cause is rather the very same structural force which is causing climate change: global capitalism. Thus Hasakah's once-abundant waters fell into crisis, and Sudan's Nuba and Nubian communities were displaced, not because of climate change but for capital accumulation through agricultural exports; conversely, the relative water security which prevails within Israel, or indeed Britain, is a function of food and virtual water imports made possible by economic power; and climate change is at root being driven by capitalism's unquenchable thirst for cheap, fungible and mobile energy resources to feed ever-expanding commodity production, consumption, competition and accumulation. So long as these basic principles and imperatives remain unchanged – or, put simply, as long as our global system remains a capitalist one – then forms of water-related expropriation, dispossession, inequality, vulnerability and degradation such as those detailed in previous chapters will follow. This, we suggest, is irrespective of whether global temperatures are somehow kept to within the Paris Agreement target of 1.5°C. Capitalism currently ranks well above climate as a cause of water insecurity – and while this could well change in future, it will not be because capitalism has suddenly become conducive to water security for all.

Indeed and second, the strong 'disaster capitalist' tendencies of contemporary capitalism are such that climate change may well become an important facilitator, even motor, of capitalist reproduction, deepening water insecurity dynamics still further. This is not just a future possibility. Just as discourses of water 'scarcity' and 'crisis' are already being mobilised to extend corporate and oligarchic power, so by the same token have climate adaptation and mitigation become big business – driving the creation of carbon markets, climate offsetting schemes, adaptation finance, new infrastructure investments and more – with innumerable negative water security implications. There are, for instance, clear links between financial speculation on international carbon markets on the one hand, and the practices of water-related green grabbing discussed at the end of Chapter 6, on the other. Extended discussion of the political economy of 'climate capitalism' is beyond what can be attempted here. But what is clear is that the implications of these changes are far from straightforwardly positive.[39]

---

[39] Klein, *Shock Doctrine*; R. Fletcher, 'Capitalizing on chaos: climate change and disaster capitalism', *Ephemera*, 12:1/2 (2021), 97–112; Dunlap and Fairhead, 'The militarisation

Third, just as Covid-19 and the responses to it have affected politics, economies and livelihoods worldwide, including patterns of water (in)security, so by the same token are other shocks likely to do similar in future, including shocks associated with climate change. For, while eco-determinist narratives of local drought-induced crises are, as explored in Chapter 3, demonstrably misleading, the same does not apply to how droughts or other environmental shocks may reverberate transnationally. The global supply of food staples is dependent on production in just a narrow band of countries, making global supplies highly vulnerable to droughts and other such weather-related shocks. In turn, there exist distinct risks of severe compound and cascading effects, in which local weather-related disruptions are amplified by, say, rising fuel prices, and amplified still further by futures markets, leading to significant price rises on international agricultural commodity markets and, within some countries at least, physical shortages – or in which the relative contributions of poor rains, unseasonal frosts, health emergencies, fuel price rises, financial speculation and state mismanagement are contested and unclear. Indeed, globally reverberating compound shocks like this and their internationally complex and uneven consequences will arguably be one of the most oft-recurring faces of our climate change–affected future.[40]

Last, it is evident that the future landscape of water (in)security will be significantly shaped by future patterns of fossil fuel extraction and use, albeit in ways which are contradictory and in aggregate far from clear. Just as fossil fuels are a crucial factor in contemporary hydro-social relations, as illustrated extensively above, so it follows that if and when there is a meaningful energy transition – if and when oil, gas and coal start being left, unexploited, in the ground – then the political ecology of water will shift in fundamental ways as well. The widespread use of solar and wind power may recast and reduce state power in water provision and generate new-found forms of local autonomy, albeit with some potentially negative consequences (emerging evidence from Afghanistan, Yemen and elsewhere, for instance, suggests that the adoption of photovoltaic pumps in conflict zones may be spurring ever-more problematic levels of groundwater mining). Alternatively, or in different contexts, transitions to low-carbon energy sources may bring

---

and marketisation of nature'; P. Newell and M. Paterson, *Climate Capitalism: Global Warming and the Transformation of the Global Economy* (Cambridge University Press, 2010).

[40] G. Pescaroli and D. Alexander, 'Understanding compound, interconnected, interacting and cascading risks: a holistic framework', *Risk Analysis*, 38:11 (2018), 2245–57; C. Raymond et al., 'Understanding and managing connected extreme events', *Nature Climate Change*, 10:7 (2020), 611–21; K. M. Adams et al., *Climate Change, Trade and Global Food Security: A Global Assessment of Transboundary Climate Risks in Agricultural Commodity Flows* (Stockholm Environment Institute, 2021).

increased dependency on large-scale and even transcontinental electricity networks, thus introducing new layers of infrastructural dependency and disruption potential to water provision. At the same time, reduced export revenues for oil and gas producer states may bring to an end the pattern of unsustainable hydrocarbon-driven virtual water exporting detailed above. These and any number of other water-related changes may follow from a shift away from fossil fuels. And this is without even touching on the broader 'power shifts' and attendant conflicts that will be necessary for low-carbon transition – a subject to which we return in Chapter 10.[41]

For the present, however, most of this seems a distant prospect. Globally, fossil fuel demand continues to rise. Renewables are adding to national energy mixes, but not yet at the expense of oil or gas. Notwithstanding the many 'net zero' declarations, global mitigation efforts are not yet being 'ratcheted' significantly upwards, as the Paris Agreement framework requires. And there has been little change so far in the behaviour of oil or gas producer states. The three great powers, in their different ways, each aspire to using fossil fuels for geopolitical and geoeconomic power projection. The fossil fuel–military power nexus remains tight, with the US military, most notably, being the single biggest consumer of hydrocarbons on the planet. New entrants like Cyprus and Israel are expanding gas production apace. And questions of water resource protection and security remain distinctly secondary. For all the noise about the need, for instance, to reverse Lake Chad's 'decline', the bigger untold story is that all four of its basin states have explored for and produced oil around it. Just two years on from nominating Lake Chad for world heritage status, the Chadian regime withdrew its application to UNESCO after signing production-sharing agreements with several oil companies for renewed production around the lake, despite warnings of a new Niger Delta in the making.[42] Whatever the future, water is clearly subordinate to fossil fuels within the political ecology of early twenty-first-century global capitalism.

---

[41] D. Mansfield, 'On the frontiers of development: illicit poppy and the transformation of the deserts of southwest Afghanistan', *Journal of Illicit Economies and Development*, 1:3 (2019), 330–45; Conflict and Environment Observatory, *Groundwater Depletion Clouds Yemen's Solar Revolution* (2021); P. Newell, *Power Shift: The Global Political Economy of Energy Transitions* (Cambridge University Press, 2021).

[42] Selby, 'The Trump presidency'; I. Stoddard et al., 'Three decades of climate mitigation: why haven't we bent the global emissions curve?', *Annual Review of Environment and Resources*, 46:1 (2021), 653–89; Belcher et al., 'Hidden carbon costs'; J. T. Omenma, 'Untold story of Boko Haram insurgency: the Lake Chad oil and gas connection', *Politics and Religion*, 13 (2020), 180–213; I. Anyaogu, 'NNPC moves to resume oil drilling in Lake Chad', *Business Day* (15/05/2017); M. Gouby, 'Chad halts lake's world heritage status request over oil exploration', *Guardian* (24/09/2020).

10 Conclusions

Reimagining Environmental Security

In his 1978 work *Orientalism*, Edward Said famously argued that modern 'Western' knowledge of the Middle East is partly scientific, rooted in a raft of post-Enlightenment procedures for collecting, validating and representing the truth of the region and its peoples, and partly imaginative, reflective of assorted fears, fantasies, stereotypes, self-images and idealisations. These two faces of Orientalism, Said argued, were and remain not just interconnected but indivisible, coursing together through even the greatest works of European culture, philosophy and social science. All such works are at least implicitly ordered, according to Said, around 'imaginative geographies' of 'the East', 'the West' and the supposed differences and hierarchies between them. When it comes to knowledge of the Middle East, the scholarly, the imaginative and also, of course, the political are bound together and mutually constitutive, inextricably joined at the hip.[1]

It has been an underlying premise of this book that something very similar applies to mainstream contemporary international or 'Western' discourse on water, climate and more broadly environmental security; and, also in line with Said, that to properly understand water and climate (in)security requires an approach which is simultaneously attentive to empirics, imaginative in its frameworks and comparisons, and political in intent. Thus even the most rigorous works of quantitative scholarship on the links between water scarcity or climatic variations and violent conflict are indebted – though not reducible – to inherited Eurocentric myths about colonised peoples and their environments, as discussed in Chapter 4. The entire quantitative research programme on climate change and conflict rests, barely acknowledged, on analogical reasoning, as discussed in Chapter 1. The global climate crisis is, as some of the more reflective works on it have emphasised, 'among other things, a crisis of the

---

[1] Said, *Orientalism*, Introduction, ch. 1.2.

imagination'. Both the 'imagined communities' analysed in the foregoing pages, and water systems configured to sustain them, only exist because they were imagined, fabricated, dreamed into being. And imaginative premises and especially geographies are omnipresent within water and climate security debates.[2] This is not to suggest at all that evidence counts for nothing. However, more fine-grained data sets alone will not, we believe, radically improve our understanding of the conflict, security or vulnerability implications of water crises and anthropogenic climate change. What is needed instead, we suggest – and what we have sought to develop through this book – is a concretely grounded reimagining of environmental security.

This concluding chapter summarises and synthesises the results of this endeavour on four levels: with regard to our five cases; with regard to what these cases, and the similarities and differences between them, suggest about the relations between water and (in)security; in terms of the broader significance of our analysis for understanding ecological politics and the study thereof; and in terms of what, by extension, this might tell us about the likely future conflict and security implications of climate change. Through the course of the chapter we thus move from the particular to the general, from the concrete to the abstract and from the past to the future. And we end with a brief nod towards alternatives – and by inverting the problem with which we began.

Our Divided Environments Revisited

Let us first consider our five cases in turn. Sudan is regularly imagined and represented in mainstream international policy and academic discourse as a land of climate and water scarcity–induced insecurity, instability and conflict – between farmers and herders, over the waters of the Nile, in Darfur. Yet the chapters above suggest very much to the contrary. There is no solid evidence that human-induced climate change has been a meaningful factor in Sudan's recent conflicts. The Sahara is not expanding. The Sahel has in recent decades witnessed greening. And both Sudan and South Sudan have relatively abundant water resources, more than any of the other 'divided environments' considered here. Sudan has not

---

[2] Mann and Wainwright, *Climate Leviathan*, 197; Ghosh, *Great Derangement*, 9; B. Korf, 'The imaginative geographies of climate wars', *Procedia: Social and Behavioral Sciences*, 14 (2011), 35–9; K. Yusoff and J. Gabrys, 'Climate change and the imagination', *WIREs Climate Change*, 2:4 (2011), 516–34; B. Anderson, *Imagined Communities: Reflections on the Origin and Spread of Nationalism* (Verso, 1983); H. Boast, *Hydrofictions: Water, Power, and Politics in Israeli and Palestinian Literature* (Edinburgh University Press, 2020).

even used up its quota under the 1959 Nile Waters Agreement. Indeed, if there is one physical hydrological property which has repeatedly been a factor in conflict and insecurity in Sudan, it is not scarcity but this relative abundance of water resources. Thus the Sudan was colonised in large measure thanks to British interests in securing the plentiful waters of the Nile: there would have been no slaughter at Omdurman without them. South Kordofan became such a focus of post-independence agrarian development and subsequently a key civil war battleground and site of genocidal actions against the local Nuba population on the back of the region's rich rain-fed soils. The 2003–5 war in Darfur took place following a period of frontier expansion, groundwater development and in-migration, not scarcity-induced collapse. Moreover, both mega-dam construction and the country's various projects of large-scale irrigation development have repeatedly been accompanied by large-scale displacement and conflict. Across each of these different spaces and forms of water-related strife, interests in exploiting abundant resources – and turning them into commodities to be exported, as virtual water, on international markets – have been the central, recurring motifs.

British colonial rule bequeathed Sudan a bifurcated state and bifurcated political ecology, divided between a modernising core around Khartoum and various dependent, subordinate and locally governed peripheries. It has been this structure and its reproduction – allied to internal resource geographies, plus a deeply unequal global economic order that positions Sudan as a producer of primary commodities – which lies at the root of the major forms and patterns of water insecurity found across the country. Thus within its core riverine regions, peoples considered obstacles to hydraulic development have been removed, their communities destroyed. Within the vast periphery which is now South Sudan, deep poverty as well as war have left most people with extremely limited access to safe water, despite relatively good rains and abundant water resources; ten years of South Sudanese independence have barely changed this at all. Both here and across Sudan's 'semi-periphery', British-invented social hierarchies and subsequent challenges to them have led to bitter 'tribal' conflicts over access to land and water resources (which in Western discourse are often, however, misrepresented as conflicts between 'farmers' and 'herders'). Across this semi-periphery, projects of incorporation and modernisation, launched within the contexts of spiralling national debt and economic liberalisation, have involved the violent expropriation of water resources by riverine elites and their allies. And in Darfur and elsewhere, Khartoum-led counter-insurgency warfare has had dire water, sanitation and associated health consequences. Overall, water-related insecurity has been and remains an endemic feature of life in the Sudans, and control and appropriation of water a key motivation for and cause of violent conflict. Water, in sum, has been central to the political ecology and conflict history of modern Sudan – not, however, primarily because of its scarcity, but because of its relative abundance within the twin contexts of deep poverty and a high level of economic dependence upon agricultural production and exports.

Lake Chad, likewise, is often characterised today as a prima facie site of climate-induced conflict and fragility, home to both a disappearing lake and, by no coincidence, a vicious sectarian movement and regional humanitarian and security crisis. Yet the evidence on these links is derisory – far weaker, indeed, than in the other cases investigated here. Lake Chad is not currently shrinking, let along disappearing. Its extent has always fluctuated. There is no evidence of anthropogenic climate change impacts on lake size. There is no evidence of scarcity-induced outmigration from the lake region. And the supposed links between a changing lake and the rise of Boko Haram are entirely speculative, being not only without evidential basis but contrary to the fact that the lake has, over recent decades, been relatively stable. By contrast, there is plentiful evidence that the Lake Chad basin has been decisively shaped and transformed by local human agency – by its history of colonial and post-colonial marginalisation; by catch-up economic development, especially dam building and irrigated wheat production; and most recently by war-induced insecurity and displacement. Thus recent irrigation development has resulted in a steep decline in river flows and limited the lake's recovery from the droughts of the 1970s–1980s, despite relatively good rains. This irrigation development has been accompanied, simultaneously, by a horizontal expansion of agriculture, increasingly concentrated land ownership, and displacement and conflict – which, in combination with its wider frontier attributes and its history of marginalisation, have no doubt been key factors behind the Lake Chad crisis. And this security and humanitarian crisis has in turn transformed local water geographies, creating new war-related water insecurities and, perhaps also, reducing irrigation and increasing groundwater and river flows into the lake itself.

Syria, too, is often represented as a land of water scarcity- and climate change–induced conflict; and indeed the onset of Syria's terrible civil war was preceded by both a deepening long-term water crisis and a severe multi-year drought which can, with some qualifications, be plausibly linked to climate change. By the early 2000s, Syria's groundwater resources, in particular, were so depleted that production of strategic crops was in secular decline, with out-migration from the worst affected rural areas occurring as a result. This was not, however, primary because of drought, still less natural water scarcity: the extreme drought of 2006/7–2008/9 struck well after the start of the country's agrarian crisis, and Syria naturally has relatively abundant water resources. What caused the country's turn of the century water crisis, rather, was the model of agrarian-led catch-up development pursued by the Ba'ath party under Hafez Al-Assad, compounded by the shock therapy–style liberalisation policies pursued by his son and heir. Under Ba'ath party rule, Syria became one of the world's leading net exporters of virtual water. Oil rents, subsidies, loans and cheap labour were all mobilised to support the super-exploitation of water resources in the name of 'self-sufficiency' – even if, in actuality, much of this water was exported. From 2000, in addition, a combination of economic liberalisation and declining oil rents meant that the props underpinning this extant system were suddenly withdrawn, leaving the country with both highly depleted water resources and a structural political economic crisis. The subsequent civil war has of course been both a humanitarian catastrophe and generated new forms and degrees of water-related vulnerability and insecurity.

Unlike Syria, Sudan or Lake Chad, Cyprus has never been seriously represented as a site of environmental scarcity conflict – partly, no doubt, beyond whatever good reasons, because such representations are rarely applied to corners of the Earth racialised as White. Moreover, unlike these other cases Cyprus is not home to significant water access, supply or sanitation insecurities, despite possessing a per capita water availability lower than Syria's and despite the problems of water storage posed by its island topography and small scale. The reasons for this are not hard to discern. On the one hand, Cyprus is a large net importer of virtual water, with large tourist and offshore service industries that can readily support the large-scale purchase of virtual water-rich agricultural and industrial commodities from abroad, reducing pressure on local water resources in the process. And, on the other hand, hydraulic infrastructure developments – the building of dams, desalination plants and, most recently, the floating Turkey–northern Cyprus water pipeline – have not only kept ahead of rising demand but given both the Republic of Cyprus and the Turkish-occupied North engineered water supply overcapacity. These water infrastructures, moreover, have been engineered and configured to reflect and reproduce Cyprus' political and military divide. Cyprus may not be a site of serious water insecurities, but it is a divided environment nonetheless.

Israel–Palestine, finally, combines many of the features of these other cases, while also bringing in new ones. Scarcity narratives are often invoked here – with both Israel and the Palestinian territories regularly being characterised as lands of 'absolute water scarcity', and this scarcity often being portrayed as both a cause of conflict and an obstacle to peace.

Yet, at least of Israel, the dominant contemporary representation is rather different from this, being not only that Israel faces extreme physical geographical and hydrological constraints but that it has managed to overcome these through sound management and scientific and technical ingenuity; Israel, as McKenzie Funk summarises this view, has pursued an 'Enlightenment answer to water scarcity', not bowing before nature but learning how to master it.[3] And indeed, such claims are apposite in more ways than one. For, in addition to developing a highly regulated and integrated national water system and being a global innovator in drip irrigation, seawater desalination and wastewater reuse technologies, Israel and with it Palestine also illustrate the darker side of the Enlightenment's bequest – that which follows from Kant's anthropology, Hegel's geography and the racialised differentiation, division and subjugation of the non-European world which followed. That water became so central to Zionist thought and practice was thanks to a combination of Orientalist fantasy and water's practical utility to the twin projects of settling the land and dispossessing the Palestinians. That Israel found itself with such plentiful water resources upon independence was mainly because the large majority of its native Palestinian population, which until 1948 had accounted for half of all water consumption, had been displaced without prospect of return, and their lands expropriated. That Israel has ever since been able to retain control of the Jordan River and regional groundwater resources is largely because of the 1967 war and subsequent occupations (of both the West Bank and Golan Heights) as well as, later, Israeli–Palestinian peace terms which essentially reproduced the political and hydro-political status quo that these military means had established. Moreover, that West Bank Palestinians continue today to face such severe water access constraints and Gaza Palestinians a combined water quality and supply crisis, while Israel has a large water surplus, are two sides of the same coin, core ethno-nationalist state-building having proceeded hand in glove with discrimination and exclusion at the frontier. And even with all this, Israel's present-day water surplus would not exist were it not for its food and other imports, Israel being, like Cyprus, a large net importer of virtual water. Perhaps Israel does indeed provide an 'Enlightenment answer to water scarcity'. But if so, this is a very different 'Enlightenment' from that which is conventionally imagined – an issue to which we return more fully below.

---

[3] Funk, *Windfall*, 86.

Ten Theses on Water (In)Security

Before doing so, however, we need to complement our overview of the specific features of each of our cases with a parallel reflection on their commonalities and on what these might tell us about the nature of water (in)security today. The central premise of mainstream public, policy and academic discourse on water security, as illustrated extensively in the first three chapters of this book, is that worsening scarcities in water-poor areas of the world are already important contributors to social vulnerabilities and political instability, and are likely to become even more so as pressures on water resources rise, including those associated with climate change. On this view, it is the brute incompatibilities between physical geographical realities and environmental limits on the one hand, and ever-rising population growth on the other, which establish the basic contours of water-related insecurity – historically, at present, and into the future. Droughts and other environmental shocks, it is typically supposed, often come on top, exacerbating long-standing or latent resource pressures and suddenly compelling local populations to either migrate en masse or fight it out over dwindling supplies. Other 'intervening variables' and 'factors' may also play a role, but it is scarcity – understood in essentially naturalistic terms – which, from this perspective, is the foundational cause of all water-related strife. In Thomas Homer-Dixon's beautifully pithy formulation, it is 'environmental scarcity' which, on this view, 'causes conflict'.[4]

This book has on one level been an extended critique of, and set of counter-arguments to, this oft-challenged but all too persistent vein of reasoning. We can boil down our case to a series of ten points. To start with, it is clear that neither the natural and physical geography of water resources, nor still less climate change, can tell us much about contemporary patterns of water consumption, water supply, water resource degradation, water resource control or water-related vulnerabilities; that in relation to water, neither nature, scarcity nor climate change impacts is destiny. Today, irrigated agriculture can flourish in marginal semi-arid soils and even deep into the desert. Cities that would have been abandoned long ago had they been forced to rely on local water reserves continue to flourish, nonetheless. Countries categorised as suffering from 'severe' or even 'absolute' water scarcity can in practice be lands of water plenty. And some national economies have become close to invulnerable to drought. Conversely, dire shortages of both domestic and irrigation supplies stalk territories with abundant rains. Groundwaters are degraded even – or

---

[4] Homer-Dixon, 'Environmental scarcities', 39; Homer-Dixon, *Environment, Scarcity and Violence*, 93.

perhaps especially – in areas with the richest of endowments. And polluted, untreated and highly saline waters are found in water-rich and water-poor countries alike. Water conveyance lines criss-cross mountains, seas and deserts. Meandering rivers are streamlined if not destroyed. Water circulates, in embedded or virtual form, throughout the global economy. And new water resources and dynamics of consumption are continually being invented, reinvented, produced and performed. Viewed thus, the idea that nature or climate change, and associated 'scarcities', are the appropriate baselines for understanding contemporary water (in)securities is not just anachronistic – it is fundamentally flawed.

Second, though, it needs to be stressed that this does not mean that water issues are unimportant; quite the contrary. The metabolic relations between water and society leave their traces everywhere. They leave them in the thousands of mega-dams which have been key to transforming the world's watercourses – and implementing all manner of socio-technical reforms, in turn – over the last century in particular. They are evident from space, in the form of concrete mega-structures, irrigation geographies, expanding cities, reclaimed coasts and wetlands, drying endorheic lakes, and declining groundwater levels. The metabolism between water and society is evident, at a very different scale, in the chemical composition of today's river flows, with their traces of shit, fertilisers, pesticides, hormones, antibiotics, industrial effluents and more. Water flows through and helps reproduce hierarchies of nation, race, class, region and gender. Sufficiency of clean water is a sine qua non of health, livelihoods and human security. Moreover, contrary to the views of many liberal and constructivist commentators – who, building on their critiques of eco-determinist water wars discourse, often deny the existence of links between water and conflict altogether – water is regularly a reason for or site of conflict, social mobilisation, land grabbing, large-scale violence, displacement and/or destruction, as has been amply demonstrated in the chapters above. In all of these respects, water matters deeply. The central failing of mainstream scarcity discourse on water is not, then, its exaggerations, however flagrant these may often be (limits have been transgressed! war and collapse are imminent!); the more fundamental problem is one of mischaracterisation, misrepresentation and misattribution.

Third, and more positively stated, in today's world water security is not a gift of nature but an essentially political economic and socio-technical accomplishment. Consider the range of socio-technical intermediaries which commonly exist between water resources and their eventual consumption. Dams, reservoirs, canals, pipelines and tubewells, which erase both the natural geographical and the temporal distances between resources and their users, annihilating hydro-social time and space. Irrigation pipes, greenhouses, flood defences, domestic plumbing systems, wastewater collection systems and meters, for controlling and rationalising water use. Water treatment and desalination technologies, through which resources can be created anew. Ample and affordable energy supplies, without which none of the above would operate. Food staple imports, on which all water-secure countries depend. Borders, which limit and define circulations of water, food staples, energy supplies, technologies and people. State institutions, which regulate, police and frequently subsidise water use and development. Expertise, to which all of these other intermediaries are in one way or another indebted. And the availability of capital, the importance of which is almost too obvious to note. It is possession of, or access to, these and other socio-technical intermediaries, much more than the direct possession of plentiful water resources, which is the key to the achievement of contemporary water security. Water security, stated simply, is much less a function of natural availability than of political and economic capacities.

Conversely, and fourth, water insecurity today is less a function of hydrology than of historically constituted patterns of subordination, marginalisation, vulnerability and powerlessness. This is so in two distinct ways. On the one hand, water insecurity is typically rooted in a series of direct 'capacity deficits', in absences or shortfalls of the range of capacities listed above: limited infrastructural development, weak state institutions, high fuel prices and so on. But more fundamentally, water security and insecurity are also mutually constitutive, structurally bound to one another and joined at the hip. Thus water typically flows, following Reisner and Allan, not only towards money and power – but simultaneously, and as a necessary corollary, away from those spaces and social groups that have much less. Globally, water flows, like so much ecologically unequal exchange, overwhelmingly from South to North, including from states and regions that are destroying their water resources in the service of rich states already in receipt of abundant rains. Regionally, water is typically dominated by 'hydro-hegemons', who – thanks to their military, political and economic capabilities and irrespective of riparian position – can monopolise trans-boundary resources at the expense of their less powerful neighbours.[5] And internally, patterns of access to and control of water resources and systems routinely reflect and entrench political and economic hierarchies, along multiple dimensions and at multiple scales.

---

[5] Reisner, *Cadillac Desert*, 12; Allan, *Virtual Water*, 19; Zeitoun and Warner, 'Hydro-hegemony'.

It follows from this, fifth, that political and hydro-political projects such as those considered in the second half of this book are always differentiating and hierarchy-creating or -reinforcing, enhancing the water security of some while directly or indirectly visiting insecurity upon others. Thus the racial categorisations explored in Chapter 4 not only imagined into being a series of peoples, but simultaneously defined some of them as capable and worthy of progressive development and others as not, including in relation to water. The infrastructural works described in Chapter 5, which were all pursued or at least justified as facilitating such development, in turn each had their victims, those deemed irrelevant to the marches of Nation and History. The projects of frontier development and internal colonisation analysed in Chapter 6 can each be understood as oriented, whether as a primary or incidental objective, to increasing some peoples' control of or access to water resources at the expense of others – what may be summarised, following Harvey, as practices of 'accumulation through dispossession'. Modern warfare as explored in Chapter 7 is simultaneously highly destructive and highly productive in its hydro-social consequences, bringing death, displacement and water vulnerability to some while enhancing the water security of others, and thus effectively redistributing or reinforcing hierarchies of water control. Contemporary peace processes, as interrogated in Chapter 8, have above all served to co-opt or delegitimise radical challenges and through that consolidate extant patterns of domination, subordination and control, including in relation to water. Even 'efficiency', as discussed in the previous chapter, has differentiating and distributive consequences. It is evident that across all of these spaces the same basic pattern recurs. Why? Because, like insecurity in general or, to give but one specific analogy, like starvation, water insecurity is a transitive, relational phenomenon, deriving not from there not being enough of the resource in question (water or food) or from capacity deficits, but at root from what we do to, and what we inflict upon, one another. Only in the most rhetorical of senses does it make sense to suppose, following Ban Ki-moon, that '[w]ater security is not an issue of rich or poor, North or South.'[6]

At the most general level, sixth, what underpins all this are the two intersecting master logics of modern world politics: capitalism and the state system. Under capitalism water security and insecurity alike become mediated by a welter of machines and technologies. They become mediated also by access to energy and capital. Water becomes a resource and commodity to be consumed, ordered and turned into

---

[6] Harvey, 'The "new" imperialism'; de Waal, 'The end of famine?', 185; WEF, *Water Security*, 2.

capital. Waterscapes are reconfigured, exploited and degraded as never before, to these ends. The inherent dynamism of capitalism – and the competitive state projects that both shape and are shaped by it – ensure the constant transformation, intensification and general ratchetting up of these productive, consumptive and destructive processes. New internal frontiers are repeatedly opened up for hydraulic development, appropriation and accumulation. Ideologies of 'national' development legitimise and drive forward this constant revolution. Inter-capitalist competition – between firms, corporations and states alike, and even between small farmers suddenly dependent on and compelled to enter the market – does likewise. And local water security becomes dependent on abstract and impersonal political economic forces and agendas and a structurally uneven, indeed discriminatory, global economy. Whatever the particularities of politics and place, from Hasakah and the West Bank to Merowe and the shores of Lake Chad, the same structural logics apply.

Several things follow from this. So, seventh, water is much more the dependent than the independent variable in the contemporary water–politics relationship; contemporary patterns of water use, water resource-making, water (in)security and water-related conflict are much more consequences of larger political and economic forces and agendas, and their histories, than the other way around. Thus, it is the overall structure of the global economy that basically shapes virtual water flows, much more than anything to do with water specifically. It is historically constructed political boundaries and identities which determine that irrigated agriculture flourishes on one side of the Israel–Gaza and Turkey–Hasakah borders, but not on the other. It is the overall structure of, and interests underpinning, recent peace processes that have essentially defined these peace processes' water politics. The recent or ongoing wars in Syria, Gaza, the Sudans and Lake Chad have reshaped patterns of water availability and (in)security much, much more – and more devastatingly – than they have been caused by them. National political projects from Ba'athism to Labour Zionism have reconfigured local water geographies in their own ideological images. And recent water-related migrations have been responses to technological changes and economic opportunities more than functions of drought. Admittedly, some caution is required on this point, given that ecology, and with it water, are not externalities but are internal to politics and political economy. It would be wrong to suggest that 'water' is only a mirror to 'politics'. Yet it is also abundantly clear that both water itself and the water sectors built around it are, in today's capitalist world order, essentially subordinate to broader state-building and development logics.

Eighth, thanks to technological, infrastructural and economic development, the economic, political and in turn security significance of water is in key respects in historical decline. As discussed in Chapter 2, the central reasons for this are clear: that agriculture, by far and away the number one water-consuming sector, is of declining relative economic importance, contributing ever-decreasing proportions of income, trade, employment and state revenues (both globally and within the vast majority of individual states); that farming interests, and with it water, have as a result slipped well down most state and elite agendas; and that the costs of producing and providing water are also in historical decline. Water is not nearly as important to the political economy of our divided environments as it was 50 or 100 years ago – and this crucial fact guides patterns of politics, war-making and peacemaking alike. Whereas in 1919 Chaim Weizmann could write that '[s]o far as the northern boundary is concerned the guiding consideration with us has been economic, and "economic" in this connection means "water supply"', in contemporary Israel – and the same applies to our other cases too – 'economic' means nothing of the sort.[7] Whereas in the 1950s when the vast majority of Asia's population were peasants, agricultural development and thus water were key US development and counter-insurgency priorities, including within the Green Revolution battle against communism, they are well down the list of priorities today. And whereas 50 and, still more, 100 years ago drought was a major cause of famine across the global South, cases of large-scale famine have both waned and become much more closely linked to war. It is this dynamic, above all, which explains why there have been no modern inter-state 'water wars'. Clearly, there are also counter-tendencies – most notably, the ever-increasing pressures on natural water resources and their consequently widespread degradation. Crucially, however, these contradictory tendencies are not mutually exclusive. Water resources are simultaneously the foundation of life on Earth, unprecedentedly degraded and less and less politically and economically valued or significant – in Marxist terms water is of incomparable 'use value' but declining 'exchange value' – which is not only a troubling and paradoxical state of affairs, but arguably one of the main causes of the current quiet march towards global water crisis.

Ninth, because the dominant economic and political interests in water resources relate to their exploitation, capture, control and consumption, it follows that contemporary water-related insecurities and violent conflicts are much more closely correlated with relative abundance – where

---

[7] Weizmann letter (18/09/1919), in L. Stein (ed.), *The Letters and Papers of Chaim Weizmann* (Oxford University Press, 1968), quoted in Wolf, *Hydropolitics*, 22.

water is available in good quantities to exploit – than with natural resource scarcity. Across our cases, the opportunities afforded by water abundance have informed everything from the Anglo–Egyptian colonisation of Sudan and the Zionist movement's settlement of Palestine; to the frontier development and internal colonisation of north-east Nigeria, north-east Syria, South Kordofan and Darfur; to dam building, irrigation and electricity production schemes from the Euphrates and Jordan to the Nile and Komadougou-Yobe Rivers. In each of these contexts, natural plenitude mediated by technological and economic change have been invitations to development, leading in turn to expropriation, livelihood destruction, displacement and conflict, both within and between states. In most if not all of these contexts, relative plenitude has also been an invitation to over-exploitation: from Gaza and Hasakah to South Kordofan, the over-abstraction and degradation of water resources has been most marked in areas of erstwhile plenty. Moreover, it is indicatively Sudan, with the highest per capita water availability (at least until South Sudan's independence), which has experienced the most acute and recurring water-related violence. Of course, these dynamics are not unrelated to scarcity – if for no other reason than that 'scarcity' and 'abundance' are relational concepts, as discussed in Chapter 2. However, to explain them as functions of 'scarcity' would be plain wrong. Sudden or secular decreases in the availability of water resources – which is what 'scarcity' is usually taken to mean in discussions of water – are neither sufficient nor necessary conditions for water-related violence. Most of the water-related disputes, dislocations and acts of violence considered in this book were barely caused by pressures on water resources at all. Population growth, while no doubt a background factor across our cases, cannot explain the Merowe Dam or northern Cyprus pipeline, the Jordan waters crisis or the frontier dispossession of Nuba communities in South Kordofan. And drought, as discussed in Chapter 3, cannot explain the civil wars in Darfur, Syria or the region round Lake Chad. By contrast, local water abundances and the political economic interests associated with them have, just like abundances of high-value resources like oil, diamonds or coltan, had direct and signal consequences for conflict. The political ecology of resources under capitalism always revolves around the construction or exploitation of local abundances and their transformation into, and circulation as, capital. Although rarely considered in these terms, and often even missing from political ecology-informed analyses of 'wars of plunder', water is no exception to this general rule.[8]

That scarcity narratives remain so dominant within mainstream water security discourse is, then, and by way of a final point, less a function of

---

[8] P. Le Billon, *Wars of Plunder: Conflicts, Profits and the Politics of Resources* (Hurst, 2012).

their explanatory adequacy than of the manifold preconceptions, interests and agendas ready to line up behind them. Contemporary scarcity narratives build upon a long legacy of racialised colonial thinking. They reflect the influence of neoclassical economics, in which scarcity is mythologised as the basic underpinning of and reason for economic exchange. They can attain extra scientific credibility through earth science models that reduce environmental problems to physical ecosystem dynamics while ignoring politics, political economy, war, history and human agency. They appeal to post-colonial regimes and elites, keen to represent themselves as victims of geography, circumstance and, more recently, climate change. They appeal also to international organisations and international NGOs keen to stay – or appear to stay – outside of politics, while showcasing their sensitivity to climate change. Scarcity narratives distract attention from the contradictions of development and from local patterns and causes of immiseration. They supply fodder for elite hero narratives, stories of those who manage to overcome natural constraints through force of their own immanent brilliance. In an age of disaster capitalism, they can construct and open up new opportunities for profit-making. They furnish striking storylines and simple images that are easily understood by deadline-driven journalists and policymakers. They are happily taken up by defence planning and national security agencies, whose job is to imagine looming threats over which they might be 'forced' to intervene. Scarcity narratives are open to right-wing and even eco-fascist readings, in which the spectres of ecological collapse and environmental refugees demand more masculine border policing, or even race war. At the same time they appeal to many (Northern) liberal and red–green audiences, ever-ready to soak up accounts of climate change–driven doom. And they are sometimes reproduced by Southern environmental and indigenous activists, keen to use them to draw attention to local political and environmental injustices. Traversing left and right, North and South, and liberal and authoritarian politics, the genealogy of contemporary scarcity narratives is complex indeed. Overall, however, both scarcity discourse generally, and its climatic variant specifically, are very much discourses of the powerful, with the latter being especially indebted to US military futurology, as discussed in Chapter 1, and both being variants of what Guha and Martinez-Alier label 'full stomach environmentalism' – and being fundamentally misleading as a result.[9]

---

[9] Guha and Martinez-Alier, *Varieties of Environmentalism*; M. Bookchin, *Toward an Ecological Society* (Black Rose, 1980), 26, 137.

Theorising Ecological Politics

These conclusions may be restated in theoretical idiom, in relation to the three traditions of thought around which most of our analysis has been framed. So, first, it is clear that Malthusianism and environmental and geographical determinism – or what we have labelled 'eco-determinist' thought – need confining to the dustbin of history. At best a zombie legacy of the time before industrial capitalism and fossil fuel–driven economies, and at worst a front for colonialism, racism, authoritarianism and militarism, eco-determinism in truth tells us little about the nature and causes of environment-related conflicts and insecurities today. This is not to say that physical geography and population, or for that matter environmental shocks, are of no import whatsoever; this is very far from being the case. It is to say, however, that the theoretical privileging of geography and demography as the master variables for understanding contemporary eco–social relations is misplaced. Today, we live in a world where local natural resource constraints matter less than they ever have to social relations and patterns of conflict and insecurity. From Malthus, Huntington, Ehrlich and Diamond to Gaza, Syria and South Sudan, eco-determinism is both an intellectual dead end and politically deeply problematic.

Now, some may read this as implying an embrace of liberalism, given that mainstream discussions of security and the environment often characterise it as the only other option available: if not Malthusian pessimism, then liberal Prometheanism it must be. Yet this would be in error, for one of the recurring threads of the foregoing pages has been that, alongside eco-determinism, the liberal tradition of politics and political economy possesses serial shortcomings too. A cursory review will suffice. So, *pace* its liberal proponents, the international virtual water trade is organised less around a logic of allocation efficiency, than around hierarchies of affluence and power. Contrary to liberal premises, improved 'drop per crop' ratios do not save water but facilitate its ever-more intensive and extensive use. Contrary to functionalist reasoning, technical cooperation over water cannot drive wider peacemaking between riparian rivals. Contrary to contemporary liberal narratives, economic development, including of water resources, and economic reforms inspired by neo-liberalism often cause conflict. And contrary to liberal idealisations, conflict and war are not only destructive but intensely productive too; they do not just amount to 'development in reverse'. Moreover, liberalism is only in the most superficial sense eco-determinism's antithesis. For, as discussed in Chapter 4, eco-determinist motifs permeate right through liberal idealist and Enlightenment thought, where they are typically mobilised to explain hierarchies between races and peoples and their supposedly differential capacities for reason, freedom and progress. Malthus himself was an early neo-liberal, committed to cutting welfare from the undeserving poor. And contemporary liberal elites and institutions are more than comfortable in deploying the languages of 'scarcity', 'carrying capacity' and environmentally driven 'collapse'. Illustratively, the co-authors of first major US defence planning report on climate security risks – which, as discussed in Chapters 1 and 2, was framed in strikingly Malthusian terms – had only four years previously co-authored a very different piece of futurology, a techno-optimist vision of a 'long boom' lasting through to 2020.[10] And why should this surprise us? As the chief ideological accomplice to capitalism, liberalism is simultaneously mired in its own contradictions and inattentive to the real material contradictions which both course through and structure modern global socio-ecological relations.

On political ecology, finally, we have both much and little to say – much because this entire book has been inspired by political ecological thought, but little because we have not particularly endeavoured to advance it theoretically. Many of the themes explored in the preceding chapters – the violence of capitalist development, the patterns of displacement and dispossession that ensue, the centrality of the state within these processes, their simultaneously social and ecological character, the importance of both local specificities and colonial legacies, the critique of scarcity narratives and more – will have been all too familiar to those versed in the tradition. Indeed, some readers may view our approach to political ecology as under-theorised and also bemoan the poor coverage of certain themes, including gender and labour. And they would in all likelihood be right.

Two theoretical points do merit highlighting, however. One, first broached in Chapter 1, is that we have sought through this book to emphasise the significance of the international within modern socio-ecological relations, and to make a case, however under-theorised, for an 'international political ecology' approach to the subject. Political ecologists have, for the most part, played scant regard to the international, focusing instead on how global structures and imperatives affect specific places, environments and communities, and operating, in effect, with what may be thought of as a 'global to local' ontology. Within IR scholarship too, and especially on its critical wings, inter-state paradigms for

---

[10] T. Homer-Dixon, 'On the threshold: environmental changes as causes of acute conflict', *International Security*, 16:2 (1991), 99–100; Collier, 'Civil war'; Schwartz and Randall, *Abrupt Climate Change Scenario*; P. Schwartz et al., *The Long Boom: A Vision for the Coming Age of Prosperity* (Perseus, 1999).

understanding world politics have lost their erstwhile hegemony, displaced by globalisation and global governance frameworks and dismissed by critics as irretrievably anachronistic. Yet, to our minds, both tendencies are unfortunate. For, whatever the limitations of traditional IR conceptions of the international, many of its characteristic features – inter-state power differentials, geopolitical alignments, foreign policy strategies and practices of competition and emulation – remain central to our contemporary capitalist world order, with consequences which reverberate through patterns of war, peace, state-building, nationalism, development and, not least, ecology.[11] The patterns of water development, water politics and water security and insecurity analysed in the preceding chapters all bear the marks of such international dynamics and cannot be explained as functions of global structures and processes alone. Moreover, at the risk of stating the obvious, international competition is one key reason why global action on climate change has been so limited. Just as the twentieth century's communist revolutions foundered in the face of geopolitics – as well as their own inherent shortcomings – so any transition towards sustainability is likely to be constrained by international forces too. Political ecologists would, in our view, do well to take note.

Last, it is worth stressing that our approach to political ecology is one in which contradictions loom large. Contradictory tendencies run like a thread throughout the previous chapters: in relation to hydraulic development, associated with relative water security for some but destruction, dispossession and displacement for others; in relation to war, associated variously with destruction, state-building and environmental recovery; in relation to eco-determinism, liberalism's ostensible antithesis but actual alter ego; in relation to spaces of abundance turned into ones of scarcity; in relation to peace, where war continues; in relation to the mutually constitutive connections between water security and water vulnerability; in relation to climate security narratives, on which authoritarian leaders, national security think tanks and eco-socialist activists can concur; and much else besides. Opposites, as we read them, typically go together, while core modern distinctions – to wit, nature–society, state–market, scarcity–abundance, construction–destruction, colonial–post-colonial,

---

[11] R. B. J. Walker, *Inside/Outside: International Relations as Political Theory* (Cambridge University Press, 1993); R. W. Cox with T. J. Sinclair, *Approaches to World Order* (Cambridge University Press, 1996); A. Linklater, *The Transformation of Political Community: Ethical Foundations of the Post-Westphalian Era* (Polity, 1998); J. Rosenberg, 'Globalization theory: a post mortem', *International Politics*, 42:1 (2005), 2–74; J. Selby, 'Engaging Foucault: discourse, liberal governance and the limits of Foucauldian IR', *International Relations*, 21:3 (2007), 324–45; Selby, 'The myth of liberal peace-building'.

war–peace, security–insecurity, progress–regress, freedom–determination, real–imagined – are dialectically related and often dissolve upon close inspection. Industrial capitalism, as interpreted by Marx, was a progressive force which liberated the peasantry while simultaneously generating both new forms of subjugation through the compulsions of the market, and the 'second contradiction' of unprecedented ecological destruction; the Enlightenment, as interpreted by Said, was a humanistic accomplishment which simultaneously rested upon racist stereotypes, as well as colonial power.[12] Modernity is suffused with, perhaps defined by, such antinomies. And the politics of water, as well as climate, provides clear evidence in point.

The Coming Landscape of Climate Change and Conflict

What, though, might all this tell us about twenty-first-century climate–conflict relations, our initial and still primary research concern? There are, to be sure, huge uncertainties. Quantitative research offers little clear guidance, being beset both by contradictory findings and methodological commitments which limit its capacity to shed light on the unprecedented. Case studies of recent 'climate conflicts' are even less revealing, as illustrated in Chapter 3 in relation to Darfur, Syria and Lake Chad. And we cannot simply assume that climate change will contribute to security crises, as environment-centric and eco-collapse narratives are want to do. Given this, our premise throughout this book has been that a different methodological strategy is required, specifically one that employs imaginative analogical reasoning alongside historical evidence and which tries to develop insights about future climate change politics through analogical analysis of the political ecology of water. Such a strategy is not without its limitations and dangers, given that no analogue is perfect and the imagination is, in any case, a source of both dreams and nightmares. But proceed we must.

Before that, however, a few words are required on how we connect past, present and future. In his landmark first essay on environmental security, Thomas Homer-Dixon comments that Malthusians have too often 'cried wolf' but then, just a few pages later, observes that 'as time passes their analysis may become ever more compelling'. What we witness here is a common strategy within eco-determinist thought, one in which, as Nicholas Hildyard has described it, 'the future is yet another country – an

---

[12] Marx and Engels, *Communist Manifesto*; J. O'Connor, 'Capitalism, nature socialism: a theoretical introduction', *Capitalism Nature Socialism*, 1:1 (1998), 11–38; E. Said, *Humanism and Democratic Criticism* (Palgrave, 2004); Said, *Orientalism*.

as of yet unoccupied political space where the "noise" of political economy can be blanked out' and Malthusian formulae are 'granted an explanatory power that they no longer enjoy when applied to the past and the present'.[13] By contrast, our premise here is that the future needs to be read as a continuation from, not a rupture with, what has gone before it. In this section we thus assume a continuation of those broad political economic and socio-ecological dynamics identified in previous chapters, merely adding accelerating and worsening climate and environmental change into the mix. Put differently, we reason both analogically from water to climate change and inductively from the past and present to the future. And on this basis we offer five main predictive conclusions.

First, claims to the effect that the environmental effects of climate change will trigger 'floods of climate refugees', 'drive political instability', be 'the biggest trigger of genocide' and cause widespread 'collapse' as the twenty-first century unfolds – or even contribute in significant ways to conflict and insecurity – are highly implausible and unlikely to come to pass. This is for three main reasons: because the extreme climate scenarios on which most such claims depend are themselves unlikely; because capitalism has furnished modern societies with astounding abilities to transcend environmental barriers and scarcities, and adapt to environmental variations; and because both large-scale violent conflict, and the states, projects and elites responsible for most of it, are even less bound by and responsive to environmental imperatives. For, just as contemporary water supplies are much more dependent on infrastructures of water provision, and the energy, expertise and capital they require, than on stable or regular rains, and just as environmental scarcities of water do not, in any significant way, translate into conflict, so the same pattern holds in relation to other environmental resources and problems too. Today, an estimated 110 million people already live below the high-tide line. Millions more live and work in conditions which would be intolerable without air conditioning. Storm deaths are thankfully limited by early warning and pre-emptive evacuation systems. There is no clear upward trend in weather disaster–related displacement.[14] Global food supplies are rising well ahead of demand. Droughts do not typically translate into famine. Even the Darfuri farmer depends more on cheap diesel and

---

[13] Homer-Dixon, 'On the threshold', 100, 104; N. Hildyard, '"Scarcity" as political strategy: reflections on three hanging children', in L. Mehta (ed.), *The Limits to Scarcity: Contesting the Politics of Allocation* (Earthscan, 2010), 154–5; Hartmann, 'Converging on disaster', 772.

[14] S. A. Kulp and B. H. Strauss, 'New elevation data triple estimates of global vulnerability to sea-level rise and coastal flooding', *Nature Communications* 10:1 (2019), 4844; Selby and Daoust, *Rapid Evidence Assessment*, 47.

remittances than on good rains. And in terms of direct interests and experience, for today's globally and financially oriented economic and political elites, heatwaves, floods, storms, rising average temperatures, rural water shortages and poor harvests are frankly insignificant, distant concerns.

Given all this, it is simply not the case that a one metre rise in global sea levels (which is the maximum generally anticipated by 2100) will translate into hundreds of millions being displaced, indeed this is highly unlikely: those models which suggest this reach this conclusion only by ignoring adaptation possibilities, whereas with the inclusion of even relatively simple adaptations such as beach replenishment and dyke construction the problem of sea level rise–induced environmental displacement 'almost disappears'. It is not the case, as the World Bank claims, that climate change impacts on sea levels, water availability and crop yields could force 216 million people to migrate by 2050: once again, this finding is achieved only by ignoring practices of adaptation which, as extensively demonstrated in the previous chapters, are a normal and far from exceptional part of modern human existence. And it is not the case that there exists a 'human climate niche' defined by average temperature of 29°C, which, as the planet heats up, will compel billions to migrate: arguments to this effect rest upon the same asocial and zoological misconceptions of the human condition that were shown to be flawed in Chapters 2 and 4.[15] Moreover, given the centrality of 'climate migration' within mainstream climate conflict thinking – typically presented as the key social variable linking environmental scarcities and political crisis – it is hard to see how the various forms of climate change–induced conflict, instability and collapse imagined within such thinking could arise.

Second, while climate change will amplify certain patterns of environmental scarcity, degradation, insecurity and vulnerability, this will rarely be as their primary cause; their more fundamental causes are likely to remain as they are today. Frontier crises from north-east Syria to northeast Nigeria are being caused by unsustainable and precarious agrarian development strategies – not climate change. The water crisis in the West Bank is a function of Israeli policies – not climate change. Recent collapses in water provision from Syria to Gaza to Lake Chad are consequences of war – not climate change. Today's 'global groundwater crisis'

---

[15] R. J. Nicholls et al., 'Sea-level rise and its possible impacts given a "beyond 4°C world" in the twenty-first century', *Philosophical Transactions of the Royal Society A*, 369 (2011), 173; V. Clement et al., *Groundswell Part II: Acting on Internal Climate Migration* (World Bank, 2021); C. Xu et al., 'Future of the human climate niche', *Proceedings of the National Academy of Sciences*, 117:21 (2020), 11350–55; Selby and Daoust, *Rapid Evidence Assessment*.

is being caused essentially by agricultural intensification – not climate change. Today's global biodiversity collapse is being driven by the same – not primarily climate change. Communities in the Ganges Delta are migrating en masse to Bangladesh's cities because of changes in river flow arising from upstream dam building, deforestation and agricultural modernisation, allied with loss of livelihood opportunities – not sea level rise induced by climate change. And even sea level rise is, according to some studies, so far a consequence more of groundwater exploitation than of climate change.[16]

Of course, as the planet warms this pattern will almost certainly shift. Climate change will result in ever-more extreme heat events, with dire consequences for the most vulnerable. Long-term changes in precipitation will exacerbate existing problems of water resource degradation, at least in places. There may be more frequent and more intense floods, droughts and storms. Much land will not be protected from the rising seas. And the possibility of more extreme changes than the current scientific consensus suggests cannot be ruled out. Crucially, however, most such climate change impacts will be refracted through existing patterns of degradation and vulnerability which, given that the latter are structured by political and political economic dynamics – by conflicts, divisions, hierarchies and capitalist accumulation imperatives – means that future climate insecurities will in all likelihood be structured by the same. Stated differently, on most issues climate change will only be a secondary cause of vulnerability, if that: climate change may make matters slightly (or periodically more than slightly) worse for populations displaced by war, living under occupation or compelled to eke out a living in precarious frontier spaces, but it will not be the primary cause of their vulnerability.

By contrast, and third, political and development dynamics enabled or legitimised by climate change are likely, in the coming decades, to have direct and significant conflict and security effects. This problem, to be fair, is now widely recognised both by climate researchers and within Northern climate security and development policy communities, the dangers of 'maladaptation' to climate change being regularly identified as an important challenge and risk. Yet the dangers, as well as complexities, run far deeper than such formulations typically allow. For just as water resource and infrastructure development has, over the course of the last century created new-found levels of water security alongside

---

[16] J. S. Famiglietti, 'The global groundwater crisis', *Nature Climate Change*, 4:11 (2014), 945–8; Y. N. Pokhrel, 'Model estimates of sea-level change due to anthropogenic impacts on terrestrial water storage', *Nature Geoscience*, 5 (2012), 389–92.

destruction, dislocation and conflict – within core regions and socio-ecological frontiers alike – so similar dynamics are likely to be associated with the response to climate change. New mega-dams, sea walls, water conveyance systems, flood protection infrastructures, land investments and livelihood strategies – all necessitated or legitimated as adaptations to climate change, and presented as 'climate-smart' solutions by international financiers and consultants, and equally interested local elites – may generate widespread local displacement, forced relocations, social tensions, political mobilisation and, potentially, violence. New mitigation investments – in which land is set aside for reforestation, biofuels production, wind turbines, solar panels and more – may have similar effects, competing for already-used land, labour and water. New frontier spaces opened up by actual or projected temperature rises may, especially in Northern latitudes, feed speculative booms relating to potential agricultural development, resource (including oil and gas) extraction, international shipping lanes and military interests; both indigenous land expropriation and militarisation are likely to follow. Meanwhile, 'climate change' will increasingly function as a socio-ecological master narrative, deployed to explain – and explain away – all manner of environmental problems and through that to deflect attention away from their bigger causes. Overall, the anticipated threat of climate change will become both a key motor of deepening capitalist development and part of the discursive armoury of the powerful, with exploitation, exclusion and immiseration inevitably in tow. Instructive in this regard are dynamics in southern coastal Bangladesh where, within the context of fears and narratives of looming 'uninhabitability' promoted by local elites and international donors alike, farmers are intentionally flooding their lands and turning from rice agriculture to shrimp aquaculture – an adaptation which is simultaneously ecologically devastating, congruent with Bangladesh's export-oriented 'blue growth' strategy, and a factor in out-migration as demand for labour plummets.[17] It is here, we suggest, rather than in model projections of people being displaced by environmental forces alone, wherein lies the more accurate picture of the climate change–related insecurities to come.

---

[17] A. K. Magnan et al., 'Addressing the risk of maladaptation to climate change', *WIREs Climate Change*, 7:5 (2016), 646–5; Eriksen et al., 'Reframing adaptation'; Eriksen et al., 'Adaptation interventions'; A. Tubi and J. Williams, 'Beyond binary outcomes in climate adaptation: the illustrative case of desalination', *WIREs Climate Change*, 12:2 (2021), e695; K. Paprocki, 'Threatening dystopias: development and adaptation regimes in Bangladesh', *Annals of the American Association of Geographers*, 108:4 (2018), 955–73; K. Paprocki, 'All that is solid melts into the bay: anticipatory ruination and climate change adaptation', *Antipode*, 51:1 (2019), 295–315.

Fourth, decarbonisation and the transition to a low-carbon global order are likely to have profound conflict and security repercussions. We acknowledge that this point is asserted here more than fully substantiated by our preceding analysis; further development of this point is evidently required. And yet, even within the foregoing pages the importance of fossil fuels to both water security and prevailing patterns of economic and political power has been one of the central, recurring threads. If we add to this three self-evident truths – that the burning of fossil fuels is the primary cause of climate change; that most remaining fossil fuels will need to be left, unexploited, in the ground if planetary warming is to be halted (and very soon if the Paris targets of 1.5 or 2°C are to be achieved); and that global fossil fuel exploitation and consumption are still continuing to rise – then it is not difficult to see that serious conflict lies in store. At best, the renewables revolution and development of other energy sources will transform the economics of fossil fuel use, such that coal, oil and gas become globally unimportant 'stranded assets'; however, even with this the ramifications will be huge, as the three great powers and some of the poorest post-colonial states, and many other countries in between, find one of the major foundations of their political economies and systems of rule pulled away. And at present – when there is no meaningful discussion of supporting poor fossil fuel producer states to make such a transition, and when parliamentarians of even the richest, and in many respects most 'green', Northern fossil fuel producer state, Norway, want the country to continue drilling for oil and gas for the next 300 years – this looks wildly optimistic.[18]

In turn, fifth, the question of climate change will surely become an ever-more important site and source of political conflict, in ways which could very easily involve large-scale violence. Here, our water–climate analogy becomes severely strained, not least because climate change has the potential to fundamentally transform and limit the basis for life on Earth, especially beyond 2100, and because fossil fuels are much more structurally important to our capitalist order than water: while water is internal to the metabolic relations of capitalism, fossil fuels are much more profoundly so. Yet the coming dynamics are not hard to see. Already, social attitudes to climate change are deeply fractured along lines of nation, race, gender, class and party: according to one recent study in the United States, the issue of climate change is more politically

---

[18] C. McGlade and P. Ekins, 'The geographical distribution of fossil fuels unused when limiting global warming to 2 °C', *Nature*, 517 (2015), 187–90; A. Allawi and F. Birol, 'Without help for oil-producing countries, net zero by 2050 is a distant dream', *Guardian* (01/09/2021); J. Henley, 'Norway votes – but is Europe's biggest oil giant ready to go green?', *Guardian* (12/09/2021).

polarising than abortion. Internationally, long-standing North–South and other divisions remain acute. And these various social and international divisions will inevitably grow as the effects of climate change become more tangible, as the negative consequences of adaptation become more evident and as pressures mount to reduce, and then halt, fossil fuel production. There is likely to be, at minimum, an intensification of the politics of 'Blockadia', as Naomi Klein labels it, as well as a continuing upsurge in conservative, paramilitary and White nationalist violence against environmental defenders; a rise in fossil fuel nationalism, especially among Southern producer states; and, quite possibly, a rise of revolutionary and counter-revolutionary movements and regimes and entirely new understandings of or approaches to the political, motivated centrally by the question of climate change. It is in these and other such political respects, we suggest, rather than as a narrowly 'environmental' trigger of mass migration and 'collapse', that climate change is likely to be a key feature of the landscape of twenty-first-century conflict. Moreover, in these terms, as against the eco-determinist registers in which they are intended, claims that climate change is 'perhaps the twenty-first century's biggest foreign policy challenge' and 'potentially the greatest challenge to global stability and security' that exists may prove unwittingly prophetic.[19]

And, we would add: necessarily so. For just as the prescriptive corollaries of Malthusian thought are birth control, premature death and God; and just as the neo-liberal equivalents are the rationalities and compulsions of the market; so political ecologists' emphasis on the centrality of politics and political economy in the making of ecological crises translates, necessarily, into an advocacy of political transformation, resistance and struggle as the keys to socio-ecological change. There are already, of course, extensive literatures on these themes, including on climate, water and environmental justice and the strategies through which these might be achieved – but we cannot open up whole new areas of discussion of these issues now.[20] Suffice to note that we have at last come full circle. The fears of 'looming climate instability' with which we began this book

---

[19] Selby, 'The Trump presidency'; O. Millman, 'Climate crisis more politically polarizing than abortion for US voters, study finds', *Guardian* (22/05/2019); A. Leiserowitz et al., *Politics and Global Warming* (Yale Program on Climate Change Communication, 2019); M. Klein, *This Changes Everything: Capitalism versus the Climate* (Simon and Schuster, 2014), 253–4; Mann and Wainwright, *Climate Leviathan*; UK Cabinet Office, *National Security Strategy*, 18; Hague, 'The diplomacy of climate change'.

[20] D. McCauley and R. Heffron, 'Just transition: integrating climate, energy and environmental justice', *Energy Policy*, 119 (2018), 1–7; F. Sultana, 'Water justice: why it matters and how to achieve it', *Water International*, 43:4 (2018), 483–93; D. McGregor et al., 'Indigenous environmental justice and sustainability', *Current Opinion in Environmental Sustainability*, 43 (2020), 35–40; P. Newell et al., 'Toward transformative climate justice: an emerging research agenda', *WIREs Climate Change*, 12:6 (2021), e733.

all posit 'conflict' not just as a consequence of global environmental change but, more precisely, as an exclusively negative consequence, reflecting dominant powers' interests in the status quo. The truth, however, is that it is hard to imagine either climate change or our deepening global water crisis being meaningfully or successfully addressed without it. 'Environmental' and 'climate' conflict are in this sense less threats that must be pre-empted than practical political necessities that need to be actively embraced.

Index

Abbas, Mahmoud, 257
Abboud, Ibrahim, 119, 160
abstract space, 139, 141, 144, 157, 213
Abuja, 85, 196
abundance, 57, 142, 170, 172, 178, 182, 188, 196, 197, 206, 302, 311, 312, 316
actor network theory, 19
Adamawa, 196, 221
adaptation, 50, 55, 60, 90, 91, 93–4, 102, 110, 111, 171, 202, 204–5, 219, 237, 318
Afghanistan, 269, 298
Africa, 44, 211
  eco-determinist representations of, 61, 107, 110–12
  European colonialism, 91, 99, 104, 115, 116, 132, 133, 277
  virtual water, 275, 277, 282
  within climate security discourse, 6–7, 8, 15, 112–13, 134, 135
African Rights, 175
Afula, 164
Agricultural Development Projects (Nigeria), 198, 200
Ahmed, Suad Ibrahim, 163
Aita, Samir, 81
Akinci, Mustafa, 247
Alaköprü Dam, 148
Alatout, Samer, 62, 142
al-Dardari, Abdullah, 96
Aleppo, 78, 79, 151, 214, 217, 218, 224, 228
Alexandria, 120
Alfa Partners, 293
Algeria, 84, 95, 244, 290
Allan, Tony, 274, 276, 280, 281, 308
Alliance for a Green Revolution in Africa, 211
Al-Qaeda in the Islamic Maghreb, 244
Amazon River, 18
America, colonisation of, 45, 55, 59–60, 104, 106–7, 169–70, 206

Amri Dam Affected Peoples Committee, 161
analogical reasoning, 24–5, 300, 317, 318, 322
Anamur River, 148
Angola, 57, 244
Annan Plan (2004), 245, 263–6, 267, 268
Annan, Kofi, 240
Anthropocene, 4, 21
Arab Belt, 193, 194–5
Arab League, 47, 161, 249, 292
Arab Spring, 11, 12
Arab–Israeli conflict, 150, 250
Arab–Israeli War (of 1948-9), 128, 185, 187, 213, 219, 226, 229–32, 235–6, 248, 305
Arab–Israeli War (of 1967), 47–8, 62, 162, 181, 182, 185, 215, 219, 229, 232, 234, 305
Arab–Israeli War (of 1973), 289
Arafat, Yasser, 254, 257
Arctic Ocean, 208
Arizona, 45, 275
Asia
  climate change, 135
  eco-determinist representations of, 110, 112–13
  European colonialism, 89, 90–1, 104
  virtual water, 274, 275, 277, 291
  within US Cold War strategy, 250, 311
Assad, Bashar al-, 80, 96, 190, 193
Assad, Hafez al-, 150, 152, 289, 304
Aswan Dam, 49, 155, 159, 162
Atatürk Dam, 162
Atbara River, 155, 156, 160
Auden, W. H., 29–30

Ba'ath Dam, 227
Ba'athism, 150, 157, 190, 203, 304, 310
Baga, 197
Baga Polder project, 198, 288
Baggara, 175

325

Index

Bahr el Ghazal, 118, 224
Balikh River, 151
Baltic region, 277
Bangladesh, 320, 321
Barada River, 151
Barak, Ehud, 131, 132
Bashford, Alison, 55
Bashir, Omar al-, 96, 119, 156, 203, 259, 260
Bechtel, 293
Beck, Ulrich, 164, 243
Beckett, Margaret, 3, 23, 33
Beer Sheva, 38, 144, 188, 230
Bengal, 89, 91, 92
Ben-Gurion, David, 126, 128, 129, 230
Bentiu, 216
Bhopal, 271
Biafra, 197
Bidibidi refugee camp, 223
Bimkom, 256
biofuels, 11, 262, 321
Birecik Dam, 162
Blaikie, Piers, 18, 21, 62
Blair, Tony, 211
Blass, Simcha, 284
Blaut, James, 106
Boko Haram, 83, 84, 86, 87, 96, 201, 215, 225, 226, 244, 303
Bookchin, Murray, 203
Bor, 216
borders, 22, 27, 130, 141, 158, 194–6, 198, 205, 230, 238, 242, 243, 266, 308, 310, 313
Borno, 196, 199, 221, 225
Boutros Ghali, Boutros, 47
Brabeck-Letmathe, Peter, 272
Braudel, Fernand, 88, 92
Brazil, 2, 46
Breadbasket Strategy, 155, 156, 159, 173, 178, 286, 288
Bretton Woods system, 92
British Empire, 27, 30, 58, 91, 92, 103, 105, 114, 115, 123, 139, 277
Brock, Lothar, 249
Brookfield, Harold, 18
Brown, Samuel, 121
Bryant, Rebecca, 121
Buckle, Henry Thomas, 107
Burke, Marshall, 6, 15
Burke, Sharon, 99
Bush, George W., 259

Cabinda, 244
California, 112
Cameroon, 28, 83, 198, 222, 225, 244
Camp David Accords (1978), 253–4
Canada, 46, 234
capitalism, 19, 20, 21–3, 43–6, 56, 68, 90, 94, 133–4, 137–40, 163, 169, 207, 210, 270–99, 309–10, 312, 315, 317
Caribbean, 6, 89, 277
carrying capacity, 33, 61, 315
Center for Climate and Security, 13, 14
Center for Research on the Epidemiology of Disasters, 222
Central African Republic, 84, 244
Centre on Housing Rights and Evictions, 287
Chad, 27, 28, 74, 76, 83, 86, 198, 222, 225, 244, 299
Chad Basin Development Authority, 197
Chamberlain, Joseph, 121, 145
Chaplin, Joyce, 55
Chari-Logone River, 99, 198, 199
Chellawa Gorge Dam, 198
Chenoweth, Jonathan, 66
China, 2, 6, 11, 46, 89, 90, 92, 95, 156, 276, 277, 278, 288, 289, 292
Christchurch massacre (2019), 136
Churchill, Winston, 99, 114, 117, 123, 132
class, 19, 22, 43, 53, 60, 89, 104, 116, 123, 124, 128, 135, 137, 150, 154, 202, 207, 232, 236, 307, 322
Clausewitz, Carl von, 236, 255
Clerides, Glafcos, 264
climate change
  adaptation, 20, 167–8, 237–9, 297, 318–19, 321, 323
  and capitalism, 167–8, 207–8, 294–9, 320–1
  and migration, 3–4, 6, 12, 15, 69, 70, 72–5, 77–83, 86–7, 98, 113, 134, 135, 167, 205, 319, 320
  and race, 105, 113, 131–6
  environmental impacts, 4–5, 63–8, 100–2, 167, 207–8, 306, 319–20
  mitigation, 4, 5, 20, 167–8, 299, 316, 321, 322–4
  scientific evidence, 4
  vulnerability, 93–4, 102, 236–9, 266–9, 319–20
climate security
  methodological issues, 5, 10–18, 23–6, 317–18
  policy discourse, 1–4, 33, 69–70, 83, 87, 96–100, 205, 312–13

Index

qualitative research, 7–8, 9, 12–14, 16–17
quantitative research, 6–7, 14–16, 20, 23–4, 70, 113, 300, 317
reinterpreted, 317–18
climatisation, 96
Clyde River, 137
coal, 42, 137–8, 246, 247, 250, 282, 283, 287, 298, 322
Coastal Aquifer, 143, 185, 187
Coca-Cola, 271
Cold War, 12, 48, 61, 76, 120, 124–5, 133, 139, 150, 213, 234, 245, 248, 249, 250–1
collapse, 6, 12, 29, 54, 55, 56, 59–60, 74, 79, 108, 109, 112, 132, 187, 205, 302, 307, 313, 315, 318
Collier, Paul, 210, 233
colonialism, 60, 61, 89–91, 99, 103–34, 314
Colorado River, 45, 275
Committee of Anti Dal-Kajbar Dams, 161
Conca, Ken, 240
conflict, definition of, 25
constructivism, 16, 26, 133, 240, 307
cotton, 42, 80, 81, 115, 118, 137, 152, 153, 154, 156, 174, 177, 191, 192, 198, 224, 278
Covid-19, 296, 298
Cox, Robert, 5
critical geopolitics, 20, 212
critical realism, 19
critical war studies, 212
Cromer (Evelyn Baring), 153
Curzon, George, 208
Cypfruvex, 229
Cyprus, 9, 27–8, 103
  northern Cyprus post-1974, 146, 148–9, 162–3, 165, 168, 213, 219, 229, 230, 233, 247–8, 266, 268, 292, 304
  peacemaking, 242, 245, 247–8, 263–6, 267, 268
  post-colonial political economy, 145–6, 147, 151, 299
  under British rule, 105, 120–5, 128, 130, 132–3
  virtual water, 279, 281, 304, 305
  water resource availability, 35–6, 39–40, 88, 94, 121–2, 130, 153, 304
  water resource development, 141, 145–9, 153, 157, 162–3, 164–5, 168, 233, 234, 235, 247–8, 265–6, 293–4, 304
Cyprus Convention (1878), 123

Damascus, 78, 79, 151, 196, 214
dams, 86, 138, 139, 145, 146, 148, 150, 151–2, 153, 155–7, 159–62, 163–4, 165–8, 198, 199, 203, 227–8, 234, 262, 288, 289, 292, 302, 303, 304, 307, 312, 320, 321
Dams Implementation Unit (Sudan), 156, 288
Danforth, John, 259
Daoudy, Marwa, 83
Dara'a, 79, 82–3
Darfur, 93, 157
  impacts of climate change, 65, 318
  within colonial state formation, 117–20, 177
Darfur war (2003-5), 220, 259
  as a climate conflict, 7, 27, 71–7, 84, 87, 96, 98, 99, 100, 102, 113, 136, 176, 301, 312, 317
  as a frontier conflict, 172, 176–80, 182, 196, 201, 206, 286, 302, 312
  impacts on water security, 215, 221, 222–3, 228, 302
Davis, Mike, 90, 91
Davutoğlu, Ahmet, 149
Dayton Accords (1995), 264
De Schutter, Olivier, 97
de Waal, Alex, 54, 93
Dead Sea, 144, 268
Déby, Idriss, 86
Delek Group, 293
Demetriades, Lellos, 247
Democratic Republic of Congo, 244
Denktaş, Rauf, 264
desalination, 42, 50, 53, 67, 144–5, 147, 165, 168, 205, 258, 265, 280, 287, 293–4, 295, 304, 305, 308
desertification, 3, 29, 76, 96, 132, 205
deserts, 45, 104, 128, 130, 170, 205, 211, 306, 307
Diamond, Jared, 55, 56, 58, 59, 62, 107, 109, 111, 112, 113, 187, 314
diamonds, 57, 312
diesel, 81, 177, 198, 204, 207, 217, 286–7, 318
Diffa, 225
Dinka, 159
disaster capitalism, 293, 313
displacement, 36, 92, 158–64, 167, 209, 219–26, 228–32, 235–6, 239
Disraeli, Benjamin, 120
Doha round, 272
Dow Chemicals, 271
drip irrigation, 50, 152, 280, 282–6, 295, 305

drought, 3, 7, 11, 34, 43, 69–102, 145, 156, 190, 191, 194, 197, 199, 201, 204, 290, 293, 295, 298, 303, 306, 310, 311, 312, 318, 320
Du Bois, W. E. B., 106
Dulles, John Foster, 245
Dunlap, Riley, 135

East Asian developmental states, 234
Easter Island, 58, 59, 62, 109, 112
eco-determinism, 54–63, 68, 76, 95, 99, 105–13, 202, 272, 314, 317
eco-fascism, 135, 313
EcoPeace, 247
efficiency, 50, 202, 272, 273, 274, 282–6, 309, 314
Egypt, 11, 12, 34, 40, 47, 48–9, 61, 62, 88, 89, 114–15, 139, 150, 153, 155, 156, 159, 162, 185, 244, 248, 249, 253, 262–3, 292, 312
Ehrlich, Paul, 47, 61, 111, 314
Eisenhower, Dwight, 249
El Niño, 89, 90
el Obeid, 118
El Paso shooting (2019), 136
electricity supplies, 148, 156, 162, 207, 216–17, 220, 227, 287, 295, 299
energy supplies, 45, 165, 216, 270, 286–7, 295, 297, 298, 308, 318
Enlightenment, 60, 106–8, 122, 132, 133, 300, 305, 314, 317
environmental Orientalism, 104, 132
environmental peacebuilding, 240–2, 267–8
environmentalism, 61, 202, 203, 313
Equatoria, 118, 224
Erdogan, Recep Tayyip, 148
Eshkol, Levi, 48, 142
Estes, Nick, 44
Ethiopia, 40, 48–9, 53, 89, 92, 155, 162, 168, 220, 292
Euphrates River, 27, 46, 49, 151, 162, 190, 227, 244, 312
Europe
  colonialism, 60, 61, 91, 105, 113, 132, 133, 139, 277
  history of famine, 89, 90–2
  self-representations, 106–9, 112, 115
  virtual water imports, 277, 278, 291
  water use, 38, 43, 46
European Coal and Steel Community, 246, 250
European Union, 6, 15, 135, 246, 247, 248, 263, 265, 283
Extinction Rebellion, 1

Fabius, Laurent, 98
Falkenmark, Malin, 35, 51
Famagusta, 120, 265
famine, 3, 34, 47, 54, 58, 62, 69, 72, 88–94, 95, 108, 134, 197, 282, 311, 318
Fanon, Frantz, 111
First World War, 91, 138
floods, 43, 94, 101
Food and Agricultural Organization, 35, 67, 190, 216, 217
forests, 3, 60, 103, 104, 121–2, 123, 129, 169, 207, 223, 225, 230, 277, 320, 321
Foucault, Michel, 19, 46, 255
France, 12, 27, 42, 85, 89, 90, 95, 98, 108, 115, 126, 130, 149, 162, 244, 250, 276
Franklin, Benjamin, 55, 108
Friedman, Thomas, 244
Friends of the Earth, 2
Friendship Dam, 233
Fromentin, Eugène, 95
frontiers, 29, 74, 141, 169–208, 281, 286, 302, 303, 305, 309, 310, 312, 319, 320, 321
Fuerth, Leon, 66
Fukuyama, Francis, 110
Funk, McKenzie, 168, 305
Fur, 228

Gaddafi, Muammar al-, 76
Galilee, 231
Galtung, Johann, 25
Garang, John, 119, 159, 260
Gates Foundation, 211
Gaza Strip, 110, 181
  impacts of war on water security, 214–15, 216–19, 223, 226, 286, 287, 310, 319
  Israeli blockade, 213, 215, 216, 217, 218
  wastewater problems, 186–7, 214–15, 216, 218
  water resources and supplies, 184–8, 196, 205, 231, 243, 254–5, 257, 258, 285, 305, 310, 312
Gaza–Jericho Agreement (1994), 254
gender, 19, 22, 61, 133, 135, 164, 203, 207, 218, 234, 236, 272, 296, 307, 315, 322
Germany, 69, 86, 155, 250, 276, 282, 292
Gezira scheme, 118, 153–4, 155, 156, 163, 177
Ghouta, 218
Gill, Stephen, 271
Gladstone, William Ewart, 120
Glasgow, 137
Gleditsch, Nils Petter, 112
Gleick, Peter, 46, 51, 77, 79, 81, 82

Global High-Level Panel on Water and Peace, 40, 43, 243
Golan Heights, 181, 194, 305
Goldstone Report, 215
Google Earth, 177
Gorbachev, Mikhail, 240
Gore, Al, 85
Graham, Stephen, 213
Grand Ethiopian Renaissance Dam, 49, 53, 162, 292
Graves, Philip, 128
great divergence, 90
Great Leap Forward, 92, 95
Greece, 122–3, 124, 125, 133, 263, 264
green grabbing, 20, 167, 297
Green Revolution, 139, 197, 211, 234, 311
Greenland, 208
Greenock, 137, 138, 139
Greenpeace, 1
groundwater, 39, 40, 43, 73, 81, 93, 94, 142, 145, 149, 151–2, 158, 162, 167, 177–8, 182, 184–7, 190–2, 195–6, 198–9, 204, 207, 217, 218, 222–3, 255, 256, 265, 271, 276, 278–9, 281, 285, 286, 290, 298, 302, 303, 305, 306, 307, 319–20
Guha, Ramachandra, 202, 313
Guterres, António, 69, 83, 240

Haas, Ernst, 246, 250
Habila, 174
Hadejia Dam, 198
Haifa, 230
Haiti, 61, 109
Hama, 214
Hamilton, Alexander, 293
Hardin, Garrett, 55, 61
Harvey, David, 56, 60, 172, 309
Hasakah, 77, 190–6, 201, 203–4, 206, 297, 310, 312
Hass, Amira, 187
Hausa-Fulani, 197
Hegel, Georg Wilhelm Friedrich, 107, 108, 110, 112, 113, 115, 305
Heglig, 261
Heilberg, Phil, 168
Hensel, Paul, 49
Herzl, Theodor, 121, 126, 130, 141, 142
Hilal, Muhamed Talab, 193
Hildyard, Nicholas, 317
Hinnebusch, Ray, 150
Histadrut, 127
Hobbes, Thomas, 1, 110, 113
Hoekstra, Arjen, 278
Holocaust, 223

Homer-Dixon, Thomas, 13, 25, 33, 56, 57, 110, 306, 317
Homs, 78, 214
Huleh Valley, 231–2
Hulme, Mike, 68
Human Development Index, 27
Huntington, Ellsworth, 61, 107, 111, 113, 314
Hurvitz, Eli, 253
Hussein, Saddam, 162
hydraulic development, 90, 137–68, 198–9, 211, 234–6, 281, 285, 287, 288, 289, 290, 292, 295, 302, 316
hydro-electric power, 48, 50, 138, 142, 143, 156, 167, 233, 243, 262, 312
hydro-hegemony, 48, 62, 308
hydro–social cycle, 26, 45, 140

Iceland, 46
IDE Technologies, 293
Idlib, 214, 218, 224
India, 2, 30, 46, 61, 89, 90, 91, 95, 120, 123, 139, 249, 271, 276, 277, 290
Indus Waters Treaty (1960), 249
infrastructural warfare, 213, 214
insecurity, definition of, 25
Integrated Water Resource Management, 271
Intergovernmental Panel on Climate Change, 8, 11, 24, 65, 66, 101
International Alert, 66
International Crisis Group, 96
International Monetary Fund, 190
international political ecology, 9, 10, 21–3, 315–16
International Relations, 12, 104, 240, 246, 315
Iran, 100, 277, 288, 290
Iraq, 15, 57, 78, 80, 150, 162, 193, 196, 227, 288, 291
Ireland, 58, 62, 90, 277
irrigation, 41, 42, 48, 49, 50, 73, 81, 86, 90, 91, 93, 118, 129, 143, 144, 145, 147–8, 150, 151–2, 153–4, 155–7, 164, 165, 166, 167, 172, 177–8, 183, 185, 190–2, 195, 197–9, 200–1, 204, 215, 217, 223–4, 225, 226, 229, 233, 243, 247, 262, 266, 278, 280, 282–6, 288–9, 294, 302, 303, 305, 306, 307, 310, 312
Islam, 103–4
Islamic State, 227, 244
Islamism, 52, 119, 148, 156, 163, 175, 259, 260

Israel, 9, 27–8
  corporate interests, 252, 293–4, 295, 299
  impacts of climate change, 65–8, 293
  impacts of war on water security, 223, 229–32, 235–6, 238, 305
  state formation, 105, 125–31, 141
  virtual water, 279–82, 305
  water conservation, 284–6
  water resource availability, 35–40, 44, 51, 53, 62, 65–8, 94, 130, 153, 164, 182, 186, 243, 258, 297, 304, 311
  water resource development, 47, 128–31, 141–5, 146, 147, 149, 151, 153, 157, 158, 161, 165, 168, 188, 202, 206, 210–12, 287, 292, 305, 310
Israel Chemicals, 293
Israel Defense Forces, 253, 256
Israel–Gaza War (2008–9), 213, 214–15, 239
Israeli Civil Administration, 183, 255
Israeli Water Authority, 38, 285
Israeli–Jordanian cooperation, 248–9, 250, 267, 268
Israeli–Jordanian Peace Treaty (1994), 248
Israeli–Palestinian conflict, 66, 110, 119, 125–31, 132–3, 181, 232, 235
Israeli–Palestinian peace process, 181, 242, 245–7, 252–8, 267, 268, 269, 305
Issar, Arie, 66

Jabalia, 215
Jaffa, 230
Janjaweed, 72, 75, 76, 215
Japan, 276
Jazira, 194, 203
Jebel Aulia Dam, 153
Jebel Marra, 73, 74, 77, 177
Jerusalem, 181, 230, 245
Jevons, William Stanley, 282–3
Jewish Agency, 127, 142, 292
Jewish National Fund, 129, 230, 292
Jisr al-Zarqa, 158
Johnston mission, 249–51
Joint Water Committee (Israeli–Palestinian), 255–8
Jones, Eric, 107
Jonglei, 215, 224, 261
Jonglei Canal, 159, 233
Jordan River, 27, 38, 46, 47, 65, 66, 67, 130, 143–4, 161, 167, 188, 215, 231–2, 248–9, 250, 255, 285, 312
Jordan Valley, 128, 143, 181–2, 183, 216, 246, 250

Jordan, Kingdom of, 38, 48, 144, 156, 183, 220, 223, 224, 226, 248, 249, 250, 268
JPMorgan Chase, 293
Juba, 39, 216, 217, 219, 221, 261, 262
Justice and Development Party (Turkey), 148, 263

Kabbara, 158
Kabul, 269
Kadugli, 173
Kajbar Dam, 160, 161, 163
Kaldor, Mary, 210
Kant, Immanuel, 106, 132, 305
Kaplan, Robert, 35, 46, 56, 59, 110–11, 112, 132
Kassala, 117, 118, 172
Kautsky, Karl, 90
Kelley, Colin, 77, 78–9, 81, 82
Kerry, John, 2, 83
Khabour River, 190–2, 195
Khartoum, 115, 117, 118, 154, 156, 160, 161, 172, 220, 302
Khashm el-Girba Dam, 155
Kiir, Salva, 260–2
Ki-moon, Ban, 3, 23, 71, 83, 96, 99, 240, 309
Kitchener, Herbert, 114, 122
Klare, Michael, 33
Klein, Naomi, 2, 98, 323
Kliot, Nurit, 47
Kobani, 214
Komadougou-Yobe River, 198, 199, 226, 288, 312
Kordofan, 118, 157
Korea, South, 27
Kosti, 117
Kurdistan Workers' Party, 203
Kurds, 97, 193–5, 196, 203–4, 214, 228
Kuwait, 44, 156
Kyrenia, 148

Lac, 225
Lake Assad, 151, 160, 162, 193, 227–8
Lake Chad Basin Commission, 244
Lake Chad region, 9, 177, 299
  as a water frontier, 171, 196–201, 206, 288, 303, 310
  as site of climate conflict, 70, 83–8, 98–9, 100, 102, 113, 167, 303, 312, 317
  during colonial period, 95, 99
  impacts of war on water security, 213, 220, 222, 225–6, 310, 319
  water resource development, 86, 168, 198–9, 268
Lake Kinneret, 38, 143, 144

Lake Nasser, 49, 155, 159, 160
Lament of Ipuwer, 88
Landes, David, 107
Languedoc, 185
Laqueur, Walter, 48
Latour, Bruno, 17
Lattakia, 223
Le Billon, Philippe, 58
Leadership Office of the Hamadab Affected People, 161
Lebanon, 141, 149, 151, 220, 223, 231, 249
Lefebvre, Henri, 139
Lemkin, Raphael, 212
Lewis, Bernard, 190
liberal functionalism, 240, 242, 246–52, 314
liberalisation, 59, 80–1, 152, 174, 178–9, 193, 252, 272, 274, 302, 304
liberalism, 110, 115, 121, 133, 138, 166, 210, 211, 242, 282, 292, 307, 314–15, 316
Libya, 73, 76, 244, 290
Litani River, 130
livestock, 73, 74, 93, 156, 172, 173, 178–80, 224, 225, 228, 262, 278, 279, 280
Livestock and Meat Marketing Corporation (Sudan), 179
Lockean heartland, 139
London, 90, 123, 153, 161, 256
Lowdermilk, Walter, 128, 143
Lowi, Miriam, 250
Lugard, Frederick, 115, 197
Luxembourg, 275

Maas, Heiko, 69, 86
Macdonald's, 244
MacMichael, Harold, 116
Macron, Emmanuel, 83
Maiduguri, 197
Makhlouf, Rami, 82
maladaptation, 320
Malakal, 216
Malm, Andreas, 137
Malta, 120
Malthus, Thomas, 47, 54, 55, 56, 58, 59, 60–1, 91, 108–9, 111, 112, 314, 315
Malthusianism, 31, 47, 54, 62, 75, 91, 99, 109, 132, 179, 187, 188, 202, 223, 237, 314, 315, 317, 323
Mamdani, Mahmood, 114, 115–16
Manchester, 137
Mandate Palestine, 62, 114, 125–30, 133, 141–3, 144, 158, 164, 181, 185, 229–32, 312

Manhattan Project, 138
Martinez-Alier, Joan, 202, 313
Marx, Karl, 44, 45, 133, 277, 295, 317
Marxism, 19, 20, 271, 311
Masalit, 228
Mason, Michael, 239
Mayan civilisation, 59–60
McCright, Aaron, 135
Mechanised Farming Corporation, 173–4
Mediterranean region, 92, 94, 101
Mediterranean Sea, 27, 143, 144, 148, 185, 279
Mekorot, 142, 145, 183, 292
Merowe Dam, 156, 160, 161, 288, 289, 310, 312
Mesaoria, 229
Messerschmid, Clemens, 67, 189
methodological nationalism, 40, 61
Mexico, 46
Mezzadra, Sandro, 171
Middle East
  climate change, 64, 65, 113
  Orientalism and, 300
  virtual water, 274, 276
  water wars discourse on, 46–7, 51, 112, 113, 244
Mihemed, Remziye, 203
militarism, 2, 5, 6, 20, 27, 50, 61, 62, 76, 96, 100, 124, 125, 138, 201, 202, 203, 209, 233, 237, 238, 250, 260, 261, 263, 288, 289, 291, 299, 313, 314, 321
Mill, John Stuart, 28
Missirian, Anouch, 6, 15
Mitchell, Timothy, 34, 35, 165
Monbiot, George, 2
Montesquieu, Baron De, 106
Moore, Jason, 169
Morphou, 229, 265–6
Morsi, Mohammed, 62
Mountain Aquifer, 182–4, 185, 229, 255–6
Mozambique, 244
Mubarak, Hosni, 244
Munif, Abdelrahman, 29–30
Myanmar, 244

Naivasha Agreement (2005), 245, 258–60, 261–3, 267
Napoleon Bonaparte, 185
Nasser, Gamal Abdel, 48, 162
National Aeronautics and Space Administration (US), 84, 85
National Congress Party (Sudan), 289
National Water Carrier, Israel, 143, 151, 287, 292

Native Administrations, 75, 116, 177, 180, 196
natural gas, 247, 287, 288, 293, 296, 298–9, 321, 322
nature, 18, 19, 23, 38, 40, 44, 45, 55, 104, 106, 109, 113, 132, 144, 305, 316
nature-exporting societies, 290
Nazareth, 164
Negev Desert, 28, 38, 128, 141, 143, 144, 188, 284, 285
Neilson, Brett, 171
neo-liberalism, 2, 20, 53, 204, 243, 271, 272, 315
Nestlé, 271–2
Netafim, 284
Netanyahu, Binyamin, 65
Netherlands, 103
New York, 156, 161, 168, 188, 293
nexus approach, 273, 274, 276, 282, 286, 294
Niamey, 196
Nicosia, 247–8, 265
Niebuhr, Reinhold, 104
Nietzsche, Friedrich, 98
Niger, 28, 83, 222, 225, 244
Niger Delta, 288, 299
Nigeria, 28, 83, 91, 92, 96, 196–201, 215, 220, 221, 222, 225–6, 244, 286, 288, 289, 291, 296, 312, 319
Nile River, 27, 29, 34, 40, 46, 47, 48–9, 50, 62, 115, 116, 118, 132, 153–5, 159, 160, 164, 173, 177, 245, 261–2, 278, 301, 302, 312
Nile Waters Agreement (1959), 48–9, 155, 302
Nimeiri, Jaafar al-, 74, 119, 155, 159, 173, 175
Nixon, Rob, 25
Noble Energy, 293
Nordås, Ragnhild, 112
North Africa
  climate change, 65, 113
  virtual water, 50, 274, 276
North America, 101, 107, 109, 113, 208, 278
Northern Rizeigat, 72, 75, 180
Nuba, 173, 174–6, 202, 297, 302, 312
Nuba Mountains, 173–5
Nuba Mountains Farmers Union, 175
Nuba Mountains General Union, 202
Nubian Alliance, 163
Nubians, 159–61, 163, 297
Nuer, 159

Nujumi, Wad al-, 89
Nyala, 65, 73, 177

Obama, Barack, 1, 23, 69, 83, 98
Obasanjo, Olusegun, 197
Öcalan, Abdullah, 203
Ofer family, 294
oil, 27, 30, 41, 46, 47, 57, 73, 155, 156, 172, 179, 190, 196, 198, 204, 217, 247, 259, 260–3, 267, 286, 288–91, 292, 296, 298–9, 304, 312, 321, 322
Omdurman, 115, 154
Omdurman, Battle of (1898), 114, 115, 123, 302
Operation Cast Lead (2008–9), 214–15, 218
Orientalism, 29, 103–4, 112, 113, 120, 121, 126, 131, 132, 140, 178, 259, 300, 305
Orontes River, 151, 215, 224, 233
Oslo Agreement (1993), 245, 252–4, 257, 259, 263
Oslo II Agreement (1995), 131, 245, 254–5, 257
Ottoman Empire, 120–1, 122–4, 127, 149–50
Oubangui River, 168

Paddock, William and Paul, 61
Pakistan, 249, 276, 277, 278
Palestine Jewish Colonization Association, 158
Palestine Liberation Organization, 131, 245, 252–4
Palestinian Authority, 235, 254–5, 257, 259
Palestinian territories (West Bank and Gaza), 9, 27–8, 203, 244, 304–5
  impacts of climate change, 65–8, 239, 319
  under autonomy, 253
  virtual water, 282
  water resource availability, 35–8, 40, 65–8, 196, 206
Palestinian Water Authority, 255, 257, 258
pan-Africanism, 76, 202
Paphos, 146
Paris, 185
Paris Agreement (2015), 4, 98, 297, 299, 322
Paris Peace Conference (1919), 130
partition, 27, 29, 119, 123, 124, 126, 133, 149, 213, 247, 249
peace processes, 148, 240–69, 309, 310

Index

peacebuilding, 240–2, 250, 258, 261–3, 265–6, 267–8
Pearce, Roy Harvey, 104
Peluso, Nancy, 29, 110
Peres, Shimon, 246
political ecology, 18–23, 29, 62, 91, 237, 271, 312, 315–17
population growth, 33, 34, 36, 47, 51, 54–6, 57, 61, 67, 69, 108–9, 111, 187, 306, 312
Port Sudan, 115, 118
post-colonialism, 16, 19
post-structuralism, 16
Prince Charles, 2
Prophecy of Neferty, 88
Purcell, Hugh Dominic, 103

Qatar, 156, 291
Qunaytirah, 194

Rabin, Yitzhak, 188, 253
race, 19, 22, 103–36, 207, 236, 287, 307, 309, 313, 315, 322
racism, 105, 106, 134, 314
Ramallah, 256
Raqqa, 214, 227
Ras al-Ayn, 192, 195
Ratzel, Friedrich, 107
realism, 104, 250, 251
rebound effect, 282–6
Red Sea–Dead Sea Canal, 268
Reisner, Marc, 45, 308
Renamo, 244
rentier state, 260, 291
Representative Concentration Pathways, 65, 101
Republican Turkish Party, 162
Rescue Nubia, 161, 163
resilience, 41, 43, 44, 60, 165, 171, 201–5, 219
Resist Kajbar Dam Committee, 161, 163
resistance, 161, 162, 171, 176, 196, 201–5
resources, 55, 56
Reuveny, Raphael, 12
revenge of geography, 35, 63–8, 102
Ritter, Karl, 107
Robinson, Cedric, 133
Rockefeller Foundation, 211
Rojava, 203–4
Roosevelt, Franklin, 138
Roseires Dam, 155, 156, 159, 289
Rosenberg, Justin, 22
Rothschild, Baron de, 127
Royal, Ségolène, 98

Russia, 2, 11, 46, 213, 216, 291
Rwanda, 109, 111
Ryan, Caitlin, 204

Sadat, Anwar, 62
Sahara Desert, 28, 84, 95, 177, 196, 205, 301
Sahel, 27, 28, 64, 69, 71–2, 84, 85, 92, 99, 173, 177, 196, 197, 205, 301
Said, Edward, 104, 132, 317
Sainath, Palagummi, 95, 100
Saudi Arabia, 30, 44, 73, 92, 156, 179, 278, 290
scarcity, 3, 4, 10, 33–5, 54, 56–8, 60–3, 110, 142, 167, 275, 293, 294, 306, 307, 312–13, 315, 316
Schlenker, Wolfram, 6, 15
Scott, James, 138
Second World War, 92, 234
securitisation, 5, 16, 62, 99, 237
security, definition of, 25
Semple, Ellen Churchill, 107
Sen, Amartya, 34
Sennar, 118
Sennar Dam, 153
Setit Dam, 156, 160
Shafir, Gershon, 252
Shagari, Shehu, 197
Shani, Uri, 256
Sharett, Moshe, 230
Sharon, Ariel, 62
Shelley, Percy, 29, 59
Shilluk, 159
Siegel, Seth, 210–12
Sierra Leone, 57, 136
Simon, Julian, 55
Sinai Desert, 130, 181, 184
Sinohydro, 289
Sloterdijk, Peter, 213
slow violence, 25
Smith, Adam, 91
Sneddon, Christopher, 139, 168, 234
soil erosion, 62, 101, 174, 199, 207, 277, 284
solar power, 216, 298, 321
South Africa, 127
South America, 90, 91, 276, 277
South Chad Irrigation Project, 198, 200, 288
South Kordofan, 172–7, 178, 180, 182, 196, 202, 206, 228, 302, 312
South Sudan, 9, 27, 173
  establishment of, 119, 259
  political economy, 260–1
  water availability and insecurities, 35–8, 39–40, 44, 48, 51, 120, 133, 168, 261–3, 287, 301, 302

333

South Sudan civil war, 92, 213, 220
  impacts on water security, 215–18, 221–3, 224–5, 233, 286
Southern African Development Community, 244
Soviet Union, 12, 92, 151, 234, 249, 289, 292
Starr, Joyce, 47
State Hydraulic Works (Turkey), 146, 233
state-building, 10, 23, 61, 125, 140, 142, 143, 153, 158, 162, 166, 170, 172, 196, 220, 227, 236, 271, 291, 305, 310, 316
Stoddard, Lothrop, 104
Stoll, Daniel, 47
Strategic Foresight Group, 243
structural violence, 25
subsidies, 53, 80–2, 143, 148, 151, 152, 154, 155, 174, 182, 183, 190, 193, 194, 200, 286, 288, 289–90, 292, 296, 304
Sudan, 9, 27, 112, 244
  first civil war, 119, 172
  Mahdist state, 89, 114, 115, 118, 132, 153, 173
  post-colonial political economy, 74–5, 119–20, 155–7, 197, 199, 258–60, 261, 286, 288–9
  second civil war, 119, 159, 172, 175–6, 213, 215, 220, 224, 225, 228, 233, 259, 302
  under British rule, 105, 114–21, 122, 124–5, 132–3, 153–5, 197, 302
  virtual water, 277–8, 281–2, 295
  water resource availability, 35–6, 40, 48–9, 71, 116–17, 188, 297, 301–2, 312
  water resource development, 118, 141, 153–8, 159–61, 163, 168, 171–81, 202–3, 262, 292, 302–3
Sudan People's Liberation Movement/Army, 76, 119, 159, 175, 215, 245, 259–60, 261
Sudan–South Sudan peace process, 242, 245, 258–63, 267
Sudd, 155, 159, 222
Suez (company), 293
Suez Canal Company, 162
Suez crisis, 162, 249
Suliman, Mohamed, 175
Sustainable Development Goals, 36, 276
swamp drainage, 128, 130, 142, 153, 158–9, 172
swamps, 45, 99, 104, 116, 121, 222
Sykes, Mark, 126

Sykes–Picot Agreement (1916), 126, 196
Syria, 9, 27–9, 47, 133, 231
  Assad regime, 2, 96, 97–8
  frontier dynamics, 171, 190–6, 201, 203–4, 206, 243, 312, 319
  impacts of climate change, 78–9
  impacts on water security, 233
  post-colonial political economy, 150–1, 152, 197, 289–90, 296, 304
  pre-war agrarian crisis, 79–82, 152–3, 190, 191–3, 204, 286, 303–4
  under French rule, 149–50
  virtual water, 278–9, 281, 294, 295, 304
  water resource availability, 35–6, 190, 195, 279, 304
  water resource development, 67, 141, 144, 149–53, 157, 159–60, 162, 190–1, 234, 249, 286, 292
  within Ottoman Empire, 149–50
Syrian civil war, 244, 269
  as a climate conflict, 7, 70, 77–84, 87–8, 94, 96, 97–8, 99, 100, 102, 113, 167, 303, 312, 317
  causes of, 82–3
  impacts on water security, 213–14, 215, 216–21, 223–4, 227–8, 233, 286, 304, 310, 319

Tabqa Dam, 151, 227, 228, 233, 234
Tahal, 292
Tartous, 223
techno-politics, 140, 158, 165, 184, 251, 286
Tel Aviv, 38, 144, 187
Tennessee Valley Authority, 138–9, 144, 234
Thailand, 12
Thom, Robert, 137–8
Tiga Dam, 198, 200
Tigris River, 49, 227
Tishrin Dam, 227
Tony Blair Institute for Global Change, 210–12
Torit, 222
trapped populations, 111, 152, 187, 226
Treaty of Guarantee (1960), 263, 264
Treaty of Lausanne (1923), 196
Troodos Mountains, 27, 146, 233
Trottier, Julie, 235
Tshuva, Yitzhak, 293
Tunisia, 11
Turkey, 46
  and Cyprus, 124, 125, 133, 146–7, 148–9, 162–3, 165, 219, 229, 233, 234, 247, 263–6, 268, 292, 304

and Syria, 78, 98, 151, 162, 190, 193–6, 214, 220, 228, 233, 243, 244, 310
Turkey–northern Cyprus water pipeline, 148–9, 168, 247, 266, 268, 292, 304, 312
Turner, Frederick Jackson, 169, 171
Tushkah, Battle of, 89

Uganda, 12, 155, 168, 197, 220, 223, 260
Ukraine, 244
Umm al-Naser, 218
United Arab Emirates, 44, 92, 156
United Kingdom, 3, 33, 42, 92, 103, 113, 137–8, 162, 187, 244, 264, 275, 276, 282, 283, 292, 297
United Nations, 1, 79, 97, 100, 161, 187, 194, 205, 214, 221, 265, 269
United Nations Children's Fund, 215
United Nations Conciliation Commission for Palestine, 230
United Nations Development Programme, 79, 193, 239, 248
United Nations Educational, Scientific and Cultural Organization, 299
United Nations Environment Programme, 71, 76, 85, 283
United Nations Security Council, 2, 3, 83, 96, 98, 243
United States, 1, 3, 6, 12, 15, 20, 24, 30, 46, 61, 92, 99, 100, 104, 112, 135, 138, 139, 190, 204, 213, 216, 228, 244, 245, 248, 249, 251, 253, 263, 264, 269, 275, 276, 277, 291, 299, 311, 313, 322
urbanisation, 52, 80, 82, 93, 123, 142, 147, 152, 182, 190, 200, 220, 237, 290

van der Pijl, Kees, 271
Varnava, Andrekos, 120, 122
Verhoeven, Harry, 96
virtual water, 42, 50, 273–82, 291, 294, 295, 297, 310, 314

Wadi Haifa, 118
Wallström, Margot, 83
Waltz, Kenneth, 12
Wang, Ranran, 275
war, approaches to, 209–12
war, impacts on water security, 209–39
Washington, DC, 156, 161
wastewater, 42, 50, 144, 186–7, 214–15, 216, 218, 247–8, 251, 258, 265, 280, 287, 305, 308

water cooperation, 48, 49, 54, 62, 240–5, 246–52, 254–5, 256, 258, 266, 267, 314
Water Development Department (Republic of Cyprus), 235
water frontiers, 170, 171, 180, 182, 206, 208
water pricing, 53, 143, 183, 217, 219, 272, 282, 293, 296
water scarcity, 35–8, 41, 49–50, 62–3, 242, 276–7, 306, 312
water security, 25, 29, 30, 40, 45, 46, 132, 139, 140, 162, 166, 210, 212, 213, 236–7, 270, 272, 273, 282, 288, 294, 295, 297, 306–13, 316, 320, 322
water stress, 35–40, 44, 67, 276
water wars, 24, 25, 27, 46–53, 62–3, 64, 68, 112, 181, 240, 276, 307, 311
Water Works Department (northern Cyprus), 233
water, properties of under capitalism, 40–6
waterscapes, 45, 140, 145, 163, 167, 310
Watt, James, 138
Watts, Michael, 29, 91, 110, 192, 201
Wau Shilluk, 222
Weizmann, Chaim, 311
West Bank, 48, 233, 285, 286, 305
  Israeli settlements, 181–4, 188, 245, 254, 257–8
  under autonomy, 254–8, 268
  water resources and supplies, 40, 53, 143, 181–8, 204, 221, 229, 231, 235, 254–8, 305, 310, 319
Westing, Arthur, 237
wheat, 11, 80, 81, 88, 152, 191, 192, 197, 198, 199, 200, 203, 204, 225, 274, 276, 289, 290, 303
wind power, 41, 298, 321
Wisconsin, 169
Wittfogel, Karl, 140
Wolf, Aaron, 247
Wolseley, Garnet, 121
Women's Committee of the Manasir Affected People, 161
World Bank, 2, 46, 152, 155, 173, 186, 190, 198, 214, 219, 319
World Commission on Dams, 161
World Economic Forum, 46, 271–2, 274, 293
World Food Programme, 216

World Health Organization, 186, 222
world systems theory, 21, 273
Worster, Donald, 169

Yangtze River, 43
Yaoundé, 196
Yarmouk River, 130, 144, 151, 224, 248
Yemen, 92, 298
Yobe, 196, 221

Yousef, Ahmed, 204
Yusuf, Mohammed, 201

Zaatari refugee camp, 223
Zaghawa, 74, 180, 228
Zimmerer, Jürgen, 13
Zimmermann, Erich, 55
Zionist movement, 62, 111, 125–31, 143, 157, 158, 182, 229, 231, 235, 248, 252, 292, 305, 310, 312

Lightning Source UK Ltd.
Milton Keynes UK
UKHW022039210922
409241UK00004B/34